樂林泉

中外园林文化展

故宫博物院 编

REJOICING IN WOODS AND SPRINGS

A Journey through Garden Cultures in China and the Wider World

Compiled by the Palace Museum

故宫出版社
THE PALACE MUSEUM PRESS

乾隆花园衍祺门北望，陈敬哲摄

乾隆花园衍祺门内假山，陈敬哲摄

乾隆花园抑斋，陈敬哲摄

乾隆花园禊赏亭，陈敬哲摄

祝辞

园林，寄托着人们对理想生活的向往与追求，是人与自然互动的核心载体。“乐林泉：中外园林文化展”以“园事”为共同的文化纽带，跨越时代与地域，让中外园林同台共话，在呈现园林空间美学的同时，深入挖掘其作为人类理想栖居地的独特意义。

故宫博物院的相关研究和展览不只呈现优秀的中国传统文化，更希望通过广泛的国际交流合作展现不同文明的多样性，同时关注时代浪潮，为现代社会的发展提供参考和借鉴。回看历史，人们在园林中建立的与自然的关系、表达的文化与精神，或许有助于我们应对当下生态环境变化等全球性挑战。

此次展览主要聚焦古典园林，而19世纪以来，城市的喧嚣和污染催生了郊野田园生活，城市公共花园也蔚然成风。园林不再由精英和贵族拥有，而是每个人都可以平等共享。与此同时，人类重新思考与自然的关系，人们欣赏自然宁静避世的美感，期待在自然中重获身心的自由。

在园林中，自然不是荒蛮的，人们可以通过培育植物、经营景观与自然建立亲密友好的关系。而人类最早的园艺，正是用双手的劳作和心灵的智慧与自然建立连接。中国园林独树一帜的山水营造，其本质也正是我们的自然观念，也就是“天人合一”的哲学。

每年有一千多万观众参观故宫这座人类文化遗产，这里的四座园林是威严宫禁的悠然之所：御花园规整对称，古木、奇石林立；慈宁宫花园宁静肃穆，为静修佳处；建福宫花园与乾隆花园集南北园林、文人园林与宫廷园林之精华。而宫苑中的花朵和绿植，则大部分依靠废弃绿色资源制成的有机肥获得持续的生机。故宫人研究园林，也养护园林，他们把对文化和自然的爱意转化为长久的守护。

“桃花流水窅然去，别有天地非人间”，古典园林中的传统智慧如何为个人家园、现代园林和“地球花园”的建设提供源头活水？希望此次展览继续激发跨文明、跨领域的开放交流与合作，共同探索绿色可持续的、以人为本的家园营造之道。

多年来，故宫博物院与国内外文博机构保持着深入的交流，本次展览就是多方合作的集中体现。乾隆花园保护项目，是与美国世界文物建筑保护基金会（WMF）持续二十多年合作的成果；建福宫花园的复建，则是在（香港）中国文物保护基金会的支持下完成的；此次展览也是我院与美国芝加哥艺术博物馆展览交换项目的终章。同时，本次展览还汇聚了美国大都会艺术博物馆、耶鲁大学英国艺术中心、法国凡尔赛宫博物馆及国家园林公共机构，以及中国国家博物馆、天津博物馆等国内外文博机构的藏品。我们也邀请到清华大学建筑学院景观学系朱育帆教授及团队参与策划，这必将写就故宫博物院百年院史上对外交流合作的精彩一页。展览得到了香港赛马会的大力支持和公益慈善研究院的独家资助。

最后，我谨向国内外的合作伙伴，以及所有为展览付出辛勤努力的同仁们致以衷心感谢，并期望“乐林泉：中外园林文化展”带给每位观众一片心灵的花园。

故宫博物院院长 王旭东

2025 年 4 月

Preface

Gardens embody people's yearning for and pursuit of an ideal life; they are one of the central ways that humanity and nature interact. Based on this shared cultural bond, 'Rejoicing in Woods and Springs: A Journey through Garden Cultures in China and the Wider World' provides a stage for dialogue between Chinese and foreign gardens across widely varied eras of history and regions of the world. Our exhibition demonstrates the aesthetics of garden space and explores its unique significance as an ideal dwelling place for humanity.

In its research work and exhibitions, the Palace Museum does not aim only to showcase the outstanding features of China's traditional culture. Through extensive international exchanges and cooperation, we also wish to present different civilizations in all their diversity. We focus on the trends of each era to provide references and inspiration for the development of modern society. A retrospective view of the ways people in history connected with nature and expressed their cultures and ideas through gardens may help us address current global challenges, such as changes in the ecological environment.

This exhibition chiefly focuses on classical gardens. From the 19th century on, the noise and pollution of urban life have inclined people to yearn for rural and pastoral life, leading to the prevailing trend of establishing urban public gardens. Gardens are no longer exclusively owned by elites and the nobility but are now equally accessible to everyone. At the same time, humanity has re-evaluated its relationship with nature. People appreciate the beauty of tranquility and seclusion in nature, hoping to regain physical and mental freedom.

In gardens, nature is not wild; through cultivating plants and building landscapes, people can establish an intimate and friendly relationship with nature. Humanity's earliest horticultural efforts employed human labor and wisdom to connect with nature, while China's unique recreation of natural landscapes in gardens embodied the traditional Chinese philosophy of 'unity between heaven and humanity'.

Every year, the Forbidden City attracts over ten million visitors, standing as a treasury of human cultural heritage. The four gardens of the Forbidden City serve as tranquil retreats within the majestic imperial complex: the Imperial Garden is meticulously arranged with a symmetrical layout, featuring centuries-old trees and exquisite rock formations; the Garden of the Palace of Compassion and Tranquility (*Cining gong*), with its serene ambience, provides an ideal space for quiet contemplation; while the Garden of the Palace of Established Happiness (*Jianfu gong*) and the Qianlong Garden embody the finest elements of both northern and southern Chinese gardens, as well as literati and imperial gardens. Flowers and green plants in these gardens flourish with nourishment from organic fertilizers made from recycled green waste. The dedicated staff at the Palace Museum not only research and preserve garden heritage but also translate their love for culture and nature into long-term stewardship of the gardens.

'Peach blossoms drift upon the flowing waters, fading into the distant horizon, where a realm beyond the mortal world unfolds.' How can the traditional wisdom found in classical gardens inspire the construction of personal homes and modern gardens, and help us turn our planet into a garden? We hope this exhibition will act as a wellspring, inspiring open dialogues and cooperation across civilizations and fields of activity, so we can jointly explore ways to create green, sustainable, and people-oriented living spaces.

Over the years, the Palace Museum has maintained in-depth exchanges with museums and cultural institutions around the world, and this exhibition showcases the achievements of many such collaborations. The project of conserving the Qianlong Garden is the result of more than 20 years of cooperation with the World Monuments Fund (WMF) in the United States, and the reconstruction of the Garden of the Palace of Established Happiness was completed with the support of the Hong Kong-based China Heritage Fund. This exhibition is also the final chapter of the exhibition exchange project between the Palace Museum and the Art Institute of Chicago. Here, we have brought together items from the collections of renowned museums and cultural institutions in China and abroad, such as the Metropolitan Museum of Art, the Yale Center for British Art, the Palace of Versailles, as well as the National Museum of China and Tianjin Museum, among others. Professor Zhu Yufan and his team from the Department of Landscape Architecture at Tsinghua University's School of Architecture participated in planning and curating.

This exhibition will surely be remembered as a remarkable example of international exchanges and cooperation in the Palace Museum's 100-year history. We would also like to acknowledge the generous support from the Hong Kong Jockey Club and the exclusive sponsorship by the Institute of Philanthropy to the exhibition.

Finally, I would like to extend my sincere gratitude to our partners in and beyond China, and all our dedicated colleagues whose tireless efforts have brought this exhibition to life. May 'Rejoicing in Woods and Springs: A Journey through Garden Cultures in China and the Wider World' inspire every visitor to nourish their own spiritual garden, enriching their spirit with the timeless beauty and illumination of nature.

Wang Xudong, Director of the Palace Museum
April 2025

乾隆花园古华轩，陈敬哲摄

祝辞

自建馆以来，芝加哥艺术博物馆一直秉持着全球合作的理念。我们深信，携手合作是实现思维交流的重要途径。虽远隔重洋，芝加哥艺术博物馆与故宫博物院的渊源仍可追溯至近百年前——故宫博物院建院伊始，我馆数位理事即捐资助力故宫的修缮工程。2018年，故宫博物院为我馆举办的“吉金鉴古：皇室与文人的青铜器收藏”特展给予鼎力支持。这场开创性展览不仅深化了与中国文博同行的友谊，更推动双方开展深度合作，促进了专业知识与艺术资源的共享。两馆秉持着共同的使命：通过跨越时空、文明与文化的艺术体验，激发探索精神，促进理念交流。

值此“乐林泉：中外园林文化展”举办之际，芝加哥艺术博物馆欣然从馆藏精品中遴选80余件艺术杰作参展。从法国印象派画作，到日本浮世绘；从波斯细密画，到意大利蚀刻版画；从古罗马玻璃器皿，到文艺复兴时期的雕塑……这些作品生动诠释了千百年来世界各地的民众如何与园共栖、寄情于园，及其享受其中的异同。芝加哥艺术博物馆与故宫博物院均坐落在园林环抱之中，这些园林与博物馆一样，既是滋养心灵的休憩之所，更是激发活力的灵感源泉；既为参观者提供独处冥思、感悟自然之美的静谧空间，又是汇聚人文想象与自然灵气的创意场所。

几个世纪以来，艺术家们不断探索园林创作，或许因其承载着生命繁衍、万物生长的永恒意象。这里千姿百态的造型、色彩搭配与空间布局，将自然界也许本难共存的元素巧妙融合，创造出独具匠心的和谐之美。博物馆恰似精神园林，汇聚不同文明瑰宝，孕育对世界日新月异的认知图景。欣逢故宫博物院百年华诞，愿我们今日播撒的交流种子，在未来岁月里绽放绚丽之花，为万千观众带来永恒的精神享受。

芝加哥艺术博物馆总裁兼艾罗斯·W. 马丁专任馆长　詹姆斯·容度

2025年4月

Preface

Global collaborations have been essential to the Art Institute of Chicago since its founding, rooted in the belief that partnerships are essential to the exchange of ideas. Despite the great distance between our institutions, the connection between the Art Institute of Chicago and the Palace Museum began nearly one hundred years ago, when several Art Institute trustees donated to the renovation of the newly established Palace Museum. In 2018 the Palace Museum provided exceptional support to the Art Institute's exhibition 'Mirroring China's Past: Emperors and Their Bronzes'. This groundbreaking presentation strengthened relationships with museum peers in China and promoted collaborations that allow us to share expertise and great works of art. Each institution embraces a mission to inspire inquiry and the exchange of ideas through an experience of art across time, civilizations, and cultures.

The Art Institute of Chicago is pleased to lend more than eighty exceptional works of art from its renowned collection to the exhibition 'Rejoicing in Woods and Springs: A Journey through Garden Cultures in China and the Wider World'. From French Impressionism to Japanese prints, from Persian miniature paintings to Italian etchings, and ancient Roman glassware to Renaissance statuettes, these works highlight the similarities and differences in how people have interacted with and perceived gardens throughout the world over the centuries. Gardens surround both the Art Institute and the Palace Museum, and much like these museums, they serve as restorative and invigorating spaces. Their beauty encourages solitary moments of reflection, meditation, and wonder, while they also serve as gathering places to enjoy the creative possibilities that emerge from human imagination and the natural world.

For centuries, artists have explored the theme of the garden, perhaps drawn in by its promise of life, growth, and regeneration. The diversity of organic forms, shapes, colors, and arrangements planted in a garden highlight the beauty created by uniting elements that might not coexist in nature. Museums offer a similar opportunity, bringing objects from diverse cultures and civilizations together that spur a new and evolving vision of our world. On the occasion of the centenary of the Palace Museum, my hope is that the seeds of exchange we plant today, through these joint efforts, will continue to blossom and be a source of inspiration for millions of visitors in the years to come.

James Rondeau, President and Eloise W. Martin, Director of the Art Institute of Chicago
April 2025

乾隆花园倦勤斋通景画（局部），陈敬哲摄

宛然画境　诗意栖居

——“乐林泉：中外园林文化展”代序

一

现在想来，我与园林，应该是有些缘分的。

从出生到小学四年级，我随祖父母住在东四与宽街之间汪芝麻胡同的一处大院儿，这里是北京某高校宿舍。院子本是贯通汪芝麻胡同和南面魏家胡同的一处大宅，原来的格局，应该南部是宅院，北部是花园，后被彻底分隔开，北侧另开院门。我们约十户人家共同居住的北院，应当就是原来大宅的花园部分，当时院内尚有残存的假山和游廊。不止如此，我上的幼儿园，就在现在的恭王府花园里。当时恭王府尚未腾退修缮，为多家单位分占，更未开放，很少有人知道其名。[1] 在我幼年的记忆中，这里只有高大的屋宇房间和杂草丛生的花园。此外，上世纪七八十年代，北京市少年宫在景山公园内的北部，在没有课的下午，我经常与小伙伴一起到那里玩耍，兼购买航模材料，回家自己制作。四年级以后，我随父母居于地安门一带，与北海公园北门仅一箭之遥，也成为周末一家人常去的休闲所在。

工作以后，因为以古书画为专业的缘故，我与园林接续了前缘。因为在中国古代书画诗文之中，屡见对园林的表现。一段时期以来，我以明代吴门地区书画为研究重点，而园林则是“聊以画图写清居”的吴门绘画的重要题材之一。

多年以来，我与园林从未走远。紫禁城中的花园自不待言，那里就是我日常的工作场所；在工作出差以及休假旅行中，中外园林也多有涉及。从皇家苑囿到私家园林，从国保单位到世界文化遗产，那一座座或宏伟，或精巧，或规整，或自然的园林，其各自独特的魅力已经深植于我的记忆之中。

1　参见何晞宇：《抢救恭王府》，载《中国新闻周刊》2016 年第 3 期。

因此，当与芝加哥艺术博物馆合作的交换展主题——“莫奈的花园”摆在眼前时，随着与同事们讨论，此前头脑中存储的关于园林的点点滴滴似乎瞬间被激活。想来，大多数人的生活中或多或少都会接触到园林，也便或多或少有着关于园林的记忆。那么，为什么不勾连拓展，办一个关于中外园林文化主题的展览，让人们更多地了解古今中外对于园林的丰富表达和追求，也让园林之美触动更多的人呢？

二

园林是什么？大家好像既熟悉又陌生。虽然常见，但它的概念，内涵、外延，其哲学基础、美学表现、营造技艺……我们普通人恐怕就很难说出个子丑寅卯了，而世人对异域文明之园林的了解则更为有限。

园林，“是以自然素材为主要欣赏对象的人工游憩空间”。园林艺术是一种较为复杂的文化现象，兼具自然美与艺术美的双重属性。[2]

在历史长河中，园林伴随了人类几千年，可谓异彩纷呈。从埃及的园林、巴比伦的空中花园，到古希腊伊壁鸠鲁的庭园、古罗马庞贝的庭园；从中世纪西欧的修道院庭园、城堡庭园，到波斯、印度莫卧儿王朝的伊斯兰庭园；从文艺复兴时期意大利的台地园，到法国路易十四时期“勒·诺特尔”式的规则型园林，再到英国的自然风景园……随着时间推移，各种风格的园林之间，相互影响、渗透，不断演变。而中国园林，在世界园林史上占有重要的一席。如果按最粗简的线条梳理中国园林发展，大体如下：自商周发其端；至秦汉呈现出山水化趋势，逐步出现了以摹拟自然为标准的造园倾向，具有中国特色的山水园林艺术及其表现手法初成框架；两晋南北朝，园林艺术发生划时代变革，自然山水式的审美进一步得到推崇，造园活动升华到了艺术创作的境界；隋唐、两宋，造园活动频繁，写意山水画的意境被融入园林创作中，中国古典园林的风格和手法基本形成并趋于成熟，达到中国园林史的高峰；明清两代，造园理论和技术日趋成熟，造园实践和著述日益丰富，出现新的繁荣。[3]

园林史是专门的学问，历史上的研究专著卷帙浩繁，累积的学术成果汗牛充栋。在中外园林交流方面，西方17世纪后期就有威廉·坦帕尔的《关于伊壁鸠鲁的园林》将中国造园艺术介绍给英国造园界，并对中欧园林进行了比较和评论；18世纪威廉·钱伯斯的《东方园林论》着重介绍了中国造园艺术，并力倡在英国风景式园林中吸取

2 参见成玉宁等著：《中国园林史（20世纪以前）》，第2页，中国建筑工业出版社，2018年。

3 参见成玉宁等著：《中国园林史（20世纪以前）》，第11、87、287页，中国建筑工业出版社，2018年。

中国趣味的创作；此外，陆续还有《欧洲造园中的中国影响》《英中式园林》《18 世纪的中国与法国》等著作，都对欧洲造园中的中国影响有所涉及。中国在中外园林交流研究领域发力较晚，但在 20 世纪以来，也有大量研究成果，如陈受颐、童寯、陈志华等前辈学者的研究。而关于中外园林文化在哲学、美学方面的比较，也有很多探讨。如，认为中西园林艺术的差别主要表现在“形和意”“含蓄与直观”“封闭与开放”等等。又如“西方园林基本上是写实的、理性的，重人工、讲规律，可以说是现实主义，但失之为机械刻板；中国园林基本上是写意的、直观的，重自然、讲想象，可以说是浪漫主义的，充满着空灵剔透和辩证法。”“（中国）传统的造园艺术精神……追求的是达于宇宙天地的‘道’，这是中国园林与西方庭院在本质上的区别。”[4]

史实的梳理与形而上的辨证都在不断深入。先贤以及当代学者的研究成果，为我们认识和呈现世界园林的风貌提供了理论基础。

三

进入 21 世纪以来，故宫博物院举办了众多有关文明交流互鉴的展览，如“中国·比利时绘画 500 年”“山川菁英：中国与墨西哥古代玉石文明展”“梵天东土，并蒂莲华：公元 400—700 年印度与中国雕塑艺术展”“天下龙泉：龙泉青瓷与全球化”“茶·世界：茶文化特展”“传心之美：梵蒂冈博物馆藏中国文物展”“历史之遇：中国与西亚古代文明交流展”等，力求从不同领域展示不同文明间的交流互鉴以及异同。同时，我们还举办了如“照见天地心：中国书房的意与象”“国子文脉：历代进士文化艺术联展”等展览，都是不仅仅展示文物之美，更力图突破物质层面，努力揭示文物背后的人的精神世界。此次展览亦是如此。

但如前所述，一方面，人类园林的历史如此复杂，各种文明之间，不同历史时期之间，社会经济发展阶段不同，艺术思潮和技术的发展，自然条件、地理水文的区别等等，都会导致园林艺术的呈现结果不同。另一方面，形而上的讨论又难免见仁见智。从根本上说，我们无法也不应对不同文明、不同历史时期的园林艺术简单地比高下，论短长，其间更多的是基于人文背景和艺术旨趣的特色不同，而非优劣之分。就像有人指出，“园囿之盛，不能相兼者六：胜宏大者少幽邃，人力胜者少苍古，多水泉者艰眺望。”[5] 这一主题的多样性与复杂性，决定了一个文物展览不可能对这一专题做整体的、系统性的呈现，更何况还必须基于能够取得的实物资料。

4 参见周武忠：《理想家园：中国古典园林艺术比较》，第 8—12 页，东南大学出版社，2012 年。

5 叶广度：《中国庭园记》，第 8 页，当代中国出版社，2015 年。

园林艺术表现形式固然多种多样，但其背后的人的精神需求则是大体一致的，存在极大的共性——都是为了营造一个理想的家园。人寄托于园林的理想，必定包含着亲近自然的追求，以及生活在其中的游憩需要。

首先是对自然的向往。不论园主的身份如何，追求富有自然意趣的生活享受这一造园宗旨，大体是相同的。本展的标题“乐林泉”，来自清代李世倬的画作《皋涂精舍图》轴上乾隆帝御笔题写的诗塘。“林泉”，是中国古人对自然山水的诗意表达，而“乐”于“林泉”，则是人们赋予园林的寄托和期许。

北宋著名山水画家郭熙在其画论《林泉高致》中说：“君子之所以爱夫山水者，其旨安在？丘园养素，所常处也；泉石啸傲，所常乐也；渔樵隐逸，所常适也；猿鹤飞鸣，所常观也。尘嚣缰锁，此人情所常厌也；烟霞仙圣，此人情所常愿而不得见也。……然则林泉之志、烟霞之侣，梦寐在焉，耳目断绝，今得妙手郁然出之，不下堂筵，坐穷泉壑，猿声鸟啼依约在耳，山光水色滉漾夺目，此岂不快人意？实获我心哉！此世之所以贵夫画山之本意也。”[6]“不下堂筵，坐穷泉壑”，山水画能够起到的作用，园林也能。即便是威仪天下的一代雄主，也难免时时需要借助山水林泉清心怡神。路易十四在凡尔赛宫苑中，在满朝文武的簇拥下，他脚下的中心大道“携带着周围的花草树木和池坛雕像，缓缓地伸向远方。它绕过层层花坛，穿过片片林园，跟随大水渠进入密集的森林之中，然后徐徐升至远方的原野，消失在广阔的天空之中。”[7]乾隆皇帝，不管是在三山五园，还是在紫禁城中有限空间内的建福宫花园和乾隆花园，“一峰则太华千寻，一勺则江湖万里”的理念恐怕是根深蒂固的。至于文人园林，更是以巧思因山就水，点缀建筑，选植花木，营造景点，将自然中的胜景缩移到壶中天地中，使人“顿开尘外想，拟入画中行”。

至于人在园林中的游憩活动，既有物质享乐层面的饮酒、品茶、游戏、赏花，更多的是精神层面的观书、作画、赋诗、抚琴、谈经、博古……在本次展览中，我们提取了雅集、鉴藏、游山、静修、观花、畅音六项典型的园事，以此提纲挈领，大体可以涵盖人们在园林中的主要活动，兼顾呈现中外园林的部分典型面貌。

在展览的各个单元中，我们围绕相关园事，选取了诸多具有代表性的展品，加以烫样、模型、局部复原等辅助手段，着力呈现中外园林的场景风貌，以及人在其中的活动及其精神内涵与情感寄托。其中既有中国历代名家的法书绘画，各类工艺、家具精品，也有西方经典油画、版画、雕塑，日本浮世绘，波斯细密画等等。不论是公众耳熟能详的名家名作，还是首次出展的展品，都紧紧围绕主题进行阐释。

6 （宋）郭熙、郭思：《林泉高致》，参见《中国书画全书》第一册，第497页，上海书画出版社，1993年。

7 ［法］贝亚特利斯·索勒、［法］马修·达·文哈：《凡尔赛：宫殿、花园、特里亚农宫》，第67页，Artlys出版社，2013年。

清华大学建筑学院景观学系是国内中外园林历史研究的重镇，而故宫与清华又有着密切的合作。鉴于此次展览的特殊性，我们的策展团队，除了故宫博物院内各个领域的优秀学者外，也荣幸地邀请到朱育帆教授及其团队一起合作策划。

四

园林是心的外化，映射人们的情感、思想和心绪。

中国古典园林在秦汉、盛唐尚追求宏大的气势，中唐、两宋后更多的则是追求素雅恬淡的氛围。人们极力追求恬淡自适和心境的超脱，在咫尺山林中怡然自乐，在小天地之间修习内心、追求精神自由并实现自身人格的完善。

园林，可以满足人们对人格完善和对宇宙至理体悟的需求。李白在桃李园中写下“夫天地者，万物之逆旅也；光阴者，百代之过客也”，感叹“浮生若梦，为欢几何”，而“寒池蕉雪诗人画，午榻茶烟病叟禅”的倪瓒，其萧散孤行的人生境界已成为中国文化史上的传奇。

园林，是人们在尘世喧嚣中的隐逸林泉。可能是“惟有王城最堪隐，万人如海一身藏”的隐于朝、市，也可能是“长安车马尘吹面，谁识空山五月凉”式的隐于山野。

园林，是人们小筑卧游、诗意栖居的理想家园，是“半榻茶烟新雨后，一庭花露午晴初”的清逸安闲，也是“日日深杯酒满，朝朝小圃花开”的无拘无碍，更是畅达乐观的生活哲学。

那么，关注园林，对于我们当代又意味着什么呢？

将近一百年前，中国学者叶广度在他的《中国庭园记》中说：“真正的庭园学家，他的使命，并不是为少数特权阶级的谁家庭院造别墅，而是为大众谋幸福，使人于自然中求忘我，不踯躅于一隅，而知人生之谜，即在自己的园地中得其消息。我国庭园前所赞美的竹庐、茅舍、水榭、凉亭，此类点缀，更可知影响国人‘淡泊以明志，临静以致远’的人生观了。”[8] 今天，我们举办展览，探讨园林文化，希望公众通过展览所呈现的内容，能够了解中外园林文化的一些侧面，并透过这些有限的点和线，激发大家对于园林文化艺术的兴趣，并自行探究，从而能够从人类璀璨的园林艺术中更多地汲取营养，弘扬园林美学，提升生活智慧。同时，在全球关注绿色可持续发展的大环境下，我们也希望公众在“意尽林泉之癖，乐余园圃之间”时，由小及大，进而关注生态环境，建设我们的现代文明，建设更加美好的世界。

8 叶广度：《中国庭园记》“自序”，第 12 页，当代中国出版社，2015 年。

五

在偶尔一个闲暇周末的清晨或傍晚，家人尚未起身或者在安静地忙着自己的事情，我独自坐在客厅靠近阳台的单人沙发上，看着熹微天光中，阳台上的桂花、木槿、绿萝、虎皮掌、蝴蝶兰，以及其他几种我叫不出名字的盆栽植物，桂花盆景中的一爿拳石，会幻化成一座孤峭的山峰。它们与玻璃窗外小区里高大乔木的茂密树冠融为一体，仿佛我也有了自己的一座小花园。

这时候，我经历过的那些园林会在心中闪现。我也会想起法国作家埃里克·法伊（Eric Faye）的小说名作《长崎》中，寺岛昌子写给志村公房信中的一段话："我心里想，在世界各国的所有宪法中，应该写上那样一条，每一个人都有不受时效约束的一项权利，只要他觉得时机合适，就可以返回他往昔生活的最初地点。应该提供给他一串钥匙，能打开他度过童年时代的所有公寓、小楼、花园，允许他在这些记忆的冬宫中停留数小时。而这些房屋的新主人绝不能给这些时光的朝圣者制造任何障碍。"[9] 每一个曾经经历过的地方，都与我们生命的一个阶段或一个时间点永远联结在一起。回到某个地方，不只是审视此地，更是审视自己那一段人生。就像女主角，在那个小房间中，感受的不仅是现实中的温暖安定，更是在曾经的家中，幻想四十多年前因意外故去的父母同处一室的心灵慰籍。

如果真能实现，我最希望回到哪里？也许是汪芝麻胡同那个大宅花园改造的，带有假山、游廊的院子？我将如何去探究它们？也许什么都不做，只是静静地享受那一刻。其实，我更希望不但在空间上回到那里，更能够叠加上时间，回到幼年时的那些花园。

真的希望，展览能够带大家回到记忆中的一座花园，或者由此构建一座自己心中的花园。

故宫博物院常务副院长　娄玮

乙巳年正月十六，于紫禁城西长房

9 ［法］埃里克·法伊：《长崎》，第 97—98 页，上海译文出版社，2012 年。

Poetic Dwelling in Picturesque Scenery:

Preface to the Exhibition 'Rejoicing in Woods and Springs: A Journey through Garden Cultures in China and the Wider World'

This article reflects on the author's long-standing engagement with gardens, both personal and professional, and the motivations behind curating the exhibition 'Rejoicing in Woods and Springs: A Journey through Garden Cultures in China and the Wider World'. It provides a brief overview of the history of Chinese and foreign landscape architecture, along with key research in the field, and explores the cultural and emotional significance that gardens have carried over time. Rather than offering a comprehensive historical comparison, the exhibition focuses on how people have experienced and interacted with gardens, showcasing the distinctive features of gardens across different cultures. It seeks to deepen public appreciation of gardens as artistic and spiritual spaces, highlighting the ways they fulfill human needs beyond aesthetics. By inspiring a renewed interest in garden culture and encouraging visitors to draw insight from ancient artistic traditions, the exhibition ultimately hopes to promote garden aesthetics, nurture a thoughtful approach to life, raise awareness of ecological issues, and contribute to a more beautiful world.

Lou Wei, Executive Deputy Director of the Palace Museum
February 13, 2025, at the Western Longfang, the Forbidden City

目录

CONTENTS

展览篇

Exhibition Chapter

展览前言

在历史长河中，园林伴随了人类几千年。各个文明的各种园林风格之间，随着时间推移，相互影响，不断演变，可谓异彩纷呈。

园林的艺术表现形式固然多种多样，但其背后的人的精神需求则是大体一致的——都是为了营造一个富有自然意趣的理想家园，都必然包含着亲近自然的追求，以及生活在其中的游憩需要。

“乐林泉”，取自清代乾隆皇帝为李世倬《皋涂精舍图》轴题写的诗塘。“林泉”，是中国古人对自然山水的诗意表达，而“乐”于“林泉”，则是人们赋予园林的寄托和期许。

人在园林中的游憩活动，包含了物质享乐和精神层面的内容。我们提取了雅集、鉴藏、游山、静修、观花、畅音六项典型园事，大体涵盖人们在园林中的主要活动，兼顾呈现中外园林的部分典型面貌，力求使公众更多地了解古今中外对于园林的丰富表达，以及人对园林的精神追求。

园林，映射人们的情感、思想和心绪。希望大家在繁忙喧嚣的现代生活中，能够找到自己平和安恬的心灵的花园。

Exhibition Introduction

Over thousands of years, gardens have accompanied human civilization on its historical journey. Out of the world's many civilizations, many styles of garden have emerged, developing and influencing each other, ceaselessly unfolding in a dazzling array.

Although gardens vary widely in design and aesthetics, the underlying human impulse remains the same a desire to create an ideal sanctuary rich in natural charm. This impulse speaks to both a longing to be close to nature and the need for leisure and reflection within garden spaces.

Our exhibition title *Le Lin Quan* (Rejoicing in Woods and Springs) comes from an inscription by the Qianlong Emperor of the Qing Dynasty for the hanging scroll painting Gaotu Vihara Pavilion by Li Shizhuo. In Chinese tradition, the term *Lin Quan* (woods and springs) captures an idyllic vision of natural landscapes in which people dream of 'rejoicing'.

Human activities within garden spaces encompass both physical pleasures and spiritual pursuits. This exhibition explores the relationship between gardens and human life through six thematic sections: Gardens and Enlightenment, Gardens and Antiquarianism, Gardens and Nature, Gardens and Retreats, Gardens and Inspiration, and Gardens and Recreation. These themes highlight six key activities within gardens, literary gatherings, art collection, scenic explorations, meditative retreats, appreciation of flowers, and musical enjoyment which broadly reflect the ways people have engaged with gardens, while also presenting the distinctive features of gardens across different cultures. Through this exhibition we hope to deepen public understanding of gardens as rich modes of expression, as well as the spiritual needs with which people approach them.

Gardens mirror the emotions, thoughts, and inner sentiments of those who create and inhabit them. Amidst life's clamor, may you discover within these gardens a moment of joy, and your own tranquil sanctuary for the soul.

雅集

宗白华先生曾有言，晋人向外发现了自然，向内发现了自己的深情。山林之中，蜿蜒水滨，那场一千六百余年前的雅集，在中国文人心中投下永恒的涟漪。觞咏之所融入园林，风雅名篇俯拾皆是。

客从桃源游，爱此玉山好。

清文引佳酌，玄览穷幽讨。

当我们感怀兰亭、西园、玉山等雅集中高士们留下的诗文时，丝绸之路西行的路上也回响着萨迪、哈菲兹等波斯诗人的篇章。他们在象征天堂的花园，藉由美酒、花鸟、山水与静夜，探讨爱、自然与生命。不同文明、不同时代的雅士在此汇聚，园林成为孕育美与哲思之地。

文明中的优雅，自此永续。

Gardens and Enlightenment

The philosopher Zong Baihua once said, ‘the people of the Jin Dynasty found beauty in the natural landscapes around them, and as they turned inward, they uncovered the depths of their hearts’. Over 1,600 years ago, the famous Orchid Pavilion gathering took place amid woods and springs, leaving a lasting impression on Chinese literati. The settings of such gatherings went on to influence garden designs and inspire literary masterworks. As Gu Jin of the Ming Dynasty wrote in his poem,

Guests wander from the Peach Blossom Spring,
Enchanted by the splendor of Jade Mountain.
With elegant writings, they toast with fine wine,
And in deep reflection, they ponder life’s mysteries.

As we reminisce on the literary legacy left by gatherings at sites such as the Orchid Pavilion, the Western Garden, and Jade Mountain, the celebrated works of Persian poets like Saadi and Hafez also reverberate along the Silk Road... In their poetry, set against gardens that symbolize paradise, they explore universal themes of love, nature, and the meaning of life through images of wine, birds, flowers, mountains, water, and tranquil nights. Across cultures and eras, scholars have gathered in gardens where the intellectual and artistic pursuits of human civilization continue to thrive.

1 乾隆花园第一进院烫样

现代
综合材质
长 96.6 厘米，宽 77.8 厘米，高 17 厘米
故宫博物院藏

烫样，是清代工匠用以表达设计的立体模型。清乾隆时期，烫样在内廷有相当广泛的应用，大到建筑、内檐装修，小到家具、器玩摆件、戏曲道具等种种设计，一般都需要先制作烫样，呈给皇帝和官员预览，确认样式之后方可营造。在表达建筑设计时，烫样不仅能反映建筑外观，往往还可以揭开屋顶，观看室内的装修与陈设，是一种具有独特表现力的建筑模型。

烫样主要是用纸张和袼褙（纸张层层裱糊所得的硬纸板）制作的，兼用少量木料，表面敷设色彩。在表现园林时，还会用蒿子、松鳞等材料制作树木山石等室外景观。许多烫样上还贴有黄色的小签，书写建筑名称、主要尺寸或修缮做法。

根据清代档案记载，乾隆三十五年（1770）十一月，在乾隆花园的设计阶段，内务府曾经制作了一座完整的花园烫样，呈给乾隆皇帝。此后，这些烫样中有一部分连同画样（设计图纸）一道被发往江南，远在江南的工匠们以这些烫样和画样为设计依据，为乾隆花园制作了一批精美的内檐装修构件，运送回京，这才有了今天乾隆花园的室内装饰。

遗憾的是，乾隆时期的烫样没有能够保存至今，而烫样的制作工艺也已失传。如今展出的这组乾隆花园烫样，是根据今日学术研究成果，使用清中期的材料与工艺复原设计制作的。其比例尺是 1:66.6，传统上称为“一寸五分样”，即以一寸五分（4.8 厘米）的长度代表一丈（3.2 米）。

乾隆花园自南向北分为四进院落，每一进院落的设计都各具特色，寄寓了乾隆皇帝对退位归政之后隐居生活的完美构想。这组烫样表现的是第一进院落，借鉴了江南园林的造景手法，布局精巧，流露出典型的文人趣味。从南面的正门衍祺门进入花园，迎面是一道障景假山，地下有五色虎皮石铺满的小路，曲径通幽。穿过两侧假山构成的屏风，曲折向北，就能到达一进院的主体建筑——古华轩。古华轩左右配置一山一水，东面是高人的叠石和承露台，西面是禊赏亭与流杯渠，背靠山石，宛然有林泉之意。整座院落将自然景观与人文意蕴融汇一体，寄寓了乾隆皇帝的隐逸追求。（刘梦雨）

1 Model of the First Courtyard of the Qianlong Garden

Contemporary
Mixed media
L.:96.6cm, W.:77.8cm, H.:17cm
Collection of the Palace Museum

Three-dimensional models known as *tangyang* were used by Qing Dynasty (1644-1911) artisans to convey design concepts to the emperor and his officials. Primarily made of paper, these models were widely used in the construction of architecture, furniture, and objects. The Qianlong Garden is divided into four courtyards from south to north. This exhibit, reconstructed based on academic research on the lost Qing Dynasty original, presents the first courtyard. The design of the courtyard draws inspiration from gardens in the Jiangnan region (to the south of the lower Yangtze River, around Hangzhou), featuring an intricate layout that reflects the refined aesthetics of literati culture. It embodies the Qianlong Emperor's aspiration for a reclusive life in harmony with nature. (Liu Mengyu)

抱厦
雲步踏跺

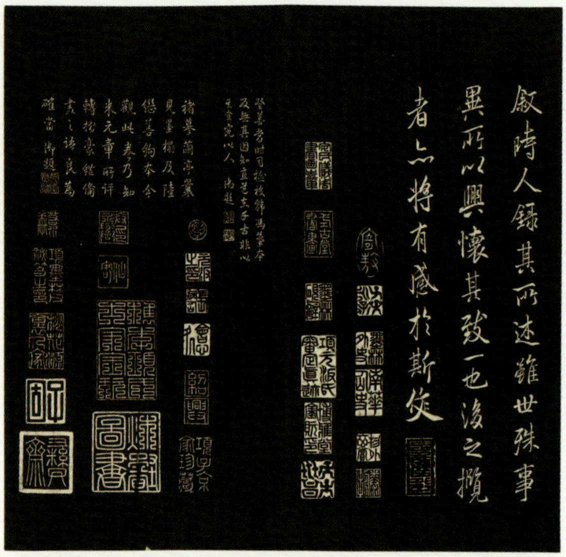
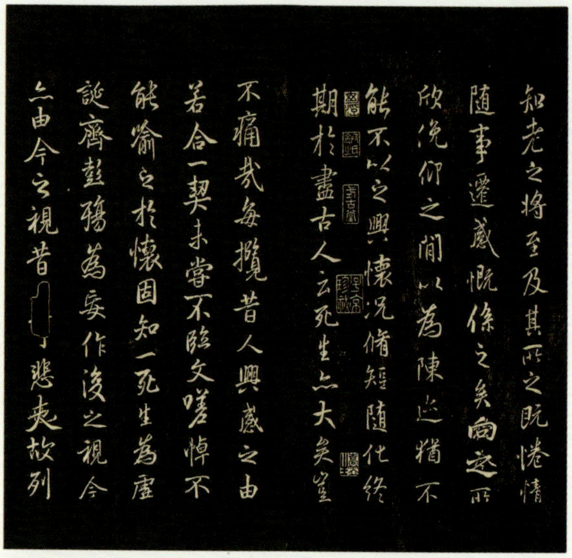
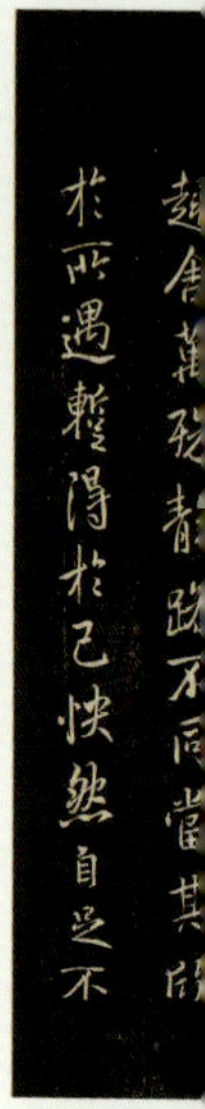
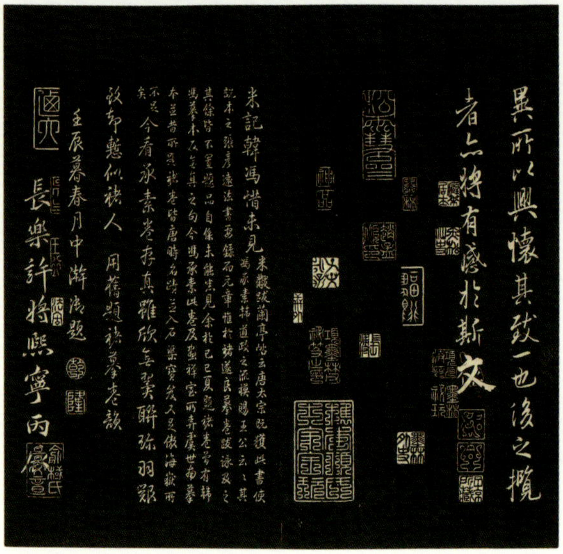
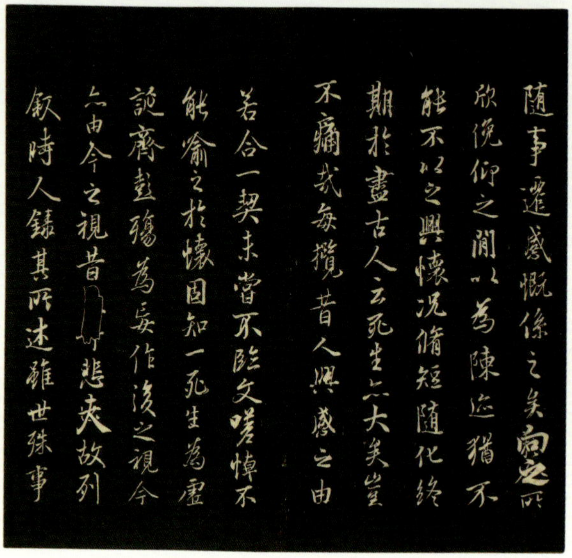
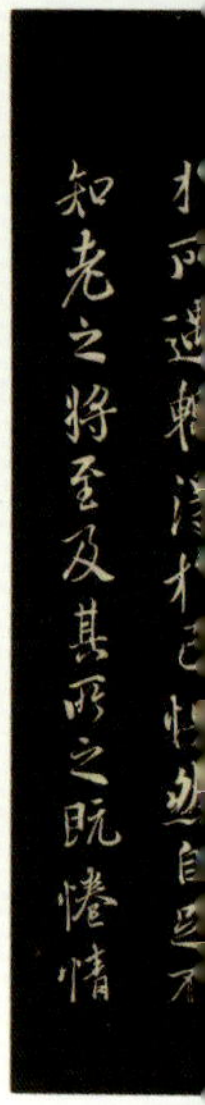

2　清拓兰亭八柱帖册

清乾隆（1736—1795）
墨纸，拓本
纵 29.8 厘米，横 34.2 厘米
故宫博物院藏

雍正帝曾仿照绍兴古兰亭，在圆明园建造了流杯亭。后其子乾隆帝命人将此亭改为重檐八角攒尖亭，并将亭内的木头柱子改为石柱。乾隆四十四年（1779），乾隆帝命人在八根石柱上分别摹刻《虞世南摹兰亭序》《褚遂良摹兰亭序》《冯承素摹兰亭序》《柳公权书兰亭诗》《戏鸿堂刻柳公权书兰亭诗》《于敏中补戏鸿堂刻柳公权书兰亭诗阙笔》《董其昌仿柳公权书兰亭诗》《乾隆帝临董其昌仿柳公权书兰亭诗》，传拓后装裱成册，共八册，即《清拓兰亭八柱帖册》。咸丰十年（1860）圆明园被焚，兰亭八柱幸未被毁，民国时被移至中山公园保存至今。

据《上谕档》《清宫陈设档》等文献记载，《清拓兰亭八柱帖册》除赏赐给皇子和大臣外，还曾用于紫禁城、颐和园、静宜园等处的室内陈设。

《清拓兰亭八柱帖册》共八册，其中第二册《褚遂良摹兰亭序》，系据《褚遂良摹兰亭帖卷》（今藏故宫博物院）摹勒上石。宋人米芾、范仲淹、王尧臣、刘泾、龚开、王申等，元人张雨、程嗣翁，明人陈敬宗，清人卞永誉并乾隆帝等的题跋、观款和印章一并保留上石。木面，托裱经折装。钤“乾隆御览之宝”“避暑

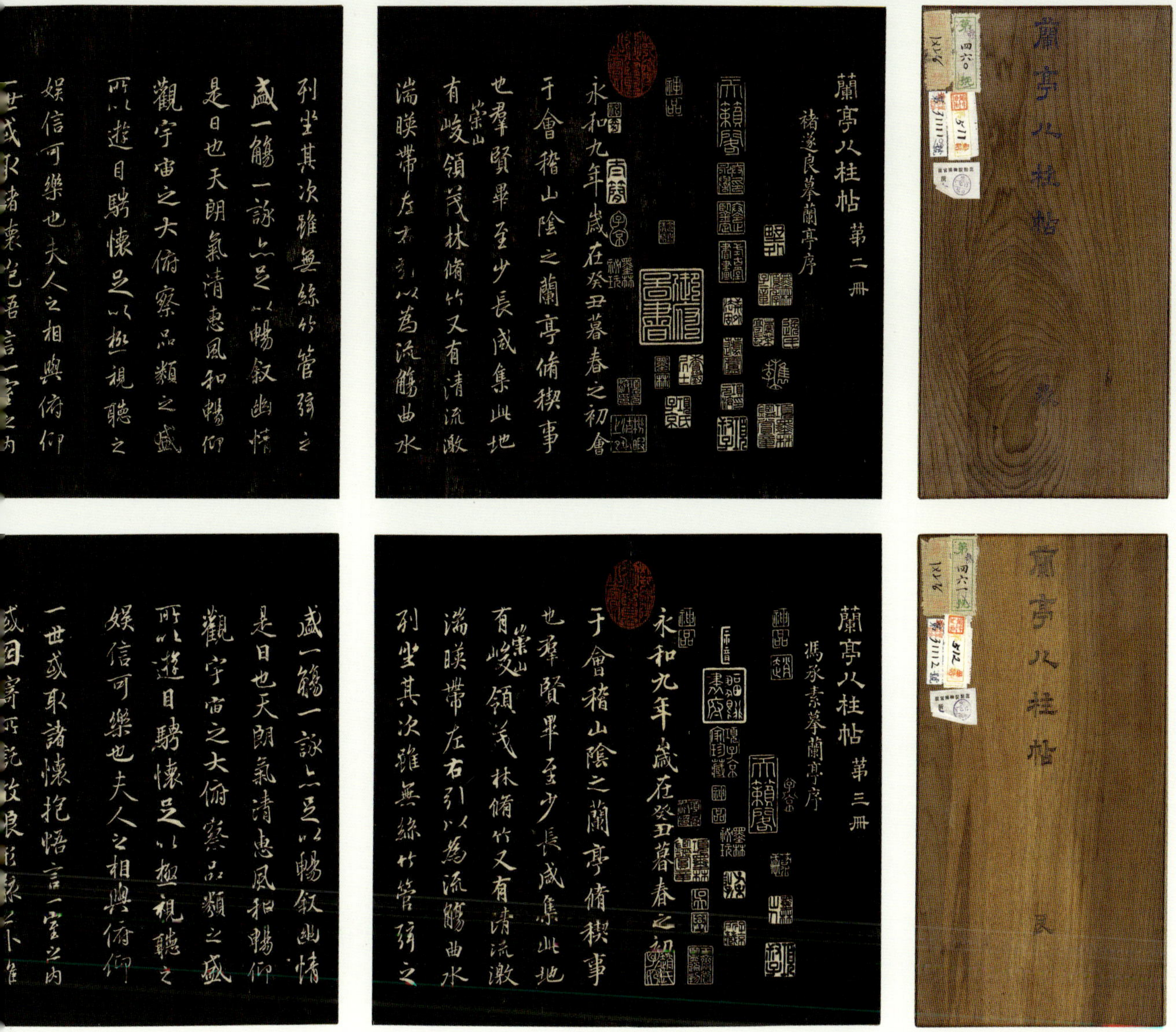

2 **Qing Rubbings of the *Eight Pillars of the Orchid Pavilion***

Qing Dynasty, Qianlong Reign (1736-1795)
Ink on paper, rubbing
L.:29.8cm, W.:34.2cm
Collection of the Palace Museum

山庄”印，此册曾旧藏于避暑山庄。

第三册《冯承素摹兰亭序》，系据《冯承素摹兰亭卷》（今藏故宫博物院）摹勒上石。宋人许将、王安礼、黄庆基、朱光裔、李之仪、李秬、王景通、张太宁、张保清、冯泽纵、王景修等，元人赵孟頫、郭天锡、鲜于枢、邓文原等，明人李廷相、项元汴、文嘉，并清乾隆帝等的题跋、观款和印章一并保留上石。木面，托裱经折装。钤“乾隆御览之宝”“避暑山庄”印，此册亦曾旧藏于避暑山庄。（黄爱民）

The Yongzheng Emperor had Liubei Pavilion built in the Old Summer Palace, modeled after the fashion of the ancient Orchid Pavilion in Shaoxing. The Qianlong Emperor later renovated the pavilion and engraved eight versions of the *Orchid Pavilion Preface* on the eight stone pillars. Rubbings of these engravings were compiled into the *Eight Pillars of the Orchid Pavilion*. During the Xianfeng reign, the Old Summer Palace was burned, but the eight pillars survived and are now preserved in Zhongshan Park. (Huang Aimin)

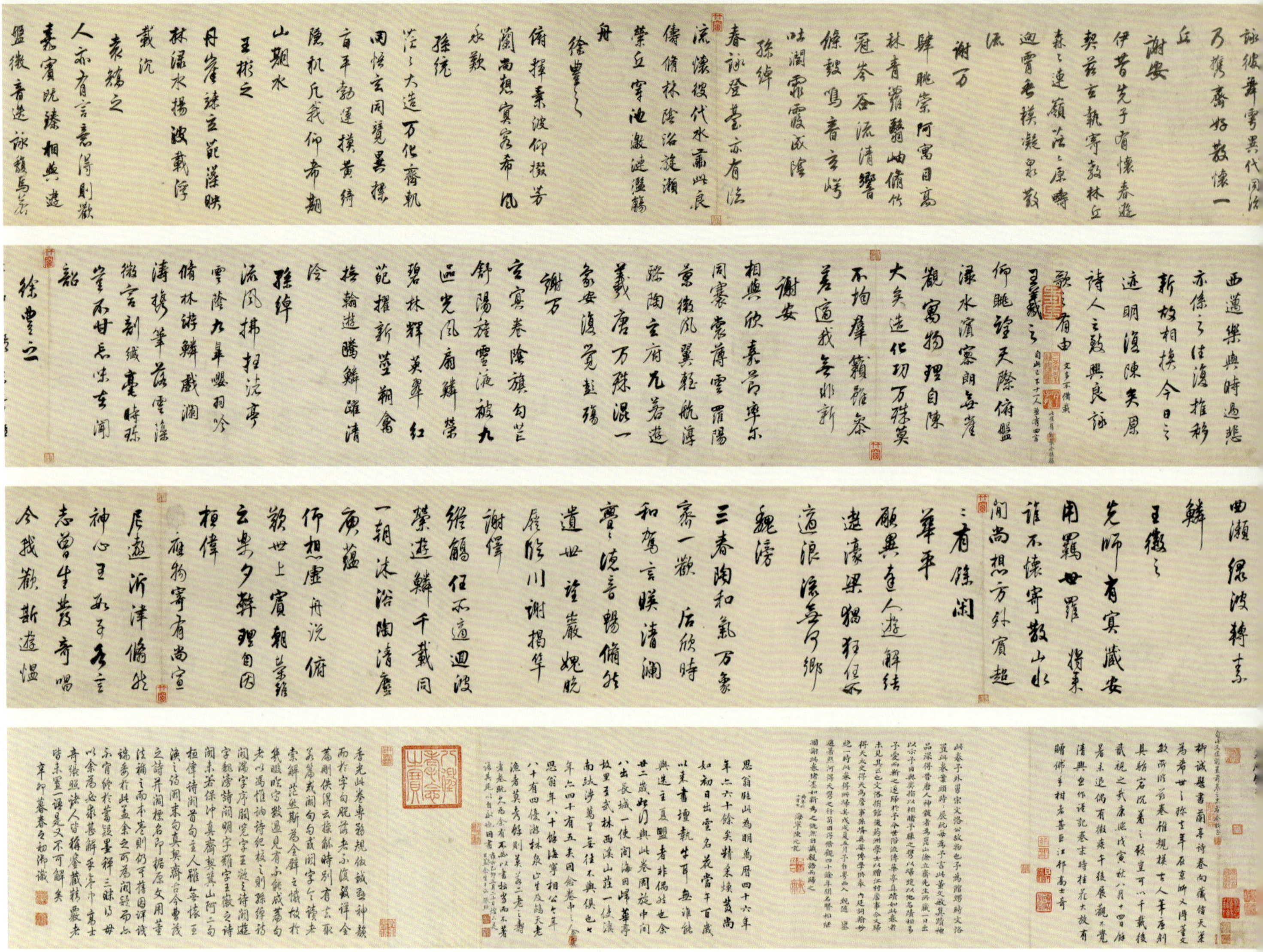

3 临柳兰亭诗卷

明万历四十六年（1618），董其昌

纸本，行书

纵 27.2 厘米，横 1070 厘米

故宫博物院藏

《临柳兰亭诗卷》为明代书法家董其昌作于明万历四十六年（1618）的行书名作，董氏时年 64 岁。诗卷内容为董其昌以自家笔法临习柳公权佳作《兰亭诗卷》。东晋永和九年（353）三月初三，书圣王羲之于会稽山阴之兰亭举办雅集，群贤毕至，赋诗吟咏。包括谢安、孙绰在内的文人雅士共作诗 30 余首，合称“兰亭诗”，名传千古。唐朝著名书家柳公权曾书录兰亭诗于长卷之上，明人董其昌复临，即本作《临柳兰亭诗卷》。

全卷书风出于己意，运笔潇洒，转折灵活，字里行间自然疏朗，顾盼有致，神完气足，展现了董氏的深厚功底和独特风格。另据卷中乾隆帝题跋，本作原被分割为两卷，陆续归入清内府后合裱一卷，得以破镜重圆，后刻入兰亭八柱。

董其昌（1555—1636），字玄宰，号思白，又号香光居士，华亭（今上海松江）人。万历十七年（1589）进士，官礼部尚书。董其昌工书善绘，慕古但不泥古。其书风拙中带秀，用墨苍润兼施、生动灵婉。绘画则平淡天真又意趣盎然，自成松江一派，主张“南北宗”论，对后世影响深远。（秦崇泰）

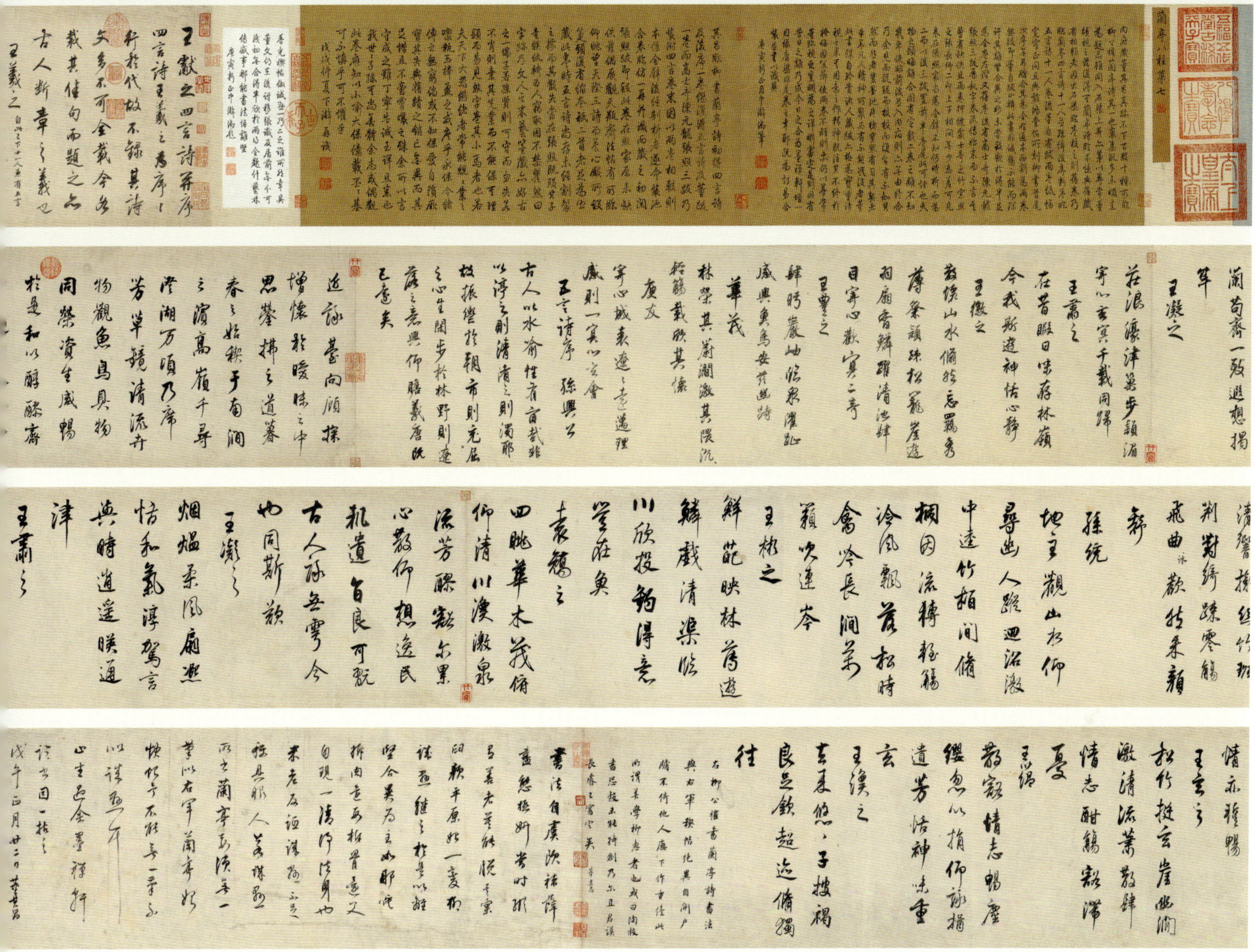

3 **After *Liu Gongquan's Orchid Pavilion Poems***

1618, Dong Qichang
Handscroll, paper, running script
L.:27.2cm, W.:1070cm
Collection of the Palace Museum

During the Eastern Jin Dynasty (317-420), the calligrapher Wang Xizhi held a gathering at the Orchid Pavilion in Shaoxing, where scholars composed the 'Orchid Pavilion Poems'. The Tang Dynasty calligrapher Liu Gongquan transcribed these poems onto a long scroll. At 64, Ming Dynasty calligrapher Dong Qichang copied this scroll, showcasing his unique style with graceful and expressive brushwork. It was subsequently engraved onto the eight pillars of the Orchid Pavilion. (Qin Chongtai)

4 描彩漆鱼纹耳杯

西汉（公元前 206—公元 8）
口径 17.5 厘米，底径 12 厘米，高 5.4 厘米
故宫博物院藏

4 **Polychrome Lacquered Ear Cup with Fish Pattern**

Western Han Dynasty (206 BCE-8 CE)
Diam.(at mouth):17.5cm, Diam.(at base):12cm, H.:5.4cm
Collection of the Palace Museum

5 描彩漆云纹耳杯

西汉（公元前 206—公元 8）
口径 16.2 厘米，底径 11.6 厘米，高 4.6 厘米
故宫博物院藏

5 **Polychrome Lacquered Ear Cup with Cloud Pattern**

Western Han Dynasty (206 BCE-8 CE)
Diam.(at mouth) :16.2cm, Diam.(at base):11.6cm, H.:4.6cm
Collection of the Palace Museum

6 描彩漆鸟纹耳杯

西汉（公元前 206—公元 8）
口径 14.5 厘米，底径 10.5 厘米，高 4.1 厘米
故宫博物院藏

6 Polychrome Lacquered Ear Cup with Bird Pattern

Western Han Dynasty (206 BCE-8 CE)
Diam.(at mouth) :14.5cm, Diam.(at base):10.5cm, H.:4.1cm
Collection of the Palace Museum

此三件耳杯均为厚木胎，椭圆形，弧壁，平底，新月式耳。一件杯内髹朱漆，彩绘一鱼，外髹赭漆。一件杯内髹朱漆，彩绘一鸟，外髹赭漆。一件杯内外髹赭漆，通身绘云气纹。1992 年出土于湖北荆州高台 28 号汉墓。

耳杯是战国至南北朝最常见的饮食器，以漆木材质居多。按汉代叫法，小者名“杯”，大者名“閜”（xiǎ）。此外还有“羽觞”的雅称。常用以饮酒，也能盛羹汤或其他食物。战国早中期流行蝶式耳，以后流行新月式耳。

曲水流觞，指中国古代园林中一种形式相对固定的景观，也可指称在上述景观中举办的文人雅集活动。规则是利用自然或人工的弯曲水道，在水中放置酒杯，令其漂流，酒杯停滞处，旁边的人便要饮酒赋诗。

已知最早设有“曲水流觞”这类景观的园林，是曹魏明帝在洛阳城建设的华林园。园中有“流杯沟”，用以宴饮群臣。此后，在历代宫廷苑囿和私家园林中，多有设置。北宋官方颁布的《营造法式》规定了“流杯渠”的做法，推动园林中的“曲水流觞”走向定型。

曲水流觞源自古代上巳节期间的娱乐活动。先秦已有水边祓禊的仪式，即通过沐浴祛除灾病。东汉形成了于三月上巳祓禊的习俗，并逐渐加入了郊游、宴饮等活动。上巳，即第一个巳日。曹魏以后，固定在三月三日这天。唐代成为官方节日。

东晋永和九年（353）三月初三，王羲之、谢安、孙绰、支遁等 30 余人在会稽山阴兰亭雅集，曲水流觞，赋诗唱和，并留下了书法史上的不朽之作《兰亭序》，成为千古佳话。当时使用的“流觞”，应当就是这种漆木耳杯。（展梦夏）

The ear cup, typically made of lacquered wood, was one of the most common dining vessels from the Warring States period to the Northern and Southern Dynasties. Also known as *yushang*, it was primarily used for drinking wine but could also hold soup or other foods. It was also the cup used in the literati game 'floating wine cups on a meandering stream'. Channels made for this game first appeared in Chinese gardens in the early 3rd century. This tradition continued in later imperial gardens and private estates. (Zhan Mengxia)

7 青釉刻花三足樽

东汉（25—220）
口径 18.5 厘米，底径 18 厘米，高 15 厘米
故宫博物院藏

樽直口，身为圆柱形筒状，腹部中间上刻有三道弦纹，下刻有内双短竖线网格纹装饰带，纹饰中心贴饰有对称的面具形铺首衔环。平底，樽底三足均铸人形，磨损较严重。外壁施青釉，釉面大部分已剥落失去光泽，内壁及底部足端无釉。

樽是古代盛酒之器，《玉篇·木部》中记载"樽，酒也"。在汉代，樽是较为常见的日常生活盛酒器，在祭祀和宴飨中起着重要作用，在墓葬中也是常用的随葬品。在 1992 年内蒙古自治区鄂托克旗凤凰山 1 号墓出土的东汉《庭院宴饮百戏图》中，描绘有在单进庭院内人们饮酒的画面，一般是席地而坐，酒樽置于席地中间，有的在下面配备托盘。饮酒器具耳杯也置于地上，用酒勺从樽中挹酒，倒入耳杯分于众人饮用。根据宴席的需要，樽有不同容量区别，成套的酒器一般由杯、勺、盘、案组成。（陈志鸿）

7 Celadon Three-legged *Zun* Vessel with Incised Pattern

Eastern Han Dynasty (25-220)
Diam.(at mouth):18.5cm, Diam.(at base):18cm, H.:15cm
Collection of the Palace Museum

A *zun* was an ancient vessel used for holding wine. During the Han Dynasty, it was a common household item. (Chen Zhihong)

8 青釉弦纹双系盘口壶

西晋（265—316）

口径 11.6 厘米，底径 9.5 厘米，高 20.3 厘米

故宫博物院藏

壶盘口，束颈，溜肩，鼓腹，下腹部渐收，平底。整体施青釉，外部施釉至下腹部，露灰色胎体，底部留有七处支烧痕。此类盘口双系壶流行于汉代至魏晋时期，可能用来盛水、盛酒。（董彦卿）

8 Celadon Double-handled Jar with Dish-shaped Mouth and String Pattern

West Jin Dynasty (265-316)

Diam.(at mouth):11.6cm, Diam.(at base):9.5cm, H.:20.3cm

Collection of the Palace Museum

This type of double-handled jar with dish-shaped mouth was popular from the Han Dynasty to the Wei and Jin periods (around the 2nd century BCE to early 5th century CE) and was likely used for holding water or wine. (Dong Yanqing)

意大利庞贝狄俄梅德别墅理想重建图

古罗马宴饮离不开以金、银、玻璃、玛瑙等珍贵材料制成的华丽酒具和丰盛的美食。主人精心布置，宾客开怀畅饮，尽享狂欢盛宴。宴饮常在别墅庭院中举办，音乐、舞蹈、戏剧等娱乐活动穿插其间，而花园中的雕塑、喷泉与植物则成为天然的布景。这样的活动不仅是社交与娱乐的场合，更是权力与财富的象征，充分展现出古罗马时期物质生活的丰盈与艺术文化的繁荣。（丁惠）

Ideal reconstruction of Villa of Diomedes in Pompeii to its original state before volcano eruption of Mount Vesuvius in 79 CE, Campania, Italy

Banquets in ancient Rome were lavish affairs, featuring luxurious drinking vessels made of gold, silver, glass, and agate, among other precious materials, alongside an abundance of exquisite food. Hosts meticulously arranged the setting, while guests indulged in merriment, revelling in a grand feast. These banquets were often held in villa courtyards, accompanied by music, dance, and theatrical performances. Sculptures, fountains, and lush greenery in the gardens served as natural backdrops. More than just social and entertainment gatherings, these events symbolized power and wealth, reflecting the opulence of material life and the flourishing artistic and cultural achievements of ancient Rome. (Ding Hui)

9 庞贝城阿克泰翁府邸的躺卧式宴厅*

约 1824，[法国] 查尔斯·弗雷德里克·夏塞里奥
钢笔和黑色墨水，水彩
纵 14.2 厘米，横 13.8 厘米
大都会艺术博物馆藏

画面复原了罗马庞贝城内阿克泰翁宅邸庭院一隅。躺卧式宴厅（Triclinium）是古罗马住宅中用于正式宴饮和社交活动的餐厅，通常配有躺卧长椅和餐桌，形成“U”形布局。宾客按社会地位高低依次侧卧于长椅上，以左肘支撑，右手取食。画面左侧柱廊是罗马建筑的标志性要素，右侧墙壁装饰有精美的壁画，描绘喷泉、植物，与院内真实的自然景物共同营造出优雅的宴饮环境。（许愿）

9 Triclinium, Excavated in the House of Actaeon, Pompeii

Ca. 1824, Charles Frédéric Chassériau (French)
Pen and black ink, watercolor
L.:14.2cm, W.:13.8cm
Lent by The Metropolitan Museum of Art, Bequest of Harry G. Sperling, 1971 (1975.131.95)

The *Triclinium* was a formal dining space in Roman homes, featuring couches and a dining table, arranged in the U-shape. Guests reclined on the couches, eating with their right hand while leaning on their left elbow. The colonnade on the left is a characteristic of Roman architecture, while the right wall is decorated with frescoes of fountains and plants, enhancing the elegant dining atmosphere. (Xu Yuan)

* 本次展览全部外方参展文物，其名称、时代、材质、工艺以及收藏和捐赠情况等信息，均依参展单位提供的原始数据处理，特此注明。

10 双耳深酒杯

公元前 410—前 400，古希腊，雅典
赤陶，黑釉压花
长 22.5 厘米，宽 16.2 厘米，高 9.2 厘米
芝加哥艺术博物馆藏

此件“双耳深酒杯”（Skyphos）是一种常见于宴会的酒器，用于盛酒或饮酒，其独特的杯形设计便于饮用者轻松握持。公元前 5 至前 4 世纪，希腊及意大利南部地区黑釉陶器制作兴盛，其中许多陶器在造型上模仿了金属器皿。（丁惠）

10 Skyphos (Drinking Cup)

410-400 BCE, Greek, Athens
Terracotta, black-glaze with impressed decoration
L.:22.5cm, W.:16.2cm, H.:9.2cm
Collection of the Art Institute of Chicago, Gift of Philip D. Armour and Charles L. Hutchinson (1889.14)

The *skyphos* (two-handled deep drinking cup) was a common banquet vessel used for holding and drinking wine in ancient Greece. Its distinctive shape enabled easy handling. During the 5th to 4th centuries BCE, black-glazed pottery flourished in Greece and southern Italy. Many of these ceramic vessels imitated the forms of metal ware. (Ding Hui)

11 玻璃壶

1 世纪末—3 世纪初，古罗马
玻璃，吹制
口径 8 厘米，底径 7.1 厘米，高 15 厘米
芝加哥艺术博物馆藏

壶身以吹制工艺成型，常置于宴席中盛放酒水或油。瓶身呈椭圆形，琥珀色玻璃透光温润，搭配优雅弯曲的手柄装饰。公元 1 世纪末，玻璃吹制技术日趋成熟，玻璃器皿成为古罗马人日常生活中的常见用品。（丁惠）

11 Jug

Late 1st-early 3rd century, Roman
Glass, blown technique
Diam.(at mouth):8cm, Diam.(at base):7.1cm, H.:15cm
Collection of the Art Institute of Chicago, Gift of Theodore W. and Frances S. Robinson (1949.1132)

This vessel was shaped using glassblowing and was commonly placed at banquets to hold wine or oil. Its elliptical body, with a translucent amber hue and smooth, warm texture, is complemented by elegantly curved handles. By the late 1st century, advancements in glassblowing led to the widespread use of glassware, making it a common household item in Roman daily life. (Ding Hui)

12 碗或杯（上）

1 世纪中期，古罗马，可能来自意大利
玻璃，铸造
口径 7.5 厘米，底径 4.1 厘米，高 7.5 厘米
芝加哥艺术博物馆藏

13 碗或杯（下）

1 世纪末—2 世纪初，古罗马
玻璃，吹制
口径 7.7 厘米，底径 4.5 厘米，高 7.7 厘米
芝加哥艺术博物馆藏

12 Bowl or Cup (up)

Mid-1st century, Roman, probably Italy
Glass, cast
Diam.(at mouth):7.5cm, Diam.(at base):4.1cm, H.:7.5cm
Collection of the Art Institute of Chicago, Gift of Theodore W. and Frances S. Robinson (1947.890)

13 Bowl or Cup (down)

Late 1st-early 2nd century, Roman
Glass, blown technique
Diam.(at mouth):7.7cm, Diam.(at base):4.5cm, H.:7.7cm
Collection of the Art Institute of Chicago, Gift of Theodore W. and Frances S. Robinson (1947.894)

此二件为用于盛放饮料与食物的玻璃杯。杯口边缘饰有突出的环形装饰，使杯形更具立体感与艺术性，杯身通体散发深邃的钴蓝色光泽。（丁惠）

This glass cup was used for holding beverages and food. The rim is adorned with a raised circular decoration, enhancing its three-dimensional form and artistic appeal. The body features a deep cobalt blue luster. (Ding Hui)

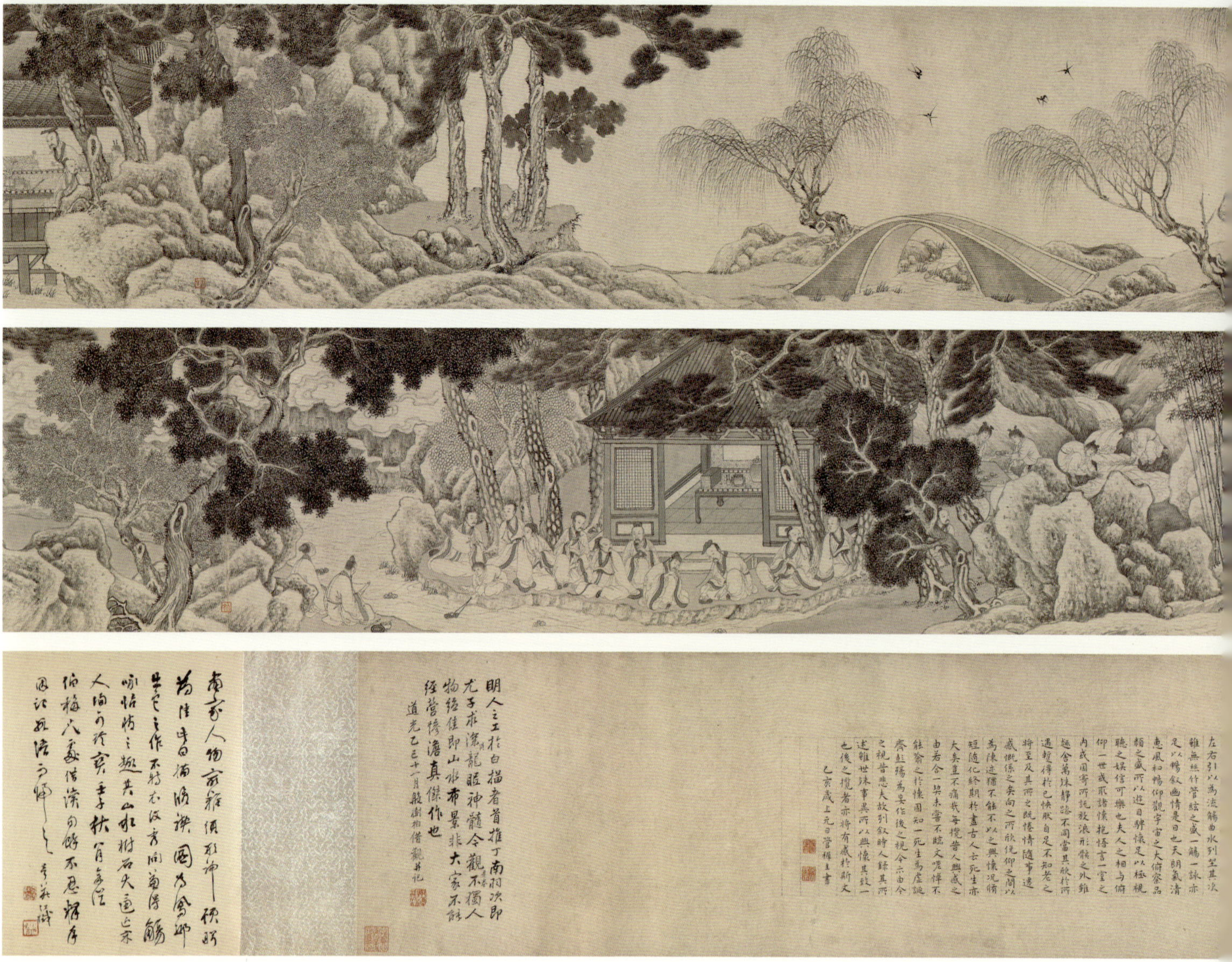

14 修禊图卷

明（1368—1644），尤求

纸本，墨笔

纵 31.1 厘米，横 520.5 厘米

故宫博物院藏

该图以岩泉间的拱桥为起点，周边的飞燕与柳树展现出一派春景，虬松翠竹点缀于园囿间。亭中坐四文士，一人展卷，一人执羽扇，一人凭栏远望，另一人回首与童子交流。河流两岸，26 位文士或坐或站，他们或赋诗，或饮酒，或沉思，或交谈，各具情态。卷尾另有两位文士正从桥上走来，身后跟着两位童子，一位背负卷轴，另一位怀抱书册。图中山石以枯墨点染皴擦，笔法松动灵秀，树木注重表现其质感。人物用白描，笔法简洁。

该卷引首有周天球书“修禊图，万历元年上巳周天球书”。后幅有管稺圭书《兰亭序》，另有壁城、殷树柏、江械题跋三段。尤求，活跃于明隆庆、万历年间，字子求，号凤丘，长洲（今江苏苏州）人，后移居太仓。工山水、人物，尤擅白描。（胥瑞頔）

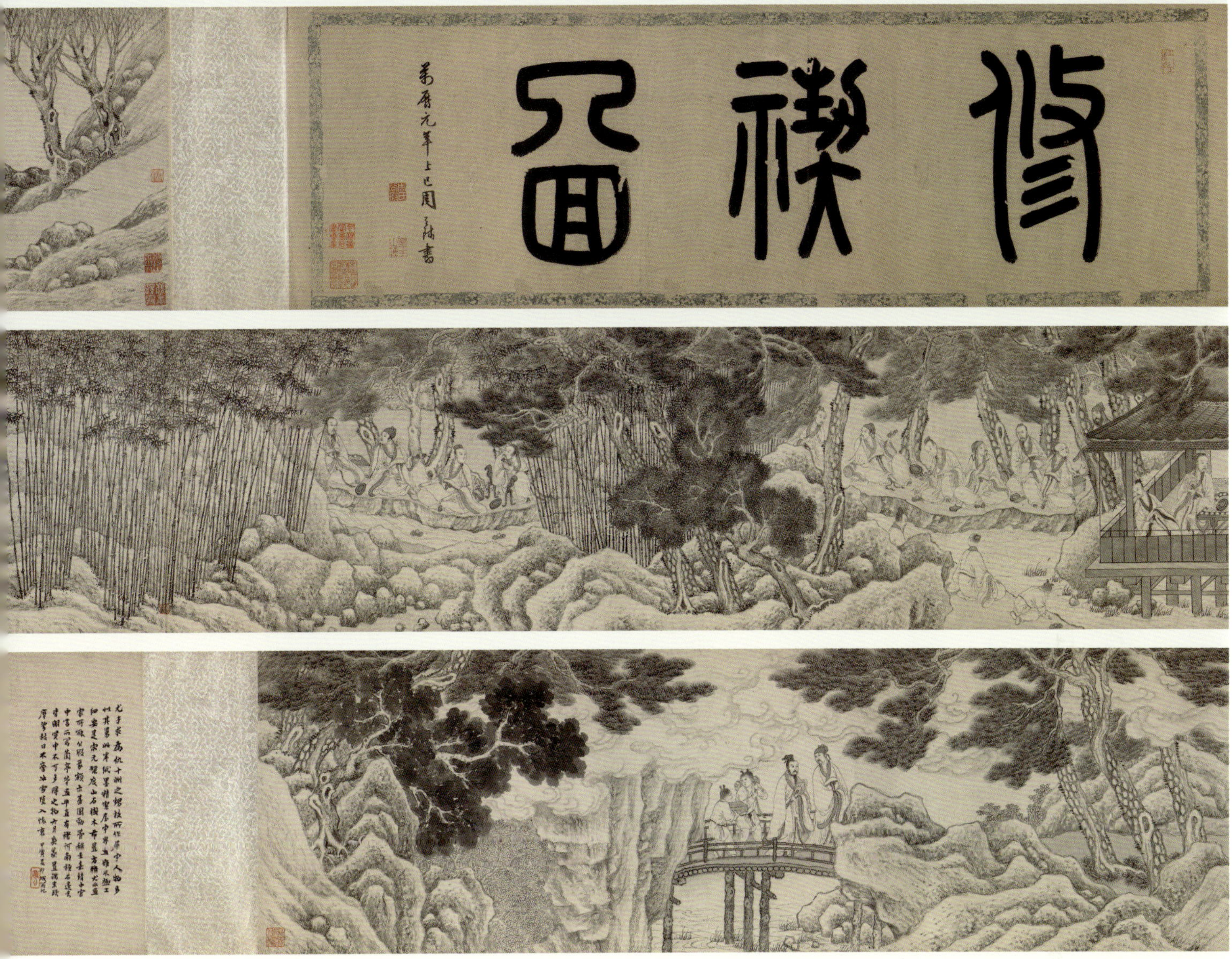

14 ***Xiuxi***

Ming Dynasty (1368-1644), You Qiu
Handscroll, ink on paper
L.:31.1cm, W.:520.5cm
Collection of the Palace Museum

Xiuxi was the annual purification ritual held on the third day of the third lunar month, when people bathed by the water to ward off misfortune. Wang Xizhi's *Orchid Pavilion Preface* records a gathering of scholars in 353 CE at the Orchid, where they floated wine cups along a winding stream. This painting depicts a spring scene in the mountains with scholars in various poses. The rocks are rendered with dry ink and the figures are drawn with simple, fine lines. (Xu Ruidi)

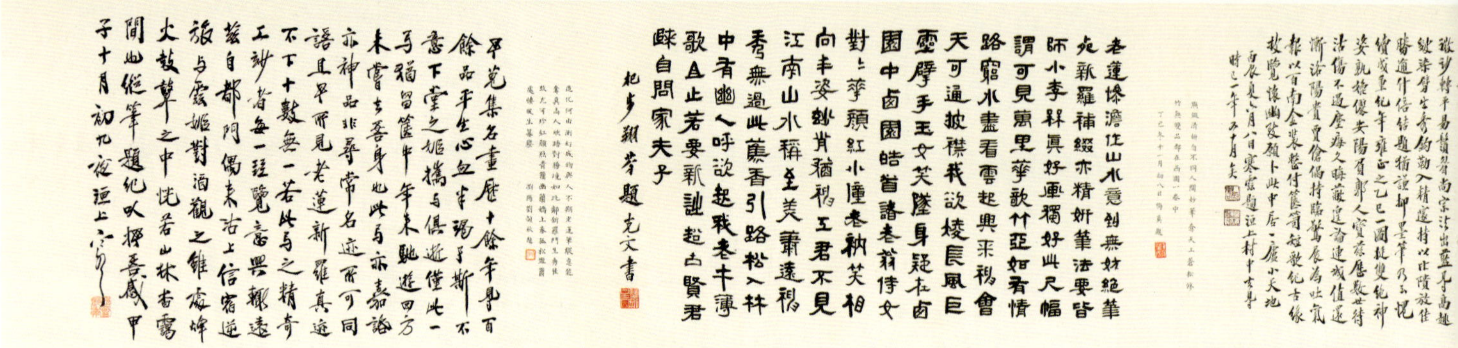

15　西园雅集图卷

明清（1368—1911），陈洪绶、华嵒

绢本，设色

纵 41.6 厘米，横 431.5 厘米

故宫博物院藏

“西园雅集”是古代画家们经常表现的题材，指的是苏轼、米芾、黄庭坚等人在驸马都尉王诜府中西园聚会之事。实际上，这一聚会从未真实发生过，其中的人员名单大多是反王安石变法的元祐党人，故对这一事件的创造很可能源于宋室南渡后贬低新法、推崇元祐政治的倾向。而传为米芾的《西园雅集图记》更有可能晚至 16 世纪才出现。但是，这一题材代表了古代文人吟咏酬唱的理想典范，因此成为被反复描摹的主题。

此图为陈洪绶及华嵒两位画家合作完成。据华嵒署款，陈洪绶完成开头至孤松的部分后，已无法继续作画，后段由华嵒补完。此画构图繁复，树石几将画面填满，山石多以虬曲弯折的墨线勾勒，辅以淡墨皴擦渲染，树叶则多为双钩，加以点染，描绘出园中绿植盎然、雾气氤氲的环境。画中以树木分隔空间，前后依次描绘了听阮、作画、观书、题壁、讲经五个场景。人物多头大身小，面部方阔，眼角上扬，造型奇崛。华嵒人物本学陈洪绶，此画中亦尽力模仿其风格。整幅画和谐统一，并无突兀之感。此图尾纸后有陈洪绶信札一通，袁克文跋文四段，刘梅真、刘韵秋跋文各一段。有“静者心多妙”“寒云子子孙孙永保”“雪夫真赏”“寒云鉴赏之珍”等鉴藏印多方。

陈洪绶（1598—1652），字章侯，号老莲，明亡后曾出家为僧，号老迟、悔迟，浙江诸暨人，晚明著名画家。其绘画技能多样，尤擅人物，笔下的人物形象常头大身小，宽袍广袖，比例与常规不合，被称为变形主义画风。华嵒（1682—1756），字秋岳，号新罗山人、东园生、布衣生等，福建上杭人，侨居杭州，后长期客居扬州，晚年归杭。《国朝画征录》评价其“善人物、山水、花鸟、草虫，皆能脱去时习而力追古法，不求妍媚，诚为近日空谷之音”。（毛翔宇）

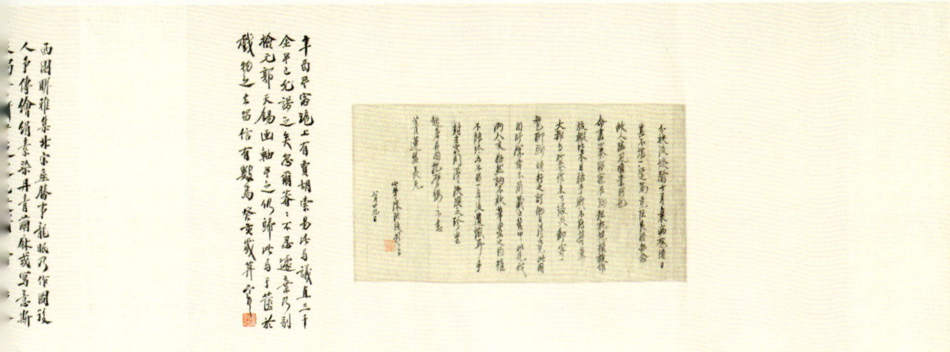

15 ***Elegant Gathering at the Western Garden***

Ming and Qing Dynasty (1368-1911), Chen Hongshou and Hua Yan
Handscroll, ink and color on silk
L.:41.6cm, W.:431.5cm
Collection of the Palace Museum

The Elegant Gathering at the Western Garden was an imaginary gathering of great scholars such as Su Shi, Mi Fu, and Huang Tingjian at the Western Garden of Prince Consort Wang Shen's estate. Since this represented an ideal image of ancient scholars engaging in poetry and music-making, it was frequently copied. An incomplete work by Chen Hongshou, this painting was later finished by Hua Yan over seventy years later. (Mao Xiangyu)

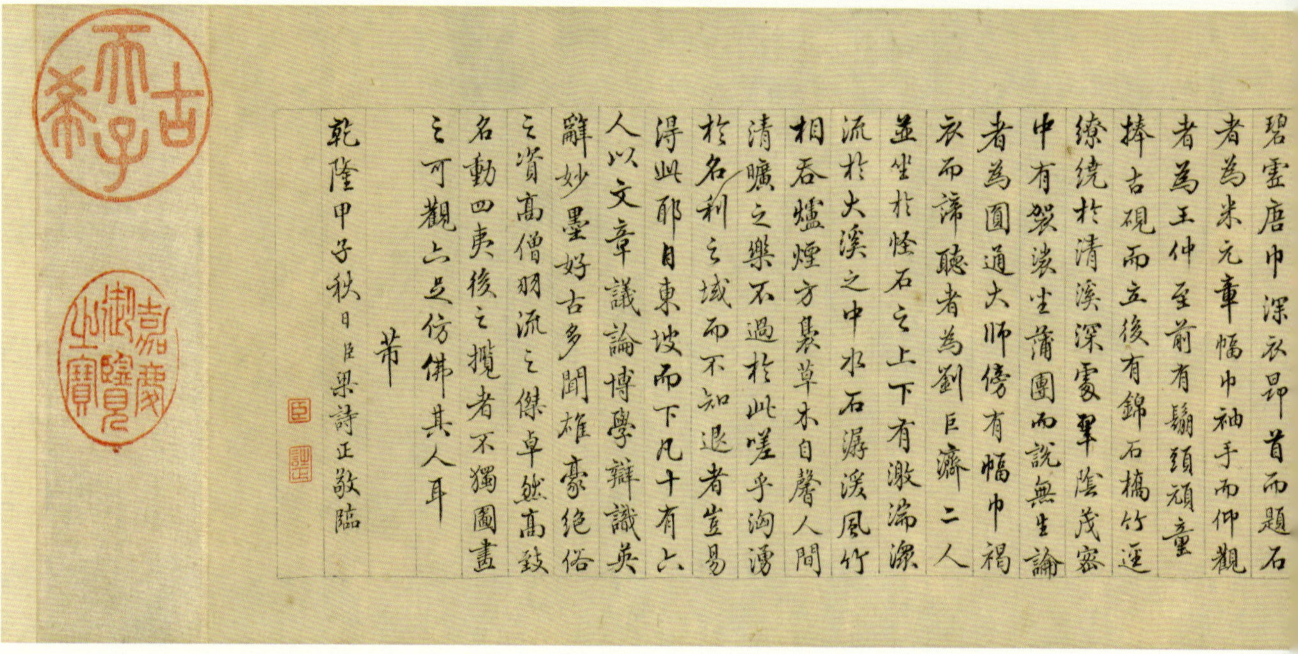

16 临米芾西园雅集图记卷

清乾隆九年（1744），梁诗正
纸本，行书
纵 15 厘米，横 59 厘米
故宫博物院藏

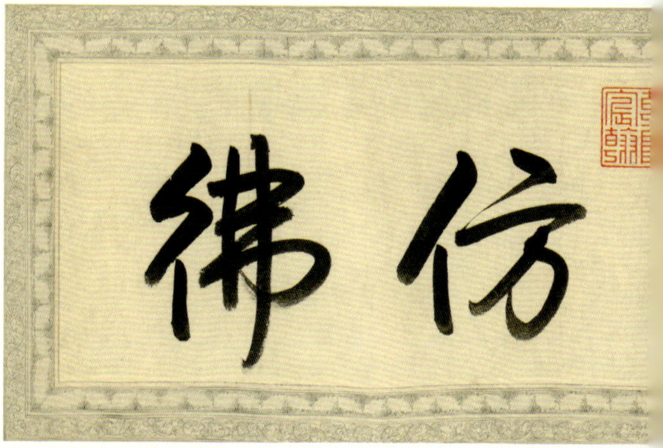

北宋驸马都尉王诜邀请苏轼、苏辙、黄庭坚、秦观、米芾、李公麟、蔡肇、刘泾、晁补之等 16 位文人名士在其府邸西园游园雅集，众人或提笔挥毫、遣兴翰墨，或品茗怡神、聚坐谈禅，或欣然拨阮、以酬知音。李公麟将此次盛集描绘成卷，米芾则作《西园雅集图记》，将人物姓名、衣冠、坐卧神态及园林环境一一记录。中国古典园林为雅集中的文士营造了清幽的环境和赏心悦目的美景，激发着他们的才思和灵感，而文士们又为园林赋予了人文意义和艺术内涵，二者互为依托、互美其美，共同造就了园林史上一次又一次的经典。

梁诗正（1697—1763），字养仲，号芗林，浙江钱塘（今杭州）人。雍正八年（1730）探花，官至东阁大学士，掌翰林院学士。多次参与修纂官书，主修《钦定叶韵汇辑》《西清古鉴》，编撰《秘殿珠林初编》、《石渠宝笈初编》等书，谥文庄。常随乾隆帝出巡，凡重要文稿，多出其手。工诗文，曾恭和乾隆帝诗作，集《矢音集》五卷。善书，初学柳公权，继参赵孟頫、文徵明，晚师颜真卿、李邕，书风媚丽婉劲。（郝炎峰）

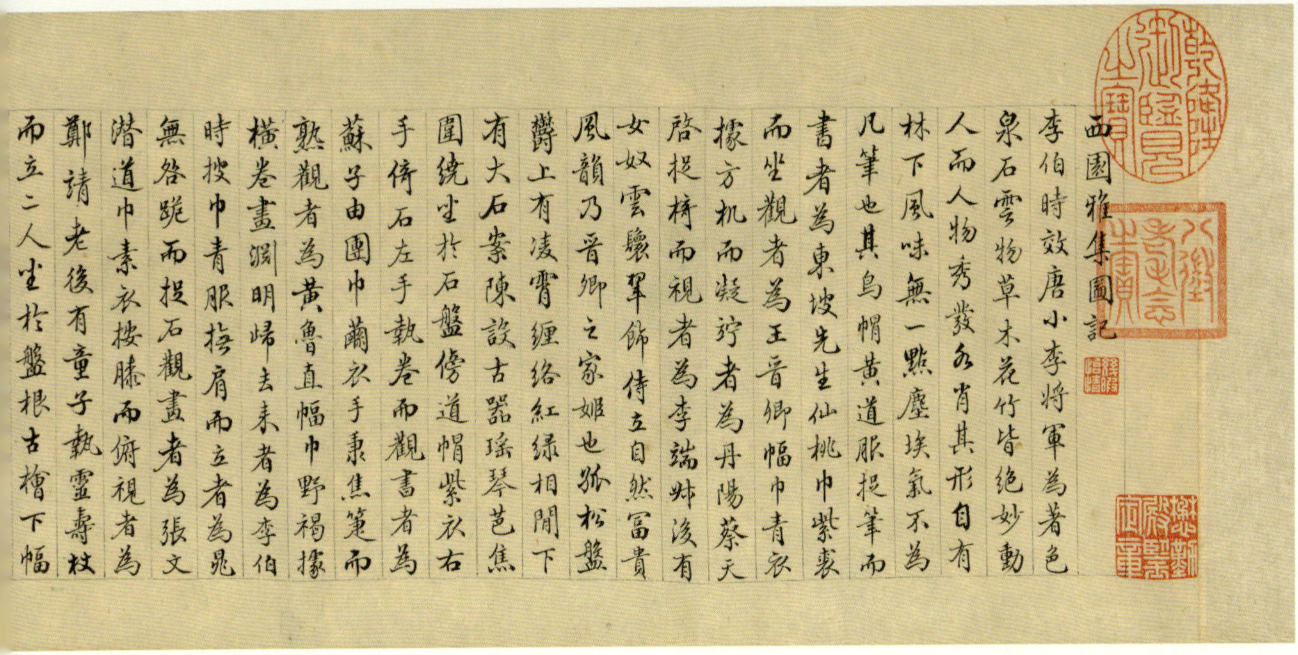

16 **After Mi Fu's *Record of the Elegant Gathering at the Western Garden***

1744, Liang Shizheng
Handscroll, paper, running script
L.:15cm, W.:59cm
Collection of the Palace Museum

The *Record of the Elegant Gathering at the Western Garden*, attributed to Song calligrapher Mi Fu, meticulously documents the individuals' names, attire, postures, expressions of the individuals, and the garden's setting. This scroll is a copy by Qing calligrapher Liang Shizheng. The brushwork is delicate and firm, with slender and vigorous strokes in a relaxed composition that reflects Mi Fu's style. (Hao Yanfeng)

17 东园图卷

明（1368—1644），文徵明
绢本，设色
纵 30.2 厘米，横 126.4 厘米
故宫博物院藏

此图为明代著名文人画家文徵明所绘，用笔秀逸，赋色淡雅。图中绘亭台楼榭，竹树湖石。游人或携琴漫步，或群聚赏文，或对坐弈棋，展现出明代文人徜徉于园林之中的闲情逸致。

款署"嘉靖庚寅秋，徵明制"，卷首书"东园记"三字图名。引首为明代书法家徐霖书"东园雅集"。卷后拖尾有湛若水《东园记》以及陈沂《太府园宴游记》并七言律诗一首，后系短跋，介绍相关创作背景。由此可知，此图描绘的是明代南京地区重要的私家园林东园。

据文献记载，东园本为明太祖赏赐开国功臣中山王徐达宅邸旁的一处小园，后由其后裔在明代中期加以翻新扩建，成为规模宏大的园林。因为与南京当时的另一座著名园林瞻园东西相望，因而称为东园。由于东园主人的好客，使其逐渐成为了当时南京地区重要的名胜，引得大量文士流连其间。正德年间，明武宗也曾来此游览。东园后来几经兴废，现为南京白鹭洲公园。（郁文韬）

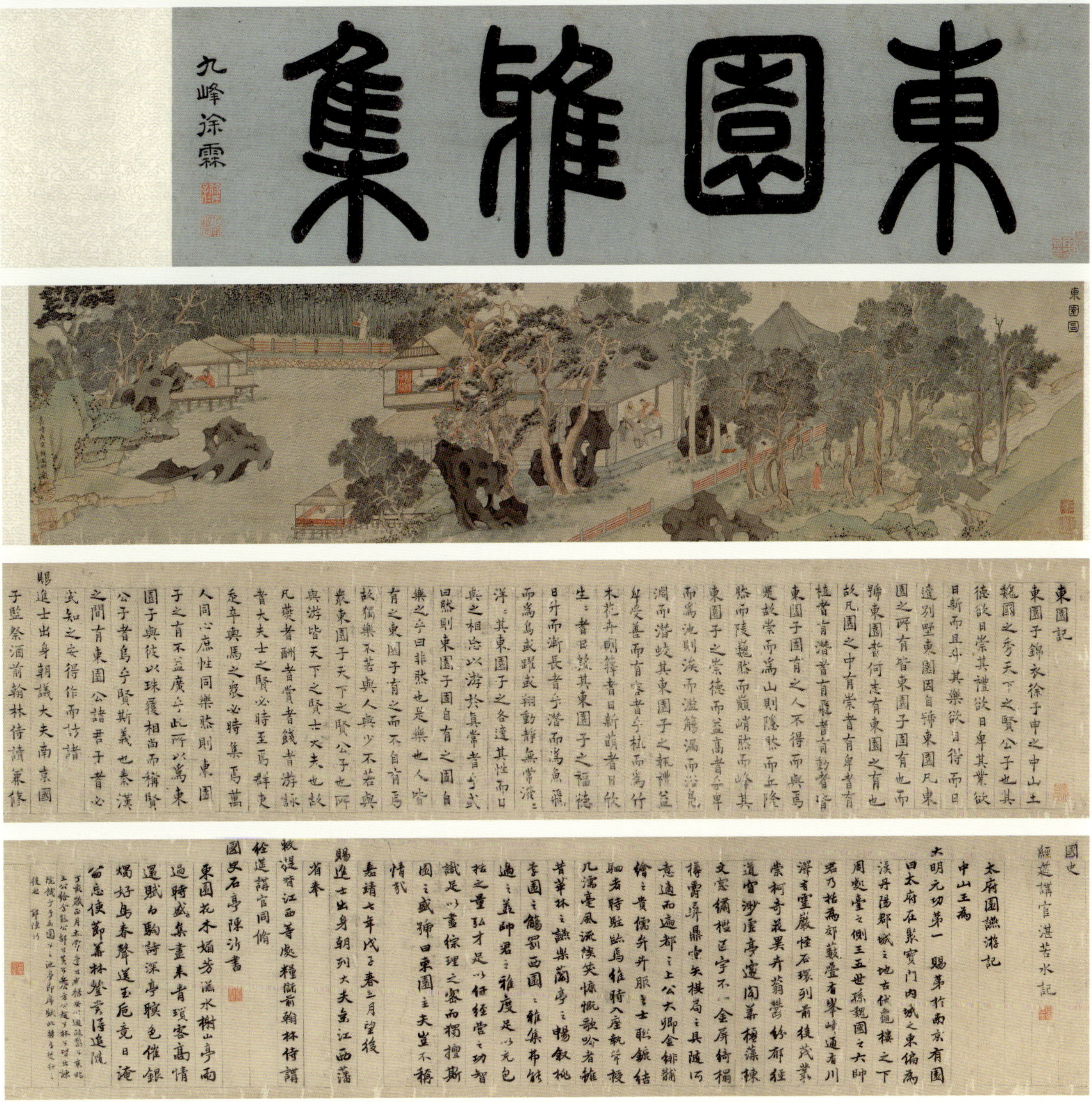

17 ***Eastern Garden***

Ming Dynasty (1368-1644) , Wen Zhengming
Handscroll, ink and color on silk
L.:30.2cm, W.:126.4cm
Collection of the Palace Museum

This painting depicts the Eastern Garden, an important private garden in Nanjing during the Ming Dynasty. The garden originally belonged to Xu Da, one of the dynasty's founding generals. The property was granted to Xu by the dyansty's first emperor, and was located next to Xu's residence. In the mid-Ming period, the garden was renovated and expanded by Xu Da's descendants. It was named the Eastern Garden because it faced the famous Zhan Garden to the west. (Yu Wentao)

波斯花园

11至15世纪，波斯文化迎来了黄金时期，花园作为尘世中"天堂"的缩影，成为绘画与文学创作的重要灵感来源。书籍插图的细密画，常描绘宴饮、奏乐、诗歌朗诵、爱情等题材，其场景多设于美丽的花园中。被誉为"波斯文坛四根柱石"的著名诗人萨迪、哈菲兹、莫拉维和菲尔多西都在诗歌中借花鸟、山水、静夜等花园意象，抒发对爱情、美与永恒的追求。（曹旭卿）

Persian Gardens

From the 11th to the 15th century, Persian culture experienced a golden age. Gardens, regarded in Islamic tradition as earthly reflections of paradise, became a significant source of inspiration for both art and literature. Persian miniature paintings in book illustrations frequently depicted scenes of banquets, music, poetry recitations, and romance, often set within beautiful garden landscapes. The renowned poets Saadi, Hafez, Rumi, and Ferdowsi–known as the 'Four Pillars of Persian Literature'–drew upon garden imagery, such as flowers and birds, mountains and rivers, and tranquil nights, to express their longing for love, beauty, and eternity. (Cao Xuqing)

18 萨迪诗集手抄本

17世纪，萨法维王朝时期，伊朗
封面：漆革
手稿：纸本，墨水、不透明水彩和金箔
纵26.1厘米，横16.7厘米
芝加哥艺术博物馆藏

萨迪（Saadi Shirazi，1210—1291），被誉为波斯古典文学最伟大的人物，其成名作《果园》(*Bustan*) 和《真境花园》(*Gulistan*) 均收录在此手抄本中。封面描绘众人在花园中载歌载舞的聚会场景，展现了萨迪作品对大自然的无限热爱与对美好生活的热切向往。（曹旭卿）

18 Manuscript of *Kulliyat* (Complete Works) by Saadi with Lacquered Cover

Safavid Dynasty (1501-1736), 17th century, Iran
Cover: lacquered leather
Manuscript: ink, opaque watercolor and gold on paper
L.:26.1cm, W.:16.7cm
Collection of the Art Institute of Chicago, Gift of Charles Zadok (1951.236)

Saadi Shirazi (1210-1291), is regarded as 'the greatest figure in classical Persian literature'. This manuscript includes his renowned works *Bustan* and *Gulistan*. The cover illustration depicts a lively gathering in a garden, where people sing and dance, reflecting Saadi's deep admiration for nature and his yearning for a joyful and fulfilling life. (Cao Xuqing)

19 **萨迪诗集内页插图**

17 世纪，萨法维王朝时期，伊朗
纸本，墨水、不透明水彩
纵 25.9 厘米，横 16 厘米
芝加哥艺术博物馆藏

此插图来自《萨迪诗集》手抄本的 Ghazaliyyat 部分，包含萨迪的爱情诗篇。左侧插图描绘了天堂花园中的宴会场景：一对皇室夫妇一边饮酒，一边欣赏音乐和舞蹈表演。乐师们演奏着“Chang”（波斯语：چنگ）和鼓。“Chang”被认为是最古老的波斯乐器，约于东汉时期经丝绸之路传入中国，称为竖箜篌。右侧插图描绘苏丹在帐篷旁接待来访者，画中花开繁盛，林木斑斓。（曹旭卿）

19 **Page from a Manuscript of *Kulliyat* (Complete Works) by Saadi**

Safavid Dynasty (1501-1736), 17th century, Iran
Ink and opaque watercolor on paper
L.:25.9cm, W.:16cm
Collection of the Art Institute of Chicago, Gift of Charles Zadok, Milwaukee (1951.236a, 1951.236b)

These illustrations come from the 'Ghazaliyyat' section of the manuscript of *Kulliyat* (Complete Works) by Saadi, which includes Saadi's love poems. In the illustration on the left, which depicts a banquet in an Islamic paradise garden, musicians are playing the chang (Persian: چنگ) and drums. The illustration on the right depicts a sultan seated by a tent receiving visitors, while flowers and lush create a scenic backdrop. (Cao Xuqing)

本页波斯文译文：

且将季节的掠夺抛诸脑后，在玫瑰园中痛饮美酒。

因这玫瑰的芳华，撑不过七日之久。

20 哈菲兹诗集

约 1540—1550，萨法维王朝时期，伊朗
封面：皮革压花，涂金、黑色颜料
内页：纸本，墨水、不透明水彩和金箔
纵 21.4 厘米，横 13 厘米
芝加哥艺术博物馆藏

哈菲兹（Khwaja Shams ud Din Muhammed Hafez-e，1315—1390）是 14 世纪波斯伟大的抒情诗人，其作品被誉为波斯抒情诗的高峰。他笔下的花园超越世俗之美，蕴含宗教与社会的多重隐喻。诗中咏叹春天、鲜花、美酒与爱情，其诗歌以精炼自然的语言和深邃丰富的修辞著称。本诗集即收录了哈菲兹的经典之作。插图为波斯细密画，描绘众人在花园中集会、奏乐、交谈和畅饮的欢乐场景，洋溢着轻松愉悦的氛围。（曹旭卿）

20 *Divan of Hafiz*

Safavid Dynasty (1501-1736), dated ca.1540-1550, Iran
Cover: bound in leather, tooled, applied gold and black pigment
Interior: ink, opaque watercolor and gold on paper
L.:21.4cm, W.:13cm
Collection of the Art Institute of Chicago, Gift of Mrs. Everett McNear (1998.202)

Hafiz (also known as Hafez, Khwaja Shams ud Din Muhammed Hafez-e, 1315-1390) was a prominent Persian lyric poet of the 14th century. The gardens in his poetry transcend worldly beauty, and are rich in religious and social symbolism. Renowned for their refined and natural language as well as profound and rich rhetoric, his verses celebrate spring, flowers, wine, and love. (Cao Xuqing)

21 花园景致

15 世纪中期，帖木儿王朝时期，伊朗，可能是设拉子
纸本，墨水、不透明水彩和金箔
纵 17.8 厘米，横 10.2 厘米
芝加哥艺术博物馆藏

这幅双页插图的画面，以右侧花树下畅饮的人物为中心展开，飞鸟落于枝头的自然意境与左侧插图中奏乐的场景相映成趣。造园离不开气候的影响，中东大部分地区炎热干燥，花木繁密多荫的庭园令人神往。因此，园林成为了细密画中的常见主题。（郭欣悦）

21 Garden Scene

Timurid Dynasty (ca. 1370-1507), mid-15th century, Iran
possibly Shiraz
Ink, opaque watercolor and gold on paper
L.:17.8cm, W.:10.2cm
Collection of the Art Institute of Chicago, Ann McNear Collection (1987.360.4)

This double-page frontispiece centers on a figure drinking under the blossoming tree, while birds perch among the branches, creating a natural ambiance. The scene complements the musical performance depicted on the left page. Garden design is influenced by climate; in the hot, arid regions of the Middle East, lush, shaded gardens have long been cherished, making them a frequent theme in miniature painting. (Guo Xinyue)

22 霍斯劳与席琳在花园中

出自尼扎米《五卷诗》
16 世纪末或 17 世纪初，萨法维王朝时期，伊朗
纸本，墨水、不透明水彩和金箔
纵 17 厘米，横 10 厘米
芝加哥艺术博物馆藏

霍斯劳与席琳（Khosrow and Shirin）的悲剧爱情故事源自著名诗人尼扎米·甘伽维（Nizami Ganjavi, 1141—1209）的创作。《五卷诗》（*Khamsa*）是一部包含五首叙事诗的诗集，每首诗中引人入胜的故事情节和优美的语言，为波斯细密画提供了丰富的背景。此幅画描绘了萨珊帝国的霍斯劳二世（590—628）和他的爱人亚美尼亚公主席琳在花园中聚会的场景。（雷浩）

22 Khosrow and Shirin in a Garden

A Scene from *the Khamsa of Nizami*
Safavid Dynasty (1501-1736), late 16th/early 17th century, Iran
Ink, opaque watercolor and gold on paper
L.:17cm, W.:10cm
Collection of the Art Institute of Chicago, Gift of Mrs. Joseph L. Valentine (1948.185)

The tragic love story of Khosrow and Shirin originates from the work of the renowned poet Nizami Ganjavi (1141-1209). *Khamsa* is a collection of narrative poems, each featuring captivating storytelling and elegant language, providing a rich backdrop for Persian miniature painting. This work depicts a garden gathering between Khosrow II (590-628) of the Sassanian Empire and his beloved, Shirin, the Armenian princess. (Lei Hao)

23 沙龙聚会

年代未知，[法国] 弗朗索瓦 · 尼古拉斯 · 巴泰勒米 · 德克沃维耶据 [瑞典] 尼古拉斯 · 拉夫伦森作品创作
蚀刻和雕刻版画
纵 41 厘米，横 50.2 厘米
大都会艺术博物馆藏

18 世纪法国巴黎贵族沙龙聚会通常在卧室或客厅举行，封闭且相对私密，缺乏与室外园林的互动，与东方宅园中建筑和园林的紧密结合形成鲜明对比。画中人物以女性为主，她们或围坐交流，或阅读书籍。背景中富丽堂皇的室内装饰、精致的家具，细致地展现了洛可可风格（Rococo）的华丽氛围。（丁惠）

23 Salon Gatherings

N.D. François Nicolas Barthélemy Dequevauviller (French), after Nicolas Lafrensen (Swedish)
Etching and engraving
L.:41cm, W.:50.2cm
Lent by The Metropolitan Museum of Art, Harris Brisbane Dick Fund, 1935 (35.100.17)

In 18th century France, salon gatherings were usually held in living rooms or bedrooms, creating a private and enclosed atmosphere, and contrasting sharply with the integration of architecture and gardens in Eastern estates. The painting mainly depicts women, either engaged in conversation or reading. The detailed furniture and interior décor in the background reflect the luxurious Rococo style. (Ding Hui)

24 蓬巴杜夫人接见画家布歇

出自《世界画报》
1861 年 6 月 29 日，[英国] 亨利 · 林顿据 [法国] 爱德华 · 热拉尔作品创作
木刻版画
纵 23.6 厘米，横 28.5 厘米
大都会艺术博物馆藏

画中描绘 18 世纪法国著名画家布歇（François Boucher）面见蓬巴杜夫人（Madame de Pompadour）的场景。蓬巴杜夫人是法国沙龙文化的代表人物之一，亦是众多艺术家及启蒙运动代表人物（如伏尔泰）的重要赞助人。（许愿）

24 Boucher presented to Madame de Pompadour

From *Le Monde Illustré*
June 29, 1861, Henry Linton (British), after Edouard Gérard (French)
Wood engraving
L.:23.6cm, W.: 28.5cm
Lent by The Metropolitan Museum of Art, Gift of Donato Esposito, 2013 (2013.525.2)

This work depicts the renowned 18th century French painter François Boucher meeting Madame de Pompadour. Madame de Pompadour was a prominent figure in French salon culture and a key patron of numerous artists and Enlightenment thinkers, including Voltaire. (Xu Yuan)

SALON DE 1861. — *Boucher présenté à madame de Pompadour*, tableau de M. E. Gérard.

25 诗歌精灵

18 世纪，[法国] 蓬巴杜夫人
蚀刻版画
纵 12.9 厘米，横 11.8 厘米
大都会艺术博物馆藏

沙龙在 18 世纪的法国也是推动艺术收藏和艺术合作的场所。蓬巴杜夫人就曾与画家布歇（Boucher）、维恩（Vien），雕刻师盖伊（Guay）合作，制作了一套版画，本件作品正是其中一幅。（郭欣悦）

25 The Genius of Poetry

18th century, Madame de Pompadour (French)
Etching
L.:12.9cm, W.: 11.8cm
Lent by The Metropolitan Museum of Art, The Elisha Whittelsey Collection, The Elisha Whittelsey Fund, 1960 (60.622.15)

Salons in 18th century France were also platforms for promoting art collecting and artistic collaboration. Madame de Pompadour collaborated with François Boucher, Joseph-Marie Vien, and gem-engraver Jacques Guay on a printmaking project, and this piece is one of the works from that series. (Guo Xinyue)

26 壁挂烛台（一对）

约 1761，法国塞弗尔瓷器厂根据［法国］让·克劳德·杜普勒西设计制作
软瓷，铜镀金
长 26.7 厘米，宽 18.4 厘米，高 43.2 厘米
大都会艺术博物馆藏

1751 年，蓬巴杜夫人成为塞弗尔瓷器厂（Sèvres Manufactory）的赞助人。此烛台由法国金匠设计，采用曲线卷轴形态的叶饰与浆果装饰，尽显洛可可风格。（丁惠）

26 Wall Sconce (a pair)

Ca. 1761, Manufactory: Sèvres Manufactory (French)
Modeler: After a design by Jean-Claude Duplessis (French)
Soft-paste porcelain, gilt bronze
L.:26.7cm, W.:18.4cm, H.:43.2cm
Lent by The Metropolitan Museum of Art, Gift of Samuel H. Kress Foundation, 1958 (58.75.65)

In 1751, Madame de Pompadour became a patron of the Sèvres Manufactory. This pair of wall sconces, designed by a French goldsmith, features leaf ornaments in curving scroll forms and berry decorations, embodying the Rococo style. (Ding Hui)

27 大观园图

清（1644—1911），佚名
纸本，设色
纵 137 厘米，横 362 厘米
中国国家博物馆藏

此幅《大观园图》采用了西洋传入的线法画，以焦点透视法描绘了亭台楼阁林立的大观园全景，营造出强烈的空间感。绘者对建筑内外空间的描绘尤为着力，无论是雕梁画栋、草木奇石，还是室内的桌椅屏风，无一不精细，连建筑匾额上的文字都清晰可辨，有牡丹亭、藕香榭、紫菱洲、蘅芜院、凹晶馆等。

画面结构打破了传统的线性叙事，将《红楼梦》小说中不同时间、空间内发生的情节场景并置于画面中。如中秋爽斋结社、螃蟹宴、湘云眠芍、四美垂钓等。各场景用建筑空间巧妙地进行了区隔，主要人物反复出现在各个场景中也不显突兀，形成了别具一格的叙事方式。人物的描绘较为呆板生硬，面部特征单一，只能凭借服饰特征区分宝玉、贾母、湘云等几个主要角色。其中最易分辨的是贾宝玉形象，他冠上有绛绒簪缨，紫色广袖长衫，外罩石青比甲，这一形象在画面中出现八次之多。

此《大观园图》1954 年由中国历史博物馆（中国国家博物馆前身之一）购于北京宣武门内大街范文斋书画店。根据尺幅形制以及入藏时未装裱且两端有残破痕迹的状态推测，该画作原本可能是某王府内的贴落，用以装饰建筑内部空间，西洋线法画尤其多用于此。（杨扬）

27 *Daguan Garden*

Qing Dynasty (1644-1911), unknown painter
Handscroll, ink and color on paper
L.:137cm, W.: 362cm
Collection of the National Museum of China

The painting is chiefly based on chapters 37 and 38 of *The Dream of the Red Chamber*. It depicts a refined gathering in which Baoyu, Daiyu, Baochai, and others form a club to recite poetry; Granny Jia hosts a grand crab feast; and an elegant literary meeting sees Daiyu triumph in a poetry and wine contest. Interwoven with these scenes are depictions of Xiangyun drunkenly reclining amid peony flowers, the 'Four Beauties' angling for fish amid ornate motifs, and a moon viewing, flute-playing scene in the Tubi Hall during the Mid-Autumn Festival. Together, these elements reveal how such cultured assemblies evolved during the Qing Dynasty as garden landscapes became even richer and more mature. (Yang Yang)

28 青花庭院仕女图高足碗

明宣德（1426—1435）
口径 15.2 厘米，足径 4.4 厘米，高 10.3 厘米
故宫博物院藏

碗敞口，深弧腹，下接中空柱形外撇高足。内、外壁均施白釉，足端无釉。外壁青花装饰，绘通景庭院仕女图。庭院中怪石耸立，花草茂盛；远处山峦起伏，天高云淡。仕女们散坐院中，或对弈，或赏画，或抚琴，诚然一幅西园雅集图。圈足外壁绘山景图，远处青山耸立云端，近处坡地树木茂盛，远山近景，错落有致。内底双圈线内署青花楷体“大明宣德年制”六字双行款。

此器造型端庄，青花色泽浓艳，系使用进口苏麻离青料所绘，釉质莹润，釉色白中闪青，款识字体端正。外壁所绘通景庭院人物图，在宣德朝御窑瓷器的装饰题材中尤少见。所绘庭院、人物、植物等纹饰生动自然，远山近景画意恬静悠然，整体构图疏密得当，笔法精湛，堪称宣德朝御窑瓷器中的上品佳器。（王振宇）

28 Blue-and-white Bowl with High Foot and Courtyard Ladies Pattern

Ming Dynasty, Xuande Reign (1426-1435)
Diam.(at mouth): 15.2cm, Diam.(at foot):4.4cm, H.:10.3cm
Collection of the Palace Museum

The exterior of this bowl is decorated with a continuous scene of ladies in a courtyard. The courtyard features towering rocks and distant rolling mountains. Ladies are scattered throughout the courtyard, either sitting opposite each other playing chess in front of a screen, appreciating paintings, or playing the *qin* zither, or reading a book, forming a serene, elegant gathering. (Wang Zhenyu)

29 青花庭院仕女图盘

明宣德（1426—1435）
口径 21.4 厘米，足径 13.6 厘米，高 4.2 厘米
故宫博物院藏

盘撇口，浅弧壁，圈足。内、外壁均施白釉，足端无釉。通体青花装饰。盘内底双圈线内绘灵芝、松树、怪石、竹子、梅花等纹饰。盘外壁绘通景庭院人物图。庭院内绘楼阁、石桥、盆景怪石、花草等景物。远处青山叠嶂，云气缥缈，画意清雅宁静。庭中绘有四位仕女，前后两位一人手握芭蕉扇缓步向前，另一人手持烛台回首后顾；中间两位仕女彼此相伴而行，其中一人搀扶头戴发饰的女子徐徐向前。结合画意，与宋代白玉蟾作《红梅》诗句相呼应，诗曰："玉妃初醉下瑶台，紫雾深深拨不开。却恐错穿桃杏径，高烧银烛照归来。"外底双圈线内署青花楷体"大明宣德年制"六字双行款。

宣德御窑瓷器所绘人物题材的装饰纹样甚少。此盘造型端庄，青花料色泽浓淡适中，釉质莹润，胎质细腻。所绘庭院仕女通景图，构图严谨，笔触生动传神，所绘仕女神态清秀婉丽，人物衣带飘逸如风。图中描绘楼阁、石桥、植物等景物细腻生动，画意玄妙，宛若仙境。此盘堪称宣德御窑青花瓷器的精品佳作。
（王振宇）

29 Blue-and-white Plate with Courtyard Ladies Pattern

Ming Dynasty, Xuande Reign (1426-1435)
Diam.(at mouth): 21.4cm, Diam.(at foot):13.6cm, H.:4.2cm
Collection of the Palace Museum

The exterior of the plate is adorned with a continuous scene of figures in a courtyard. The courtyard features pavilions, stone bridges, miniascape, and herbs and flowers, creating an atmosphere of elegance and tranquility. Within the scene, four graceful ladies are depicted strolling at leisure through the garden.
(Wang Zhenyu)

30 喷泉花园中的公主聚会

约 1770 年，印度莫卧儿时期
纸本，不透明水彩和金箔
纵 42.5 厘米，横 55.2 厘米
大都会艺术博物馆藏

画面描绘了宫廷花园的聚会场景，其中身着高腰飘逸束腰外衣（angarkhas）的女性形象，以及黄绿色的山坡，展现出北印度宫廷绘画的典型特征。水在伊斯兰园林艺术和文学中常被视为通往天堂的通道，画家通过非透视的画法使喷泉从多个角度平铺呈现于观众眼前。（曹旭卿）

30 Princesses Gather at a Fountain

Ca. 1770, Mughal period (1526-1857), India
Opaque watercolor and gold on paper
L.:42.5cm, W.:55.2cm
Lent by The Metropolitan Museum of Art, Cynthia Hazen Polsky and Leon B. Polsky Fund, 2001 (2001.421)

This painting depicts a gathering in a palace garden, featuring long-legged women in flowing, high-waisted angarkhas against yellow-green hills–characteristic elements of North Indian court painting. In Islamic gardens and literature, water is often regarded as a passage to paradise. The painter abandoned the rules of perspective to dramatically emphasize the overall layout of the fountain. (Cao Xuqing)

鉴藏

唐代《历代名画记》中就提到了“鉴识”与“收藏”。中国古代的文人在雅集中品古、鉴古，也在书房中藏古、阅古。

“四旁修竹百竿，以招清风；南面长松一株，可挂明月。老梅寒蹇，低枝入窗；芳草缛苔，周于砌下。东屋置道释二家书，西房置儒家典籍，中横几榻之外，杂置法书名绘……”

明代文人兴致盎然地在溪山迂曲处规划自己的小园书斋，伴着园林的幽静，与古物交游，与古人相接，如此风雅的园居生活羡煞今人。

在文艺复兴时期的意大利，大量出土的古代雕塑等艺术品被教皇、贵族与富商们所收藏，用于装点宅邸和花园，彰显权力与财富。18 世纪，来自东方的异域风情席卷英伦，设计师们构建出想象中的中国亭廊、宝塔等建筑，宛如一座座“飞来峰”，成为自然风景园中时尚的点睛之笔。

Gardens and Antiquarianism

As far back as the Tang Dynasty, the *Record of Famous Paintings Through the Ages* already mentioned the practices of ‘appraisal’ and ‘collection’. For ancient Chinese literati, savoring, appraising, collecting, and studying antiquities was an essential part of both social gatherings and private contemplation.

With great enthusiasm, a Ming Dynasty scholar designed his studio nestled in the garden, “Plant a hundred bamboo stalks around to beckon the breeze, and a tall pine to the south to ‘hang’ the moon, as well as an aged plum tree, with its bent branches reaching into the window. Carpet the steps with fragrant grass and lush moss. Store Taoist and Buddhist texts in the east chamber; Confucian classics in the west; the masterpieces of calligraphy and painting in the center...”

This elegant lifestyle, combining the tranquility of the garden and daily contact with antiquity, is one modern people might envy.

During the Italian Renaissance, popes, nobles, and wealthy merchants amassed collections of newly unearthed ancient sculptures and artworks, displaying them in their residences and gardens as symbols of power and prestige. In the 18th century, a fascination with the East swept across Britain, and Chinese-inspired architecture such as pavilions, corridors, and pagodas were constructed and incorporated into gardens. Bringing to mind the exotic and mystical, much like the ‘Floating Peaks’ of Hangzhou, they became fashionable highlights of English landscape gardens.

31 建福宫花园

建福宫花园，亦称西花园，位于紫禁城内廷西六宫的西北部。清乾隆七年（1742），在拆除乾西四所、五所基址上逐步建成。乾隆十五年（1750）建成延春阁、敬胜斋、凝晖堂、吉云楼、玉壶冰、积翠亭，乾隆十七年（1752）建成碧琳馆，之后又增建慧曜楼、妙莲花室等。

建福宫花园位于建福宫西北部，是建福宫组群之一部分，全园占地近 4020 平方米，整体布局精巧，融合了江南园林的婉约与北方皇家园林的恢宏。园内建筑以延春阁为中心，北有敬胜斋、吉云楼，西有碧琳馆、妙莲花室、凝晖堂，西南有玉壶冰，南有堆山及积翠亭，东有游廊，东北有静怡轩。园中建筑形式各不相同，集宫、殿、楼、阁、斋、堂、亭、轩于一体，错落有致，步移景异，配置山石树木，景色秀丽典雅，是乾隆初年在宫中开辟的一处风格独特的皇家园林，乾隆时期将珍奇文物收藏于此。

1923 年 6 月，建福宫花园连同无数珍宝化为灰烬，仅留存少量石质建筑构件。2000 年，建福宫花园复建工程启动，现全园已复建完成。此为建福宫花园焚毁前后历史照片（摄于 1923 年），并复建竣工后影像。（崔茜）

建福宫花园今貌，史宁昌摄

31 **The Palace of Established Happiness (*Jianfu gong*)**

The Palace Museum of Established Happiness, also known as the Western Garden, is located in the northwestern part of the western Six Palaces within the Inner Court of the Forbidden City. Its layout is exquisitely designed, blending the elegance of gardens in the Jiangnan region with the grandeur of northern imperial gardens. Established in the early Qianlong reign, the garden was a distinctive imperial retreat storing rare and precious artifacts. In 1923, a fire reduced the entire garden to ashes. (Cui Qian)

32 缂丝《五事箴》及龙额围屏

清乾隆（1736—1795）
长 88 厘米—133 厘米，宽 40.4 厘米—70.8 厘米
故宫博物院藏

《尚书》中《洪范》篇虚拟了箕子向周武王传授为君之道的对话，托箕子之口论述了所谓的“天地之大法”，即帝王治理国家必须遵守的九条根本大法。在这九条大法中，有一条为“敬用五事”（又称“洪范五事”），即修养身心的五个方面。而帝王修身须在貌、言、视、听、思这五个方面做到貌恭、言从、视明、听聪、思睿。之后的历代文人多创作有《五事箴》，赞颂“洪范五事”，并谈论自己的修身心得，表达对于“洪范五事”的理解，借以自勉。乾隆帝早年也写有自己的御制《五事箴》，并将其用各种媒材进行表现，包括将此文写成卷轴、刻成玉册，也包括将自己纸本书写的《五事箴》用缂丝的方式织出，作为屏心嵌于屏风之上。缂丝因为织造工艺复杂、造价高昂，且适合表现书画作品，因此将宫廷中一部分重要的纸本书画用缂丝的方式制成副本，是热爱书画的乾隆帝的常见之举。乾隆帝除了将自己的《五事箴》作为屏心镶嵌在屏风之上，还为每一箴下方配置了裙板，裙板上镶嵌缂丝织成的坐龙捧圣图案。这套缂丝御笔《五事箴》及坐龙捧圣裙板心，原嵌于乾隆帝授意制作的屏风之上，屏风则放置在建福宫的静怡轩中，用于展示皇帝的座右铭（铭座），直至乾隆十九年（1754）被皇帝下旨换下。（仇泰格）

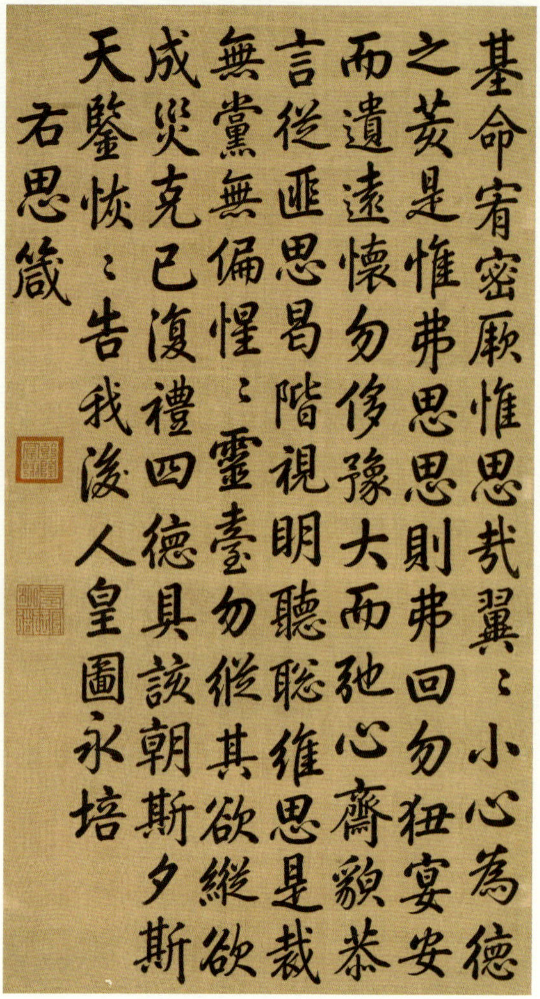

32 *Kesi* Embroidered *Five Admonitions* and Dragon-Pattern Screen Panels

Qing Dynasty, Qianlong Reign (1736-1795)
L.:88-133cm, W.:40.4-70.8cm
Collection of the Palace Museum

This *kesi*-woven imperial inscription, titled *Five Admonitions*, along with the dragon pattern panels, was originally part of a screen commissioned by the Qianlong Emperor. The *Five Admonitions* outline the moral self-cultivation of an emperor, emphasizing respectfulness in demeanor, compliance in speech, clarity in sight, keenness in listening, and wisdom in thought. The Qianlong Emperor composed his own version of the *Five Admonitions* in his early years, which he had transcribed into scrolls, inscribed onto jade tablets, and woven using the *kesi* technique. (Qiu Taige)

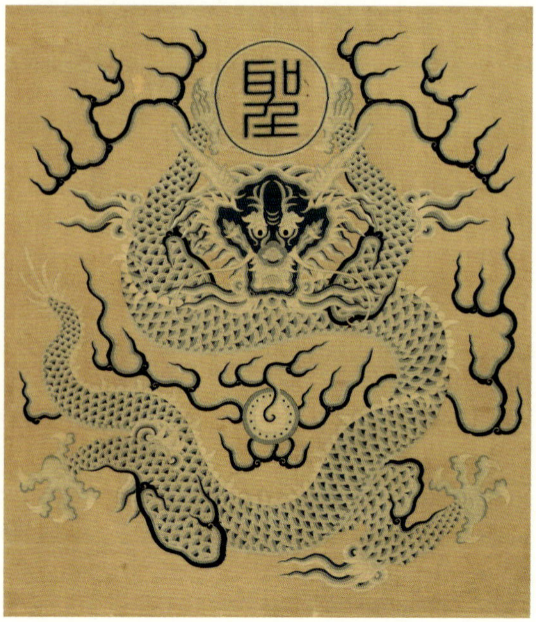

欽戒有訓抑抑威儀洪範五事始貌
終思動罔或愆敬義夾持作肅惟恭
非禮弗為顧諟明命日監在茲輯柔
爾顏友君子時屋漏有愧差之豪釐
顯卬令望夙夜是祗毋尚嚴嚴咫尺
天威毋示弛弛右史有司溫恭允塞
為德之基樹之軌範視此箴辭

右貌箴

維目司視其德曰明明有由蔽物引
欲萌伊古有訓五色目盲勿眩錦繡
勤省農耕勿翫珠玉企望賢能知人
則哲體察物情曰視有要視於無形
如水如鑑萬象自呈靜虛動燭其本
必清垂旒端冕儲神致精日月環照
無幽弗瑩克艱厥后永正玉衡

右視箴

恢恢寰宇一人是主維命靡常有德
斯祜王言如綸被乎下土無易由言
惟敬作所彝訓聿昭四方是矩卓彼
前王樹茲諫鼓柔亦不茹剛亦不吐
進彼嘉言袞職堪補聲為之律毋逞
喜怒白圭謹言玷不在鉅金人三緘
非禮勿語逆斯作乂敬茲九五

右言箴

宣聖有言非禮勿聽紹茲建極承天
永命惟精惟一敬勝義勝何以達聰
容納諫諍執兩用中旌賢表正既高
乃聽亦遠波佞五音耳亂樂必放鄭
九歌驗民八風體政亦越卷阿矢音
有咏雖隔九重而知民病雖聆萬言
而執一柄毋俾聽熒尚其居敬

右聽箴

四美具

"四美具"为建福宫花园静怡轩中珍藏晋顾恺之《女史箴图》和传为北宋李公麟《潇湘卧游图》《蜀川胜概图》《九歌图》四件名品的专室。乾隆十一年（1746），乾隆帝喜见明代董其昌品题顾从义所藏四大名卷在自己手中重聚，"千古法宝，不期而会，正复不可思议"。欣喜之余，特辟专室用以存放，并取唐代王勃《滕王阁序》"四美具，二难并"典（"四美"即良辰、美景、赏心、乐事），命名曰"四美具"，并一再赞道："乃今四美具一室，赏心乐事无伦比。"

根据档案可知，当时室内金砖墁地，六扇玻璃窗户，宝座床西向，上设靠背，靠背上立"四美具"插屏匣（内存四画）。两侧设雕龙博古柜，摆设"静怡轩四美具百什件"。其中，从"天"字屉至"阳"字屉的32个抽屉，每屉分别以千字文依次命名。其内器物种类极其丰富多样，卷轴、册页、玉器、陶瓷、铜器、珐琅器等等，共计千余件各类文玩。地上设紫檀琴桌、香几，室内悬乾隆帝御笔"四美具"匾额。屏间有"貌、言、视、听、思"《五事箴》缂丝围屏等。乾隆帝在静怡轩中打造出了一方属于他私人的收藏宝室，"四美具"也成为乾隆帝"寓意于物""清嘉托兴"的审美空间与精神园地。

乾隆帝自幼熟习汉文化，喜爱艺术，亦好博古收藏。在位60年间建立起庞大有序的清宫收藏体系。由其亲自打造的多种类型的"多宝格"及"百什件"即是其收藏体系中十分重要的一环。清宫"多宝格"通常指能单面陈设及摆放各式器物的格架，大者如墙面，小者可置于书桌炕案之上。而"百什件"则一般指盛装各式古玩器物的箱匣。据清宫档案可知，乾隆帝曾在紫禁城、圆明园等多处宫殿居所安设"多宝格"或"百什件"，并亲自挑选合适的古玩器物入藏其中。

本文后所示器物即参考"静怡轩四美具百什件"及造办处《活计档》相关所载，如"寒"字屉中的白玉鹦鹉、霁青小瓶，"来"字屉中的白玉莲花罐、掐丝珐琅海棠洗、掐丝珐琅筒子炉等等，配合格架，意象性地展示乾隆时期静怡轩"四美具"内的博古收藏。（赵赢赢、蒋艺）

The Four Beauties Collection Chamber (*Simeiju*)

The Four Beauties Collection Chamber was a special room in the Gallery of Tranquil Ease (*Jingyi xuan*) in the Garden of the Palace of Established Happiness (*Jianfu gong*), dedicated to four treasured masterpieces: *Admonitions of the Instructress to the Court Ladies* by Gu Kaizhi of the Jin Dynasty and three works attributed to Song Dynasty artist Li Gonglin (*Imaginary Tour Through Xiaoxiang*, *Magnificent Views of Shuchuan*, and *The Nine Songs*). In the 11th year of his reign in the Qing Dynasty (1746), the Qianlong Emperor was delighted to have the four renowned scrolls reunited under his possession. Overjoyed, he established a special room for them, hence the name 'Four Beauties Collection' after a phrase from Tang Dynasty poet Wang Bo.

The chamber housed a rich variety of items, including scrolls, albums, jade, ceramics, bronze, enamelware, and more, totaling more than a thousand pieces. The room also featured red sandalwood tables for the *qin* zither, incense tables, and a plaque inscribed with Qianlong's handwriting reading *The Four Beauties Collection*. Here, the Qianlong Emperor created a personal treasure room, doubling as an aesthetic space and spiritual sanctuary.

The items displayed here are based on the related historical records. Together, they symbolically showcase the antiquities in the Four Beauties Collection Chamber during the Qianlong reign. (Zhao Yingying, Jiang Yi)

建福宫花园静怡轩，"四美具"位于其西间，林京摄

33 紫檀三屉画箱插屏

清（1644—1911）
长 43 厘米，宽 16 厘米，高 47 厘米
故宫博物院藏

插屏紫檀质，夹堂内设三屉，可存贮手卷。屉面安铜纽，屏心另一侧面为玉石镶嵌的“梅竹图”。站牙雕云蝠纹，下承方盘，足外撇。（赵赢赢）

33 Red Sandalwood Table Screen with Inset Three-drawer Painting Box

Qing Dynasty (1644-1911)
L.:43cm, W.:16cm, H.:47cm
Collection of the Palace Museum

34 紫檀炕几

清（1644—1911）
长 77 厘米，宽 33.5 厘米，高 35 厘米
故宫博物院藏

炕几桌面与腿直接用粽角榫连接，以攒棂的手法做成横枨、矮佬与花牙。桌面边沿双混面起阳线，方腿。炕几整体呈现出轻妙空透之感，展示了细致考究的紫檀工。（赵赢赢）

34 Red Sandalwood Bed Table

Qing Dynasty (1644-1911)
L.:77cm, W.:33.5cm, H.:35cm
Collection of the Palace Museum

35 紫檀炕几

清（1644—1911）
长 96 厘米，宽 35.5 厘米，高 35.5 厘米
故宫博物院藏

炕几紫檀质，有束腰，束腰下托腮，方腿，内翻马蹄足。此几料厚质精，体现出清代宫廷家具庄重文雅的特点。（赵赢赢）

35 Red Sandalwood Bed Table

Qing Dynasty (1644-1911)
L.:96cm, W.:35.5cm, H.:35.5cm
Collection of the Palace Museum

36 楠木描金博古格（一对）

清（1644—1911）
长 96 厘米，宽 32 厘米，高 95.5 厘米
故宫博物院藏

博古格齐头立方式，正面及两侧面透敞。边框描金装饰折枝花纹，镶夔龙纹、云纹坐牙或托角牙。两件左右上角分别对开两扇门，门下及格底部抽屉屉面镂空，饰团螭纹卡子花拉手。（赵赢赢）

36 Catalpa Wood Gilt Curio Display Shelves (a pair)

Qing Dynasty (1644-1911)
L.:96cm, W.:32cm, H.:95.5cm
Collection of the Palace Museum

37 紫檀框嵌染牙竹石图挂屏

清乾隆（1736—1795）
长 102 厘米，宽 56 厘米
故宫博物院藏

挂屏紫檀素面双阳线边框，线间嵌绿色染牙花纹为饰。屏心在黑色地上嵌染牙朱竹一丛，青金石制出湖石，螺钿雕作灵芝。上部有绿色染牙字，为和珅书乾隆帝《题朱竹》诗一首，此屏图像当是参照乾隆帝御笔《朱竹图》制作而成。（黄剑）

37 Red Sandalwood Hanging Screen with Inlaid Dyed Ivory and Bamboo-and-rock Pattern

Qing Dynasty, Qianlong Reign (1736-1795)
L.:102cm, W.:56cm
Collection of the Palace Museum

38 碧玉嵌白玉寿字八宝缠枝莲纹委角方盒

清（1644—1911）

长 15 厘米，宽 14.5 厘米，高 7.1 厘米，通高 10.7 厘米

故宫博物院藏

方盒以碧玉嵌白玉雕制，玉质上乘无瑕。方盒由盖与身两部分组成，呈四方委角式，深壁，矮圈足。盖平顶，面镶嵌白玉，白玉面正中雕团寿字，外琢一周八宝纹，白玉盖面边缘琢一周缠枝莲纹，枝蔓层层缠绕，花瓣翻卷，舒展自如，形成强烈的立体层次效果。盖、盒身四面为碧玉，玉质坚实温润，配有紫檀木座。此类方盒或为呈香之用，或为摆件陈设，存于多宝阁中，玉质难得，多以供赏玩。（杨立为）

38 Jasper Box with Inlaid White-jade Featuring Eight Treasures and Interlocking Lotus Sprays Pattern

Qing Dynasty (1644-1911)

L.:15cm, W.:14.5cm, H.:7.1cm, H.(in total):10.7cm

Collection of the Palace Museum

39 白玉团寿字圆盒

清（1644—1911）
直径 7.3 厘米，底径 4.5 厘米，通高 3.5 厘米
故宫博物院藏

圆盒玉质上乘，为白玉质，质地匀净，光泽细润。盒圆形，为盖与身两部分组成，盖正中浮雕团寿字，身为光素。此类圆盒多为焚香用具，常盛放所焚的香面或香条（粒），多以炉、瓶、盒三事出现，配以铜铲、箸使用。多为圆形、圆筒形、方形，或大或小，一些还配有木座。（杨立为）

39 White-jade Box with '*Shou*' Character

Qing Dynasty (1644-1911)
Diam.:7.3cm, Diam.(at base):4.5cm, H.(in total):3.5cm
Collection of the Palace Museum

40 白玉三鸠罐

清乾隆（1736—1795）
口径 4.4 厘米，足径 5.5 厘米，高 6.9 厘米
故宫博物院藏

罐白玉质，圆口，收颈，鼓腹，圈足，足壁镂雕三个方形孔。罐身等距饰有三鸠鸟，作昂首展翅直立状，圆雕头部，翅膀与爪浅浮雕于器身，以阴线刻出羽毛纹理，线条简洁流畅。底部阴刻“乾隆年制”篆书款，下附镂雕木座。此器仿清宫旧藏唐代白玉三鸠罐，鸠为“不噎之鸟”，故以此寓意刚直和祝年长者不得“噎病”，是一件寓意吉祥并为陈设赏玩的仿古玉器。（刘梦媛）

40 White-jade Jar with Three Turtledoves

Qing Dynasty, Qianlong Reign (1736-1795)
Diam.(at mouth):4.4cm, Diam.(at foot):5.5cm, H.:6.9cm
Collection of the Palace Museum

41 白玉荷花式盖罐

清乾隆（1736—1795）
口径 4.1 厘米，底径 2.8 厘米，通高 6 厘米
故宫博物院藏

盖罐白玉质，花瓣形口，平底，底心浅浮雕六瓣花形。盖作倒覆重瓣荷花形，顶雕一小鸟为钮，鸟作回首状立于花上。苏轼有诗“翻空白鸟时时见，照水红蕖细细香”，此件玉器展现的正是鸟栖芙蕖的自然意趣，可赏可用。（刘梦媛）

41 White-jade Lidded Jar in Shape of Lotus

Qing Dynasty, Qianlong Reign (1736-1795)
Diam.(at mouth):4.1cm, Diam.(at base):2.8cm,
H.(in total):6cm
Collection of the Palace Museum

42 碧玉缠枝莲纹瓜棱式盖罐

清乾隆（1736—1795）
口径 8.6 厘米，底径 8.5 厘米，通高 11.5 厘米
故宫博物院藏

盖罐碧玉质，为瓜棱式圆罐，花形口，鼓腹，平底。整体运用浅浮雕工艺，罐身以瓜棱线等距离排列六组西番莲图案，间饰以茛苕叶。花形盖，子母口，盖钮镂雕瓜瓞纹，盖面浅浮雕茛苕叶纹，下附紫檀木座。此器为中式造型，纹饰模仿痕都斯坦玉器风格，体现了中外文化的交流与融合，是一件富有巧思的陈设玉器。（刘梦媛）

42 Jasper Melon-shaped Lidded Jar with Interlocking Lotus Sprays Pattern

Qing Dynasty, Qianlong Reign (1736-1795)
Diam.(at mouth):8.6cm, Diam.(at base):8.5cm,
H.(in total):11.5cm
Collection of the Palace Museum

43 掐丝珐琅缠枝莲纹象耳方瓶

清（1644—1911）
口长 5.1 厘米，口宽 4.7 厘米，足长 5.3 厘米，足宽 4.5 厘米，高 14.1 厘米
故宫博物院藏

方瓶铜胎鎏金，掐丝珐琅工艺。通体天蓝色，口沿下如意云头纹、缠枝菊瓣纹，瓶颈两侧象首耳。肩部錾刻莲瓣纹，瓶身饰缠枝莲纹、联珠纹。足底中心錾刻“景泰”款。此珐琅瓶的装饰工艺是典型的清中期特征。（邢娜）

43 Square-section Cloisonné Vase with Elephant-shaped Handles and Interlocking Lotus Sprays Pattern

Qing Dynasty (1644-1911)
L.(at mouth):5.1cm, W.(at mouth):4.7cm, L.(at foot):5.3cm, W.(at foot):4.5cm, H.:14.1cm
Collection of the Palace Museum

44 古铜彩描金双耳炉

清乾隆（1736—1795）
口径 9 厘米，足径 7.3 厘米，高 5 厘米
故宫博物院藏

炉唇口，微撇，扁弧腹，两侧对称各置一三角形耳，圈足。全器通体施仿古铜彩，釉色呈黄褐色，泛有金属光泽。外壁口沿至胫部分别以描金绘饰回纹、如意云头纹、变形螭龙纹及卷草纹。外底刻篆体“大清乾隆年制”六字三行款。（蒋艺）

44 Porcelain Censer with Two Handles with Imitating Archaistic Bronze Glaze

Qing Dynasty, Qianlong Reign (1736-1795)
Diam.(at mouth):9cm, Diam.(at foot):7.3cm, H.:5cm
Collection of the Palace Museum

45 古铜彩描金盖盒

清乾隆（1736—1795）
口径 6 厘米，足径 4.5 厘米，通高 3 厘米
故宫博物院藏

盒口微敛，斜弧腹，圈足，附平顶圆盖。通体施仿古铜彩，釉色呈黄褐色，泛有金属光泽。盖面及盒外壁皆用描金装饰，盖面中心绘一圆形开光，开光内绘花卉纹及如意云头纹，外壁绘一周朵花纹及点状纹。外底以金彩署篆体“大清乾隆年制”六字三行款。（蒋艺）

45 Porcelain Box with Imitating Archaistic Bronze Glaze

Qing Dynasty, Qianlong Reign (1736-1795)
Diam.(at mouth):6cm, Diam.(at foot):4.5cm, H.(in total):3cm
Collection of the Palace Museum

46 青花如意纹洗

清康熙（1662—1722）
口径 11 厘米，足径 12 厘米，高 5 厘米
故宫博物院藏

洗敛口，扁圆腹，腹以下向内斜收，圈足。外壁以青花绘饰一周如意云头纹。外底以青花署楷体“大清康熙年制”六字三行款。（蒋艺）

46 Blue-and-white Brush Washer with *Ruyi*-shaped Design

Qing Dynasty, Kangxi Reign (1662-1722)
Diam.(at mouth):11cm, Diam.(at foot):12cm, H.:5cm
Collection of the Palace Museum

47 掐丝珐琅海棠式洗

清（1644—1911）
长 16.6 厘米，宽 14.1 厘米，高 4.1 厘米
故宫博物院藏

洗为铜胎鎏金，掐丝珐琅工艺。海棠式，敞口，平底。洗内白色地中心饰莲纹一朵，四周围绕缠枝莲纹，外壁天蓝色地上饰缠枝莲纹。此洗小巧精致，工艺精湛，是清中期掐丝珐琅的上乘之作。（邢娜）

47 Begonia-flower-shaped Cloisonné Brush Washer

Qing Dynasty (1644-1911)
L.:16.6cm, W.:14.1cm, H.:4.1cm
Collection of the Palace Museum

48 斗彩花蝶纹罐

清雍正（1723—1735）
口径 8 厘米，足径 7 厘米，通高 13 厘米
故宫博物院藏

罐唇口，短颈，深弧腹，圈足。罐口内随形配一黄铜内胆，内胆顶面开五瓣花形孔。罐外壁以青花斗彩绘洞石花蝶图，外底无款识。此器内胆应为清宫后配，配置内胆后，可作花插陈设。（蒋艺）

48 *Doucai* Jar with Flower-and-butterfly Pattern

Qing Dynasty, Yongzheng Reign (1723-1735)
Diam.(at mouth):8cm, Diam.(at foot):7cm, H.(in total):13cm
Collection of the Palace Museum

49 白釉刻蟠螭纹兽耳活环四足盖炉

清乾隆（1736—1795）
口长 13 厘米，口宽 5.6 厘米，底径 7.5 厘米 ×3.7 厘米，
通高 13 厘米
故宫博物院藏

炉长方形，两侧对称置铺首耳，铺首下接活环。器身四角及底作圆柱形。平盖，盖顶置蟠龙形钮，龙呈团身正面状。器身凸饰弦纹，腹部凸雕夔龙纹及云头纹。通体施白釉，釉色匀润。

清乾隆时期，宫廷器作仿古之风颇盛。从档案来看，此时期的景德镇御窑厂多有仿烧前代名窑釉色及器形。此盖炉釉色仿定窑白瓷，器形则来源于古铜器。同形器物材质多样，除瓷器外，更多见玉器、珐琅等材质，并与瓶、盒等成套使用。在清代宫廷绘画中，同类造型的方炉多陈列于博古架中或书案之上，可见这类造型器物在清宫中应为陈设之用。（蒋艺）

49 White-glazed Imitation *Ding*-kiln Lidded Censer

Qing Dynasty, Qianlong Reign (1736-1795)
L.(at mouth):13cm, W.(at mouth):5.6cm,
D.(at base):7.5cm×3.7cm, H.(in total):13cm
Collection of the Palace Museum

50 炉钧釉三孔花插

清雍正（1723—1735）
足径 5.5 厘米，高 7 厘米
故宫博物院藏

花插口部为一平面，面上均匀分布三个圆形插孔，束颈，扁腹，圈足。通体施炉钧釉，釉面似有流动之感，釉色蓝中满布红色及青色斑点。外底刻“雍正年制”四字双行篆书款。

炉钧釉是雍正朝新创的一种以炉内低温烧成的仿钧釉。清《南窑笔记》载：“炉钧一种，乃炉中所烧，颜色流淌中有红点者为佳，青点次之。”据嘉庆七年（1802）十一月所立《静怡轩四美具百式件》档案，四美具百什件内曾有多件“钧釉撇口小花插”或“花插”，此件仿钧釉三孔花插或可与之对照。（蒋艺）

50 ***Lujun*-glazed Flower Receptacle**

Qing Dynasty, Yongzheng Reign (1723-1735)
Diam.(at foot):5.5cm, H.:7cm
Collection of the Palace Museum

51 冬青釉刻缠枝花纹三孔花插

清乾隆（1736—1795）
足径 6 厘米，高 7.8 厘米
故宫博物院藏

花插束颈，扁腹，圈足。口部封闭，仅留三个圆形插孔，作三角形分布。外壁通体刻花，施冬青釉。口部一周刻缠枝卷草纹，颈部、腹部刻缠枝莲纹。釉呈淡青色、微泛绿。外底署青花篆体“大清乾隆年制”六字三行款。（冀洛源）

51 **Pale-green-glazed Flower Receptacle with Veiled Incised Interlocking Sprays Pattern**

Qing Dynasty, Qianlong Reign (1736-1795)
Diam.(at foot):6cm, H.:7.8cm
Collection of the Palace Museum

52 青玉石榴鸟纹云耳盖瓶

清（1644—1911）
口长 4.6 厘米，口宽 2.6 厘米，长 18 厘米，厚 4 厘米，
通高 12 厘米
故宫博物院藏

盖瓶青玉质，玉质良好，为整块玉料镂雕而成。瓶带盖，盖钮琢为绳纹圆形，瓶肩雕有双云形耳。瓶盖、腹部光素，颇有复古之风。瓶足山石状，瓶身一侧雕一鸟，落于山石之上，另一侧镂雕一石榴，籽粒饱满，刻画精湛，栩栩如生。本件配有镂雕花卉紫檀木座。

石榴、花鸟为清代常见雕琢题材，寓意吉祥、平安、多子多福。盖瓶造型复古，乾隆时好古之风盛行，虽复古但不泥古，一些世俗题材常与之共同雕琢，为乾隆玉器题材的一大特点。（杨立为）

52 Green-jade Lidded Vase with Pomegranate and Bird Pattern

Qing Dynasty (1644-1911)
L.(at mouth):4.6cm, W.(at mouth):2.6cm, L.:18cm, Th.:4cm,
H.(in total):12cm
Collection of the Palace Museum

53 斗彩缠枝花纹蒜头瓶

清康熙（1662—1722）
口径 2 厘米，足径 7 厘米，高 21 厘米
故宫博物院藏

瓶作敛口、鼓腹形口，细长颈，斜肩，曲腹，实心高圈足，整体呈锥形。外壁施斗彩装饰，口部绘青花缠枝卷草纹，其间署楷体"大明嘉靖年制"仿款。颈、腹、足通体绘缠枝花卉装饰纹。本件造型、装饰具有清代前期的工艺特点。（冀洛源）

53 *Doucai* Garlic-head Vase with Interlocking Sprays Pattern

Qing Dynasty, Kangxi Reign (1662-1722)
Diam.(at mouth):2cm, Diam.(at foot):7cm, H.:21cm
Collection of the Palace Museum

54 青玉鹦鹉

宋（960—1279）
长 7.8 厘米，宽 2 厘米，高 4.4 厘米
故宫博物院藏

鹦鹉白玉质，玉质温润上乘，整体以圆雕加透雕技法雕琢。鹦鹉坑眼，弧形喙，长尾，爪踏如意祥云，颇具动感。此鹦鹉神态、做工俱佳，翅和尾的放射阴刻线、小坑眼、尾部刻划精细，应为宋代佩饰或把玩件。

宋代上层饲养鹦鹉较为普遍，在这一时期流传的诗歌、绘画、瓷器、玉器等作品中多有体现。传宋徽宗所绘《五色鹦鹉图》和定窑鹦鹉壶等作品的传世或出土，表明宋人对鹦鹉的喜爱。（杨立为）

54 Green-jade Parrot

Song Dynasty (960-1279)
L.:7.8cm, W.:2cm, H.:4.4cm
Collection of the Palace Museum

55 青花蕉叶缠枝花卉纹贯耳瓶

清乾隆（1736—1795）
口径 5.5 厘米，足径 7 厘米，高 20 厘米
故宫博物院藏

瓶侈口，直颈双贯耳，圆肩，斜直腹，圈足。外壁施青花装饰，笔触仿明早期铁锈斑风格，颈部绘回文，肩部及底部绘蕉叶纹，腹部通体绘缠枝花卉装饰纹。外底署青花篆体“大清乾隆年制”六字三行款。此器烧造量较多，应属宫廷内常用陈设器。（冀洛源）

55 Blue-and-white Vase with Tubular Ears, Interlocking Flowers Sprays Pattern

Qing Dynasty, Qianlong Reign (1736-1795)
Diam.(at mouth) :5.5cm, Diam.(at foot) :7cm, H.:20cm
Collection of the Palace Museum

56 青花卷草纹梅瓶

明宣德（1426—1435）
口径 2.8 厘米，足径 4.5 厘米，高 11.7 厘米
故宫博物院藏

瓶小口，短颈，丰肩，腹部渐收敛，圈足。通体青花装饰，口部绘弦纹，肩部绘如意云头，腹部绘卷草纹，近足处绘朵花纹一周。外底署青花“大明宣德年制”六字楷书外围双圈款。（董彦卿）

56 Blue-and-white *Meiping* Vase with Interlocking Flowers Sprays

Ming Dynasty, Xuande Reign (1426-1435)
Diam.(at mouth):2.8cm, Diam.(at foot):4.5cm, H.:11.7cm
Collection of the Palace Museum

57 掐丝珐琅花卉纹蒜头瓶

清（1644—1911）
口径 1.2 厘米，足径 5.5 厘米，高 15.9 厘米
故宫博物院藏

瓶铜胎鎏金，掐丝珐琅工艺。圆形，蒜头式口，长直颈。通体黑色地，饰朵花纹，口下璎珞纹，外底鎏金光素。此瓶装饰雅致，小巧玲珑，一般作为室内陈设器。（邢娜）

57 Cloisonné Garlic-head Vase with Floral Design Pattern

Qing Dynasty (1644-1911)
Diam.(at mouth):1.2cm, Diam.(at foot):5.5cm, H.:15.9cm
Collection of the Palace Museum

58 仿哥釉双耳炉

明（1368—1644）
口径 10.5 厘米，足径 7.8 厘米，高 9.2 厘米
故宫博物院藏

炉撇口，弧腹，圈足，腹两侧贴塑双耳。内外壁满施青釉，釉呈米色。釉面满布开片，形成“金丝铁线”的特征。圈足端刷酱色护胎釉。内底有三个支钉，外底有六个支钉。此炉为明人仿烧的哥窑瓷器。从耳部破损处可观察胎土呈黑色，与哥窑瓷器相同，说明此器不仅注重外形神似哥窑瓷器，更是在胎、釉原料层面追求逼真效果，足见仿制水平之高。炉外底刻乾隆帝御制《咏哥窑炉》诗：“伊谁换夕薰，香讶至今闻。制自崇鱼耳，色犹缬鳝纹。本来无火气，却似有云氲。辨见八还毕，鼻根何处分。”诗末署“乾隆丙申孟春御题”，钤“乾”“隆”二印。（单莹莹）

58 Imitation *Ge*-glazed Censer with Two Handles

Ming Dynasty (1368-1644)
Diam.(at mouth):10.5cm, Diam.(at foot):7.8cm, H.:9.2cm
Collection of the Palace Museum

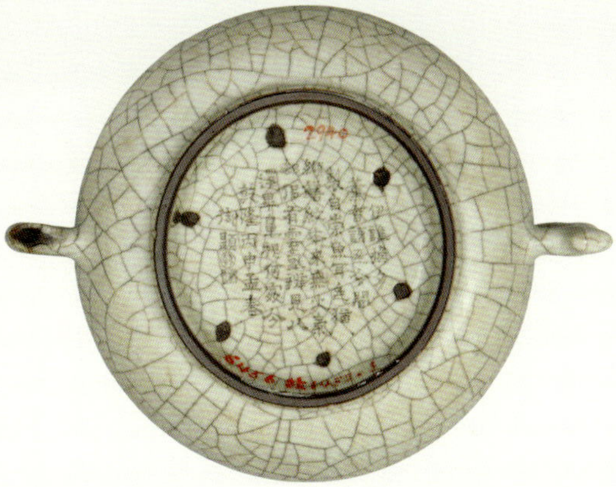

59 祭蓝釉直颈瓶

清雍正（1723—1735）
口径 2.4 厘米，足径 5.8 厘米，高 19.4 厘米
故宫博物院藏

瓶直口，直颈，圆肩，曲腹，圈足。外壁施祭蓝釉，釉色浓艳、均匀。外底署青花楷体“大清雍正年制”六字双行外围双圈款。此瓶的造型、装饰能够体现雍正朝御窑厂高水平的烧制工艺。（冀洛源）

59 Sacrificial-blue-glazed Vase with Straight Neck

Qing Dynasty, Yongzheng Reign (1723-1735)
Diam.(at mouth):2.4cm, Diam.(at foot):5.8cm, H.:19.4cm
Collection of the Palace Museum

60 冬青釉七孔鼓式花插

清雍正（1723—1735）
面径 10.5 厘米，底径 10 厘米，高 16.5 厘米
故宫博物院藏

花插敛口，曲腹，隐圈足。顶部封闭，留七个圆孔，呈双菱形排布。腹部上下各塑一周乳钉，两侧塑兽首衔环。外壁通体施冬青釉，釉色呈淡青绿色，釉层较薄。外底近足跟一周刮釉、露胎，底心署青花篆体“大清雍正年制”六字三行款。此罐造型复古，改实用器为陈设器用，乾隆、道光等朝仍有仿制。（冀洛源）

60 Pale-green-glazed Drum-shaped Flower Receptacle

Qing Dynasty, Yongzheng Reign (1723-1735)
Diam.(at surface):10.5cm, Diam.(at base):10cm, H.:16.5cm
Collection of the Palace Museum

61 宜兴窑仿哥釉玉兰式花插

明（1368—1644）

口横 16 厘米，口纵 14.7 厘米，足距 8.5 厘米，高 20.3 厘米

故宫博物院藏

花插造型借鉴自然花卉，器身主体似一朵含苞待放的玉兰花。花瓣口外撇，深腹，腹部外侧、底部塑贴花叶和枝干作为辅助装饰与底座。内、外施青釉，釉色仿宋代哥窑瓷器，釉面布满黑、金二色开片，呈现“金丝铁线”特征。外底内凹，施釉不匀，露出黄色胎。（董彦卿）

61 Imitation *Ge*-glazed Magnolia-shaped Flower Receptacle, Yixing ware

Ming Dynasty (1368-1644)

L.(at mouth):16cm, W.(at mouth):14.7cm, Distance between feet:8.5cm, H.:20.3cm

Collection of the Palace Museum

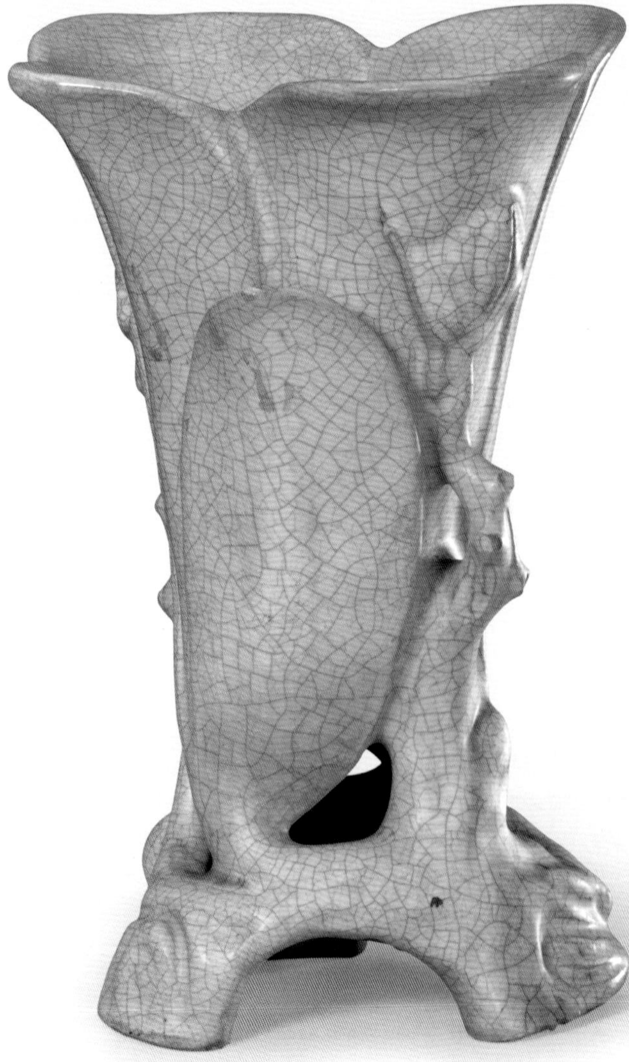

62 掐丝珐琅缠枝莲纹三足筒式炉

清乾隆（1736—1795）

口径 9.5 厘米，足距 8.4 厘米，高 9.2 厘米，通高 10.8 厘米

故宫博物院藏

炉铜胎鎏金，掐丝珐琅工艺。圆筒状，微弧壁，平底，三垂云足，通体天蓝色地。外壁饰缠枝莲纹，上下边沿如意云头纹各一周，底部饰缠枝六瓣花纹。外底中心凸起铜鎏金方片，上阴刻“乾隆年制”楷书款，另有阴刻“羊”字，并配木盖。（邢娜）

62 Three-legged Cloisonné Cylindrical Censer with Interlocking Lotus Sprays Pattern

Qing Dynasty, Qianlong Reign (1736-1795)

Diam.(at mouth):9.5cm, Distance between feet:8.4cm, H.:9.2cm, H.(in total):10.8cm

Collection of the Palace Museum

63 掐丝珐琅缠枝莲蕉叶纹出戟花觚

清乾隆（1736—1795）

口径 8.4 厘米，足径 4.8 厘米，高 13.9 厘米

故宫博物院藏

花觚铜胎鎏金，掐丝珐琅工艺。觚式插瓶，有四出戟，通体天蓝地。口沿内缠枝莲纹一周，口外壁及高足饰蕉叶纹间缠枝莲纹，蕉叶纹内填夔龙纹，腹部饰缠枝莲纹。外底中心双方框内起地阳文“乾隆年制”楷书款，配木座。清中期善做仿古器，此件为典型的仿古掐丝珐琅器。（邢娜）

63 Cloisonné *Gu* Vessel with Protruding Bars, Interlocking Lotus Sprays and Banana-leaf Pattern

Qing Dynasty, Qianlong Reign (1736-1795)

Diam.(at mouth):8.4cm, Diam.(at foot) :4.8cm, H.:13.9cm

Collection of the Palace Museum

64 青白玉乾隆帝御制诗赤壁图山子

清乾隆（1736—1795）
长 20 厘米，厚 7 厘米，通高 14.2 厘米
故宫博物院藏

山子青白玉质，细腻上乘，其上依料随形雕琢《前后赤壁图》景。上部琢山崖，中部琢江水，水中波光粼粼，芦苇荡荡，泛一篷舟，穿梭于芦苇、山崖之中。舟中一乘舟老者、二文人、一童子、一执橹老者，刻画细腻生动，再现了苏子与众人游览赤壁之景，颇有泛舟舒畅、怀古伤今之感。背面亦琢山石、江水、芦苇等，一侧山石刻有填金乾隆帝御制《题和阗玉前后赤壁图》诗："山阳舟泛山阴陆，秋水冬峰景顿移。试问丹青工幻者，可能一幅幻如斯。乾隆己亥秋御题。"山子配青玉座，玉质、刻工均上乘。座上刻山水、崖石、松柏等，与山子相得益彰，共成一景。

清代山子常以各种画作为样进行雕琢，赤壁、海屋添筹、竹林七贤、高山流水、松下童子、兰亭等典故为常见题材。此类山子常放于厅堂以供赏玩，体型大者或为书桌后的小型屏风，以增加文人情趣。多以紫檀或其他硬木为器座，本件配以上乘玉质器座，较为少见，且于其上题刻御制诗，可见乾隆帝之喜爱。（杨立为）

64 Green-white Jade Carving with Poem by the Qianlong Emperor

Qing Dynasty, Qianlong Reign (1736-1795)
L.:20cm, Th.:7cm, H.(in total):14.2cm
Collection of the Palace Museum

65 仿哥釉石榴式洗

清乾隆（1736—1795）
口径 7 厘米，底横 11.5 厘米，底纵 10.5 厘米，高 3.2 厘米
故宫博物院藏

洗敛口，矮曲腹，平底，平面及腹部造型仿石榴果实，两侧各塑出石榴叶、蒂。通体施仿哥釉，呈淡青色，乳浊质感，有黄褐色开片纹。石榴作为装饰主题或辅助纹样，广泛用于清宫各类工艺品。与此洗相类的石榴式洗另有白玉制品。（冀洛源）

65 Imitation *Ge*-glazed Brush Washer in Pomegranate Shape

Qing Dynasty, Qianlong Reign (1736-1795)
Diam.(at mouth):7cm, L.(at base):11.5cm, W.(at base):10.5cm, H.:3.2cm
Collection of the Palace Museum

66 仿哥釉葵花式三足洗

清乾隆（1736—1795）
口径 22 厘米，足距 18 厘米，高 5.5 厘米
故宫博物院藏

洗作花瓣形，窄平沿，矮曲腹，平底，三足。通体施仿哥釉，呈淡灰青色，有深褐色开片。三足微曲，足梢无釉。外底心署青花篆体“大清乾隆年制”六字三行款。（冀洛源）

66 Imitation *Ge*-glazed Mallow-petal Brush Washer with Three Legs

Qing Dynasty, Qianlong Reign (1736-1795)
Diam.(at mouth):22cm, Distance between feet:18cm, H.:5.5cm
Collection of the Palace Museum

67 青玉梅瓶

清乾隆（1736—1795）
口径 5.2 厘米，底径 8.1 厘米，高 23.3 厘米
故宫博物院藏

梅瓶青玉质，小口，短颈，丰肩，肩下渐收敛，平底，通体光素，底刻“乾隆年制”篆书款。此件仿瓷器中的梅瓶造型，因口径之小与梅之瘦骨相若而得名。梅瓶始于隋唐，宋代逐渐流行，初为盛酒具，在明清时转变为陈设器。（刘梦媛）

67 Green-jade *Meiping* Vase

Qing Dynasty, Qianlong Reign (1736-1795)
Diam.(at mouth):5.2cm, Diam.(at base):8.1cm, H.:23.3cm
Collection of the Palace Museum

68 青花凤穿花纹盖盒

明嘉靖（1522—1566）
口径 13.9 厘米，足径 9.2 厘米，高 9 厘米
故宫博物院藏

盖盒子母口，浅弧壁，圈足，内、外壁均施白釉，足端无釉，通体青花装饰。盒盖面双圈线内绘鸾凤云纹，外壁绘凤穿花纹，盒身外壁绘三组首尾相接云凤纹。外底署青花楷体“大明嘉靖年制”六字双行外围双圈款。此盖盒造型小巧，青花色泽蓝中泛紫，系使用进口回青料研磨而绘。所饰鸾凤纹华美飘逸，是嘉靖朝御窑瓷器中难得的小器大样精品。（王振宇）

68 Blue-and-white Box with Phoenix and Flowers Pattern

Ming Dynasty, Jiajing Reign (1522-1566)
Diam.(at mouth):13.9cm, Diam.(at foot):9.2cm, H.:9cm
Collection of the Palace Museum

69 仿哥釉胆式瓶

清雍正（1723—1735）
口径 3 厘米，足径 6 厘米，高 20 厘米
故宫博物院藏

瓶直口，长颈，垂腹，圈足。全器通体施仿哥釉，釉面布满黑褐色细碎开片。外底署青花篆体“大清雍正年制”六字三行款。雍正、乾隆年间，在督陶官唐英主持下，景德镇御窑厂生产了大量仿古瓷器，仿哥釉即其中一例。唐英撰于雍正十三年（1735）的《陶成纪事碑记》载：“一仿铁骨哥釉，有米色、粉青二种（俱仿内发旧器色泽）。”雍正、乾隆时的景德镇御窑厂仿哥釉器，不但仿其釉质，也仿其胎质，釉层肥厚，釉质温润。哥釉特有的“油腻酥光”及密布的“金丝铁线”片纹均仿得神似，其釉色有青灰、灰白和粉青等。（蒋艺）

69 Imitation *Ge*-glazed Gall-bladder-shaped Vase

Qing Dynasty, Yongzheng Reign (1723-1735)
Diam.(at mouth):3cm, Diam.(at foot):6cm, H.:20cm
Collection of the Palace Museum

70 剔红花卉纹三层方形委角盒

清乾隆（1736—1795）
长 20.3 厘米，宽 20.3 厘米，底座长 21.5 厘米，
底座宽 21.5 厘米，通高 20.8 厘米
故宫博物院藏

剔红盒三层，盒壁各层委角开光内雕牡丹、芙蓉、水仙、梅花等折枝花卉，开光外为莲纹。盖面锦纹上花卉纹由中心的牡丹和对角菊花、牡丹、梅花的混枝花卉构成。盒下饰锦纹底座。（邢娜）

70 Carved Red-lacquer Three-tiered Box with Floral Pattern

Qing Dynasty, Qianlong Reign (1736-1795)
L.:20.3cm, W.:20.3cm, L. (at base):21.5cm, W.(at base):21.5cm,
H.(in total):20.8cm
Collection of the Palace Museum

71 剔红“福寿宝胜”百什件提梁盒

清乾隆（1736—1795）

长 35.7 厘米，宽 19.5 厘米，底座长 40.7 厘米，底座宽 24.7 厘米，通高 29.8 厘米

故宫博物院藏

多宝盒外匣为剔红提梁盒，三层带盖。外盒浮雕缠枝花卉纹及双蝠捧寿纹，盒内髹以黑漆，每层皆配有锦匣，匣内贮藏玉器、漆器、匏器、珐琅、象牙、卷轴等各式清玩 37 件，品类多样，色彩丰富。其中多件玉器花纹及器形皆仿自商周青铜器，或将古旧玉剑饰改制为墨床，赋予新的功用，成为造型古雅的文房用具。镂雕螺钿象牙盒的器盖镂雕“卍”字纹，其上又嵌入雕刻花卉纹的螺钿，精工细作，此类镂雕应为乾隆时期广东牙匠的绝活。

造办处《活计档》中或将此类文物称为“百什件”“百事件”。雍正六年（1728）即有“养心殿后殿明间屋内桌上有陈设玉器、古玩，俱系平常之物，尔持出配作百事件，用作漆箱盛装”的记载，乾隆十八年（1753）的《活计档》中，则更详细地记述了乾隆帝参与选择与装配这套“福寿宝胜”的细节。

此套多宝盒空间设计灵活，其内所收珍玩包罗万象，材质丰富多样，反映出清宫重视文物组合、收纳、传承、求新的风尚。（于倩）

71 Carved Red Lacquer Hundred-Treasures Box with Handle

Qing Dynasty, Qianlong Reign (1736-1795)

L.:35.7cm, W.:19.5cm, L.(at base):40.7cm, W.(at base):24.7cm, H.(in total):29.8cm

Collection of the Palace Museum

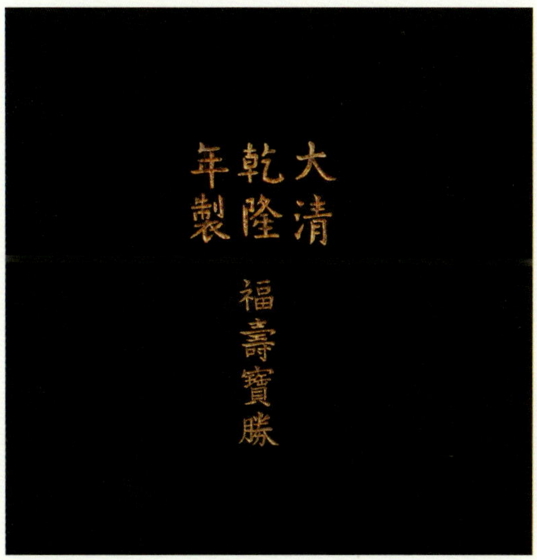

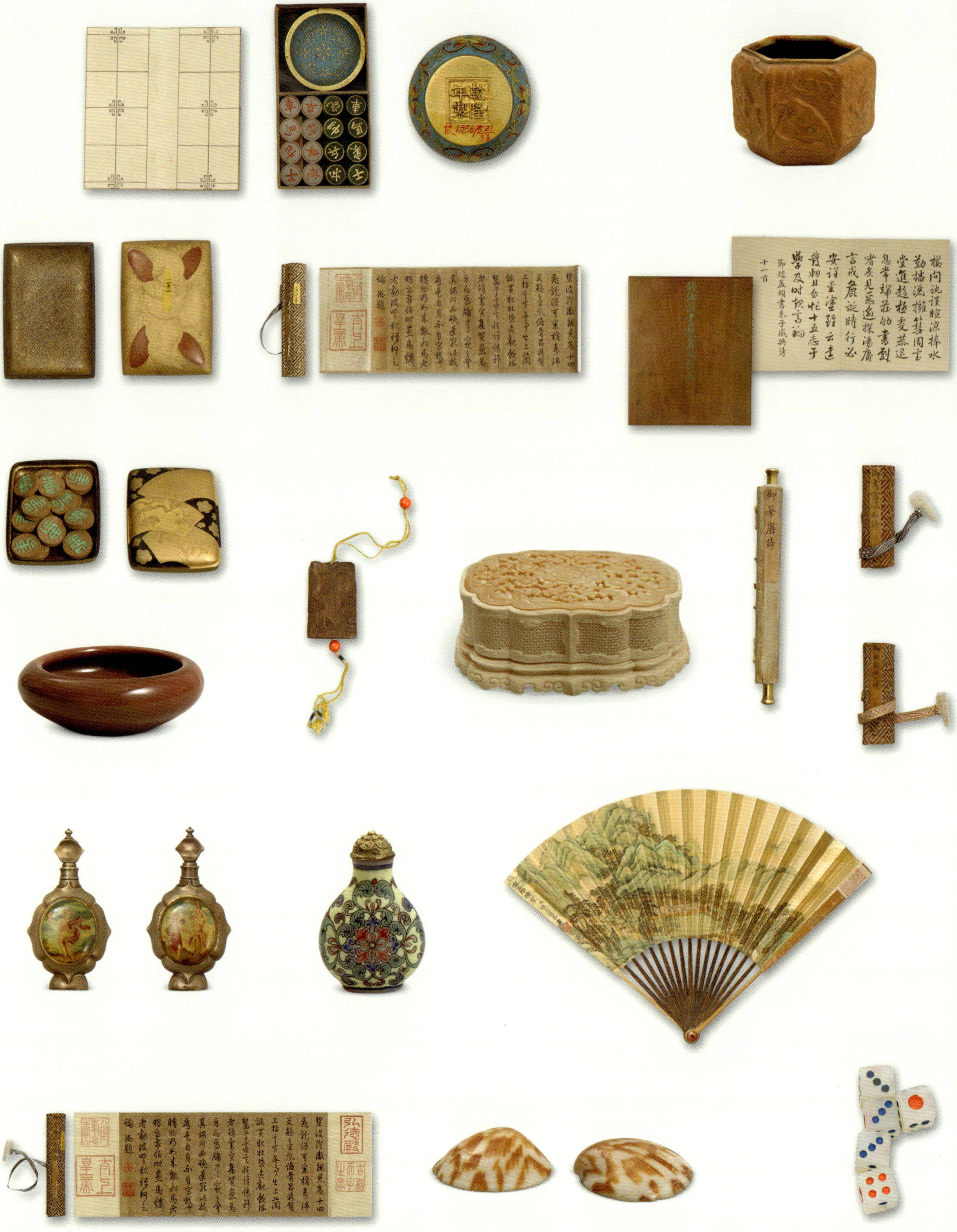

72 四美具图页

清乾隆（1736—1795），董邦达
纸本，墨笔
纵 30.7 厘米，横 57.1 厘米
天津博物馆藏

董邦达（1696—1769），字孚存、争存，号东山、非闻，浙江富阳人。雍正十一年（1733）进士，官至礼部尚书，谥文恪。董邦达的绘画技法主要来源于对古人的临摹和学习，其篆、隶得古法，山水取法元人，善用枯笔。其上追“黄王吴倪”，近学“四王”，画风介于虞山和娄东之间，乾隆帝也将其与董源和董其昌并称为“三董”。其风格延续了正统的文人山水画一脉，并且成功越过了“四王”而上追董其昌及宋元经典。

乾隆十一年（1746），乾隆帝发现明代顾从义旧藏的东晋顾恺之《女史箴图》和北宋李公麟《潇湘卧游图》《蜀川胜概图》《九歌图》四幅绘画名卷在内府重聚一堂，而这四卷被晚明著名书画鉴赏家董其昌认为是顾氏所藏绘画中最佳的四件。特在宫中静怡轩仿“三希堂”置“四美具”，令董邦达绘制了《四美具图》页和《四美具合幅图》轴（现藏台北故宫博物院），以志此事。图页与图轴在构图上有所不同，有研究者认为此图页或为《四美具合幅图》轴的母本。图页右上方有乾隆帝长题，记述绘制此图的缘起，表达其欣喜的心情。这幅《四美具图》页和《四美具合幅图》轴是典型的奉敕之作，皇帝赐下命题，词臣要扣题而作。从图页和图轴之间的区别来看，乾隆帝对于母本样稿还进行了修改，然后再由词臣来完成。此作画面严谨精细，从内容展现、构图设计到笔墨运用，都是画家经过深思熟虑的结果。以干笔淡墨反复皴擦山石，以求浑厚；以界画笔法描绘茅屋村社，工整细致。这是董邦达山居图的一大特色。（韩小赫）

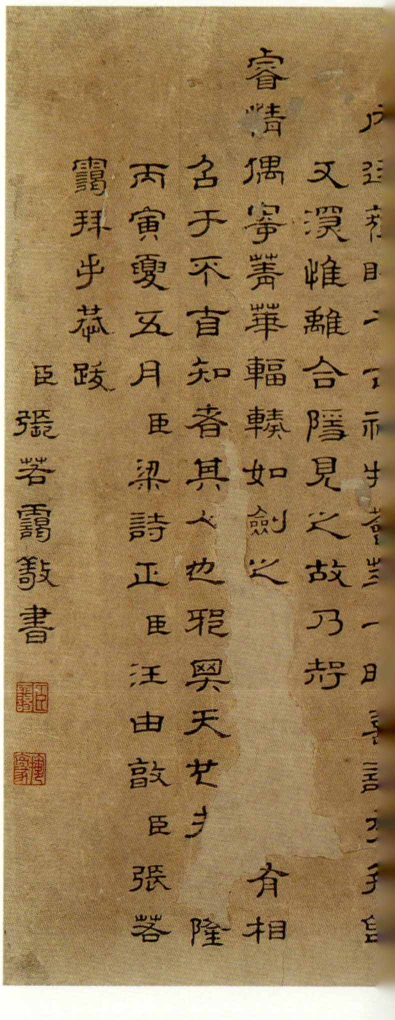

72 *The Four Beauties Collection*

Qing Dynasty, Qianlong Reign (1736-1795), Dong Bangda
Album leaf, ink on paper
L.:30.7cm, W.:57.1cm
Collection of the Tianjin Museum

In 1746, the imperial collection gathered four famous painting scrolls. To commemorate this, the Qianlong Emperor commissioned Dong Bangda to create the Album of the Four Beauties Collection and the Combined Scroll of the Four Beauties Collection. This album page is likely the prototype. Dong used a dry, light-ink technique to layer texture on the rocks and meticulously depicted thatched cottages–a hallmark of his style. (Han Xiaohe)

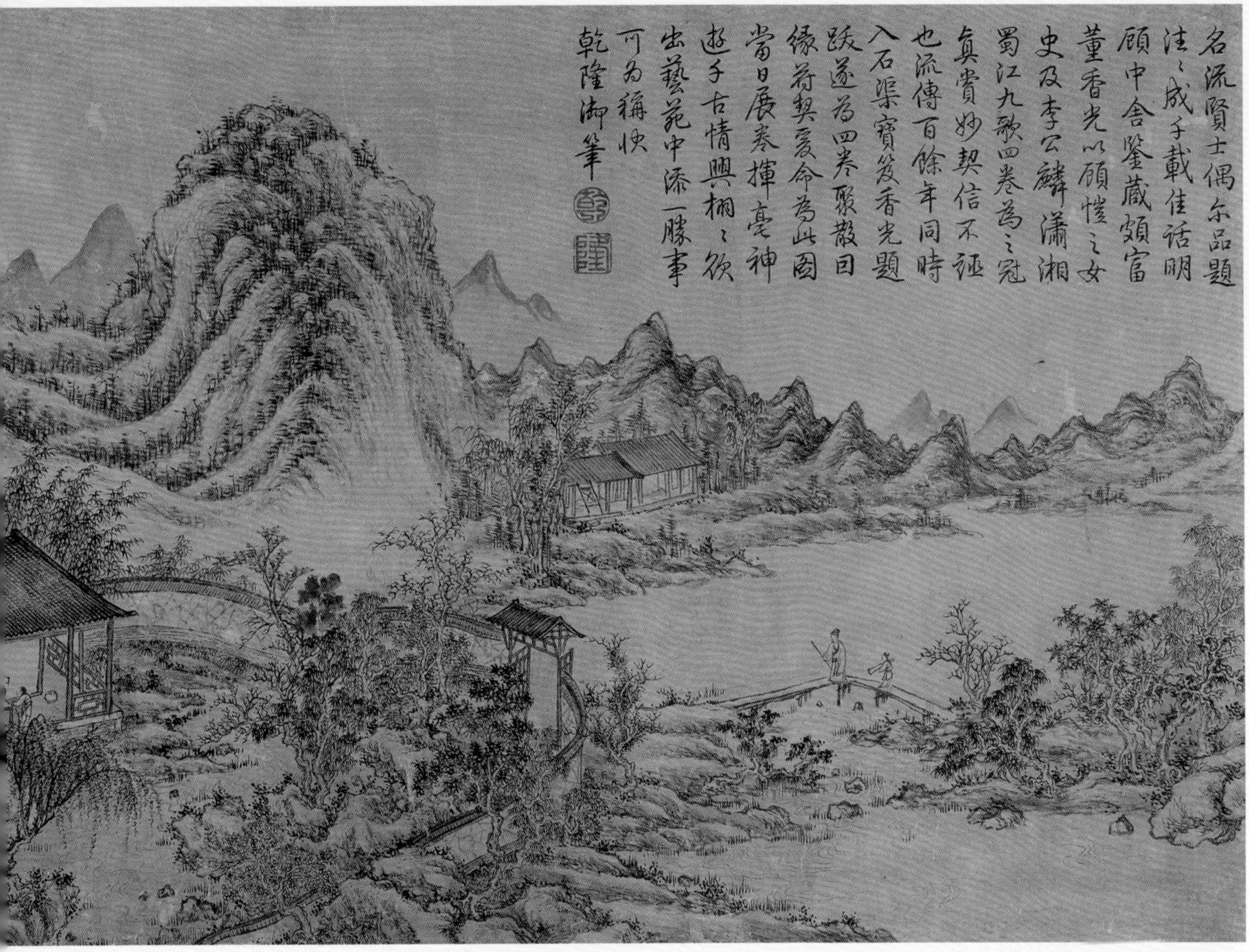

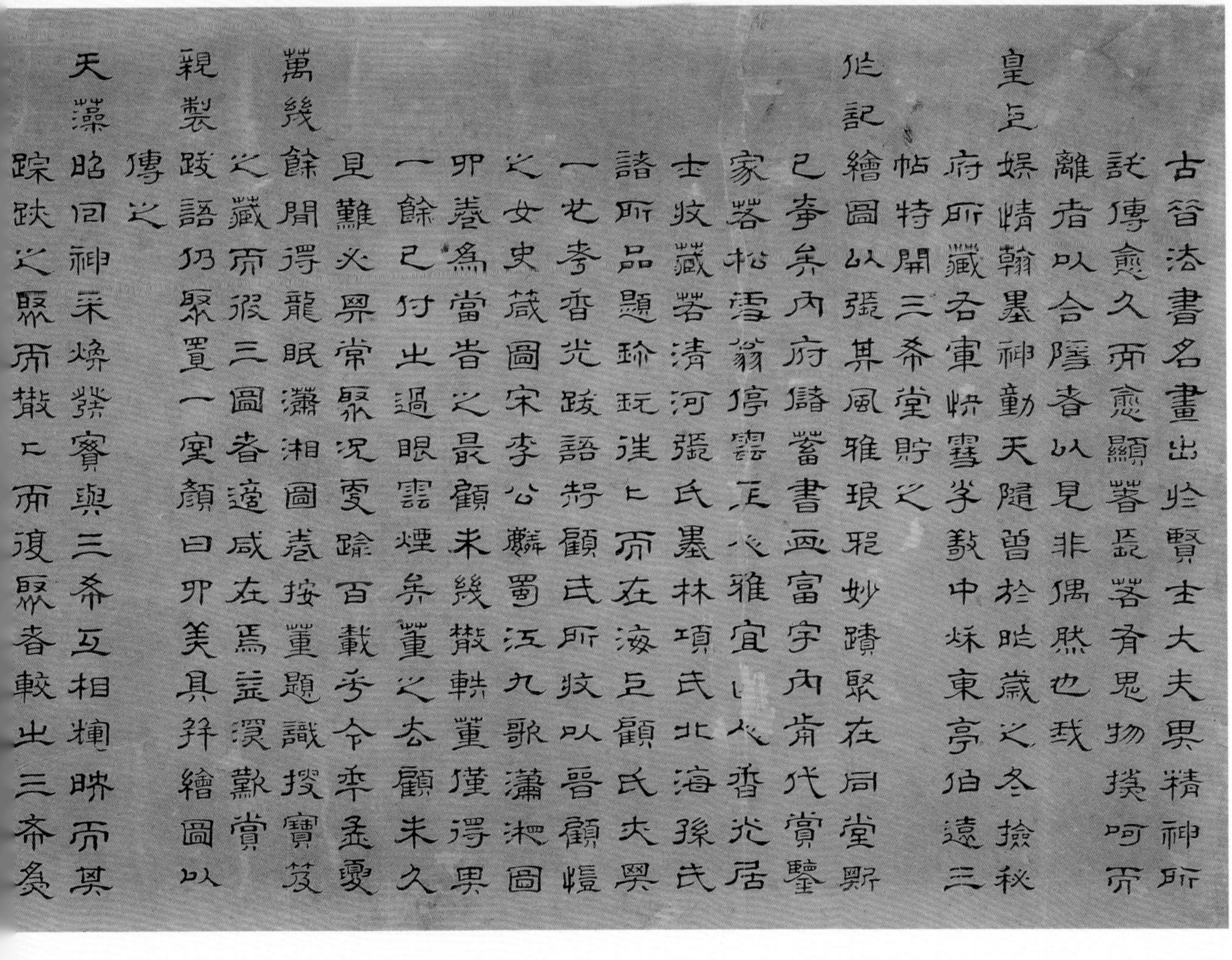
古昔法書名畫出於賢士大夫畀精神所
託傳愈久而愈顯著豈著有鬼物撝呵而
離者以合隱者以見非偶然也我
皇上
娛情翰墨神動天隨曾於睢歲之冬檢秘
府所藏名畫快雪學數中娣東亭伯遠三
帖特開三希堂貯之
御記
繪圖以張其風雅琅琊妙蹟聚在同堂夥
己事爲內府儲蓄書畫富字內肯代賞鑒
家若松雪翁停雲生之雅宜山人香光居
士收藏若清河張氏墨林項氏北海孫氏
諸所品題珍玩往々而在海上顧氏夫畀
一也李香光跋語稱顧氏所收以晉顧愷
之女史箴圖宋李公麟蜀江九歌瀟湘圖
四卷爲當昔之最顧未幾散軼董僅得畀
一餘已付出過眼雲煙弃董之太顧未久
見難必畀常聚況更踰百載予今季孟夏
萬幾
餘閒得龍眠瀟湘圖卷按董題識搜寶笈
之藏而得三圖者適咸在焉益深歎賞
親製
跋語仍聚置一室顏曰四美具并繪圖以
傳之
天藻
昭回神采煥發寶與三希互相輝映而其
蹤跡之聚而散散而復聚者較出三希爲

73 九歌图卷

清乾隆二十五年（1760），丁观鹏
纸本，墨笔
纵 45 厘米，横 738 厘米
故宫博物院藏

《九歌》是屈原所作的楚辞名篇，共 11 段，最初是先秦时期楚国所辖湖南沅、湘一带民间祭祀神灵的歌词，屈原将其改写，去除了原诗中淫祀的部分。改写后的《九歌》文采绮华，想象瑰丽，成为流传千古的经典诗歌。因诗中神灵、山鬼等特殊形象，《九歌》也成为画家描摹的题材。

乾隆内府藏有三卷李公麟款《九歌图》卷，均为墨笔，分别贮藏于静怡轩、养心殿和御书房，其中静怡轩所藏被定为上等，著录于《石渠宝笈初编》。乾隆帝喜爱李公麟此画，称“《九歌》辞采奇丽，意境惝恍，非伯时通神之笔，不能得其仿佛……几余披览，但觉灵气往来，非寻常意度所及”，并将其与顾恺之《女史箴图》并提，给予了很高的评价。乾隆帝考虑到李公麟所画《九歌图》是墨笔，但《九歌》中“青云衣、白霓裳（东君）”“孔盖翠旌（少司命）、龙堂贝阙（河伯）”之句，非五彩设色不能彰显其瑰丽，因命丁观鹏仿为此图，并加设色，意欲和李公麟一争高下。丁氏此图完成于乾隆二十五年（1760）六月。

丁观鹏仿李公麟《九歌图》卷，采用通景布局，将前九段内容布置于画幅之中，每段用山水云气分隔，使画面和谐统一。丁氏仿李公麟笔意，大体布局和人物形象、动作相似，山水点景等具体画面略有变化，这也是乾隆时期宫廷画家模仿古画的特点。与李氏《九歌图》一样，乾隆帝在丁摹画作的每段相应处也抄录有《九歌》诗句。

丁观鹏（生卒年不详），清代宫廷画家，乾隆时供奉内廷，善画释道人物，画风工整精细，并受到西画影响。丁观鹏奉敕临摹了大量古代名作。（施含牧）

73 ***The Nine Songs***

1760, Ding Guanpeng
Handscroll, ink on paper
L.:45cm, W.:738cm
Collection of the Palace Museum

In 1760, court painter Ding Guanpeng was commissioned by the emperor to create this painting after the style of *The Nine Songs* by Li Gonglin, which was housed in the Gallery of Tranquil Ease (*Jingyi xuan*). The long scroll features landscapes interwoven with divine figures, capturing their vivid expressions and dynamic postures, with flowing garments rendered naturally. The artwork also includes lines from *The Nine Songs* transcribed by the Qianlong Emperor. (Shi Hanmu)

74 临李公麟女史箴图卷

清乾隆十二年（1747），弘历
纸本，墨笔
纵 28 厘米，横 28 厘米—80 厘米
故宫博物院藏

晋惠帝时，贾后专权善妒，张华撰《女史箴》以历代贤德女子事迹教育、规劝宫中女子。“女史”指宫廷妇女，“箴”为规劝之意。传顾恺之将文中故事以图画形式描绘，使之通俗易懂。清宫旧藏两卷《女史箴图》，即大英本和现藏故宫博物院的宋摹本，二本构图接近，均保留祖本遗意。较之大英本，宋摹本画面内容完整，为白描本，入藏后被定为李公麟作。乾隆帝对此卷颇为喜爱，曾亲手临摹。

本卷即乾隆帝款摹本，体现了他的书画鉴藏活动。画卷共 12 段，首段绘“妇德尚柔，含章贞吉”，其后为“樊姬感庄”“卫女矫桓”“冯婕妤挡熊”“班婕妤辞辇”“防微虑远”“知饰其性”“出其言善”“灵监无象”“欢不可以黩”“靖恭自思”“女史司箴”，每段均题有《女史箴》文字。引首乾隆帝题“彤管贻芬”，卷尾署：“向爱李公麟《女史图》并所书《女史箴》，有顾恺之遗意。甲子秋曾手临一过，兹值时雨既沾，笔墨滋润，再成是卷。宋曾巩谓《家人》之义，归于修身，二南之化，本于文王。予之留意于此，岂惟豪翰是娱己耶。乾隆丁卯夏六月，御识。”这段文字和宋摹本上的乾隆御笔“乾隆甲子秋八月抚临一周”相吻合，知此卷依宋摹本作，绘于乾隆十二年（1747）。虽题作御笔，但并非皆出自乾隆帝手。

乾隆帝弘历（1711—1799），自幼受祖、父影响，广泛收集历代书画名作，一生钟情于诗文、书画的创作及鉴藏。其绘画技巧稚拙平实，但基于广泛的收藏和鉴赏经验，自有一种清雅趣味在其中。（邢陆楠）

彤管貽芬
乾隆宸翰

衛女矯桓
耳忘和音
志厲義高
而二主易
心

御臨李公麟女史箴圖

74 **After Li Gonglin's *Admonitions of the Instructress to the Court Ladies***

1747, Hongli (the Qianlong Emperor)
Handscroll, ink on paper
L.:28cm, W.:28cm-80cm
Collection of the Palace Museum

In the Gallery of Tranquil Ease, part of the gardens of the Palace of Established Happiness, the Qianlong Emperor designated a special room named the Four Beauties Collection Chamber (*Simeiju*) to house four paintings: *Admonitions of the Instructress to the Court Ladies, Imaginary Tour Through Xiaoxiang, Magnificent Views of Shuchuan,* and *The Nine Songs*. This scroll is an imitation by the Qianlong Emperor of a Song Dynasty copy. His artistic creations were often closely tied to his collection. (Xing Lunan)

道同隆而
不殺物無
盛而不衰
日中則昃
月滿則微
崇猶塵積
替若駭機

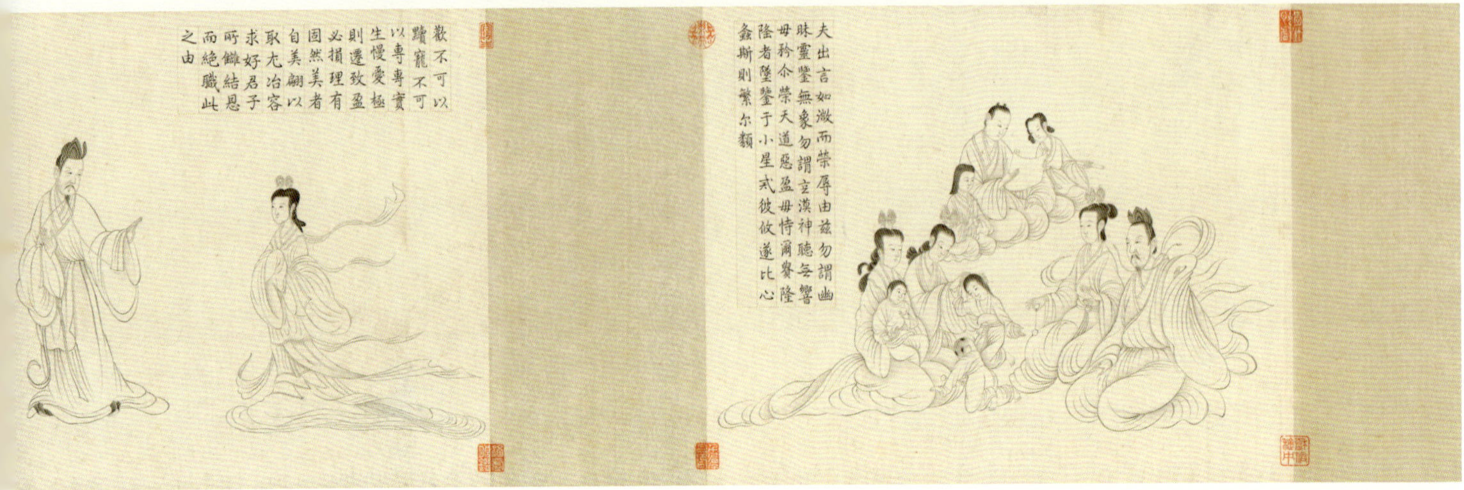
夫出言如微而榮辱由茲勿謂幽
昧靈鑒無象勿謂玄漠神聽無響
毋矜尒榮天道惡盈毋恃爾貴隆
隆者墜鑒于小星式彼攸遂比心
螽斯則繁尔類
歡不可以
瀆寵不可
以專專實
生慢愛極
則遷致盈
必損理有
固然美者
自美翩以
取尤冶容
求好君子
所讎結恩
而絕職此
之由

乾隆御臨李公麟女
史箴圖

75 墨竹图卷

元（1271—1368），李衎
纸本，墨笔
纵 32.5 厘米，横 92 厘米
故宫博物院藏

本幅无款，钤“李仲宾”“息斋”印。李衎以自然为师，对竹子的种类、形态皆穷究熟察，并形诸笔墨。图中墨竹一竿仰天直立，形如凤尾，虚空摇曳。笔墨挺劲厚重，意境萧散俊朗。引首有明代文彭隶书“君子林”。

此图《石渠宝笈初编》著录为《元人君子林卷》，全卷共收七家之作，存建福宫西偏之室，定为上等。乾隆帝将内府所藏曹知白《十八公图》、元人《君子林图》及宋元人《梅花合卷》一并收贮此处，因此定室名为“三友轩”，并在乾隆十二年（1747）敕命做檀匣合贮其中。又御制长歌，敕诸臣和韵汇刻其上。三友轩室内以松、竹、梅三友图案为装饰，室外特地遍植三友，皆为应和三卷内容。

李衎（1245—1320），字仲宾，号息斋道人，蓟丘（今北京）人。元仁宗时任吏部尚书、集贤殿大学士，晚年定居扬州。擅枯木竹石，尤善双勾设色竹及水墨竹。（聂卉）

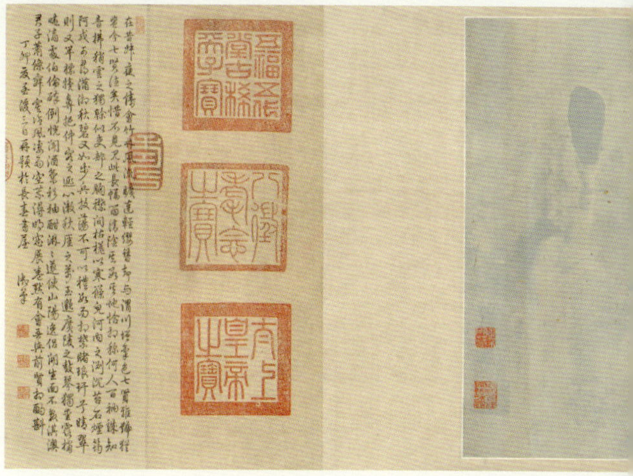

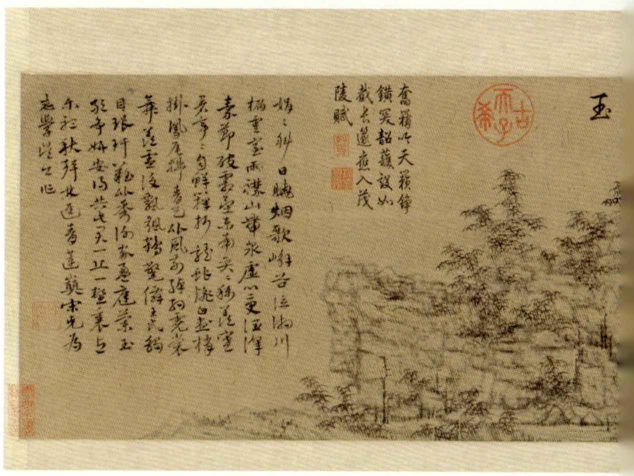

75 *Ink Bamboo*

Yuan Dynasty (1271-1368), Li Kan
Handscroll, ink on paper
L.:32.5cm, W.: 92cm
Collection of the Palace Museum

The Qianlong Emperor stored *Eighteen Scholars, Plum Blossom Scrolls by Song and Yuan Artists,* and *The Forest of Gentleman Scholars by Yuan Artists* in the Bower of the Three Friends in the Palace of Established Happiness. The Bower was adorned with motifs of these plants and the garden outside was planted with them to match the theme. This scroll bears Li Kan's seal. Known for his naturalistic bamboo paintings, he depicts a single bamboo stalk rising like a phoenix tail. (Nie Hui)

元人君子林圖神品
君子林

獨幹梢雲

瀟湘秋碧
苔石幽篁
丹丘柯九思

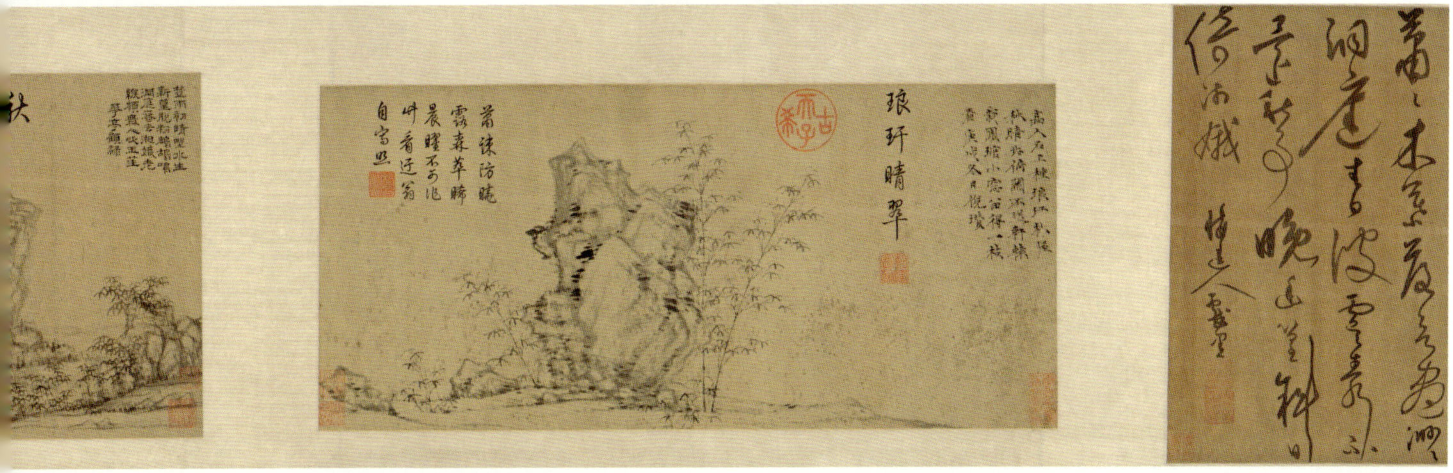
琅玕晴翠

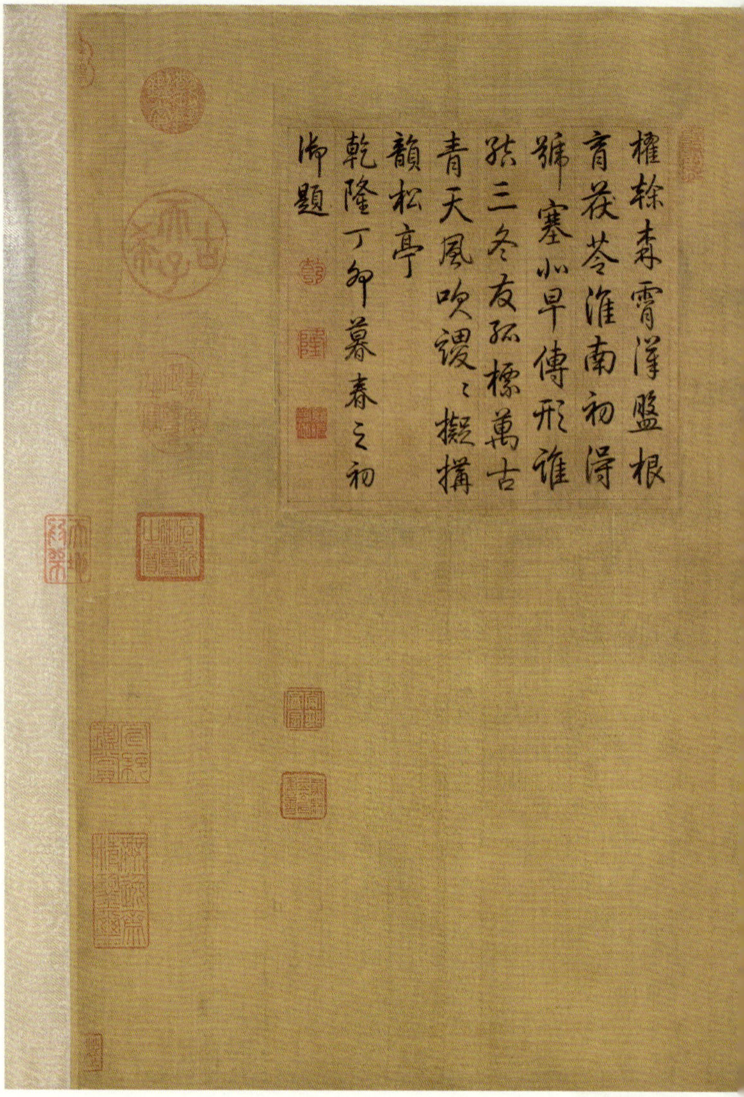

76 十八公图卷

元（1271—1368），曹知白（传）
绢本，墨笔
纵 45 厘米，横 95 厘米
故宫博物院藏

此图曾与元人《君子林图》、宋元人《梅花合卷》同贮藏于建福宫花园三友轩。“十八公”为松树之名。图绘参天巨松一株，松底部分叉，共 18 股，其中两株被砍，瘢痕清晰可见，各枝干均笔挺峭立，16 根主干之上又分出数根斜枝，彼此方向不同、姿态各异。画家以墨线勾勒，辅以皴擦渲染，并借助墨色的浓淡深浅，展示前后的叠压与空间关系，表现松针的线条短促而直挺有力，构图疏密有间，使得整树繁密不显杂乱。松根部有“曹知白画”四字款题。据徐邦达《古书画伪讹考辨》，笔墨与曹书不类，应系后添。但此图曾经元代祥哥剌吉收藏，流传有序，故仍应定为元人之作。树右侧有乾隆帝御书“十八公”三字，左侧有御制诗文：“擢干森霄汉，盘根育茯苓。淮南初得号，塞北早传形。谁结三冬友，孤标万古青。天风吹谡谡，拟构韵松亭。乾隆丁卯暮春之初御题。”

此图曾经安岐《墨缘汇观》及《石渠宝笈初编》著录。据著录，知其原有冯子振《十八公赋》及邓文原、赵岩跋文，赋及跋今均已不存。“十八公”本为一松，因共有 18 根枝干，故名。据冯子振赋文，此树在长城之北数百里，18 干中有两根已被斫，冯子振多次经过此树，令“方外客卢生”画此图。

此图鉴藏印有“皇姊图书”“都省书画之印”“真澹”“棠村审定”“蕉林玉立氏图书”“秋碧”（半印）“苍岩”（半印）“安仪周家珍藏”及清乾隆、嘉庆、宣统诸帝鉴藏印。（毛翔宇）

76 ***Eighteen Scholars***

Yuan Dynasty (1271-1368), Attributed to Cao Zhibai
Handscroll, ink on silk
L.:45cm, W.:95cm
Collection of the Palace Museum

This scroll was once stored in the Bower of the Three Friends, alongside *The Forest of Gentleman Scholars by Yuan Artists* and *Plum blossom Scrolls by Song and Yuan Artists*. *Eighteen Scholars* actually refers to a pine tree: the painting depicts a massive pine with eighteen branches. Although the inscription 'painted by Cao Zhibai' was added later, the scroll was once owned by Sengge Ragi during the Yuan Dynasty and has a well-documented provenance, confirming it as a Yuan work. (Mao Xiangyu)

77 竹院品古图页

明（1368—1644），仇英
绢本，设色
纵 41.1 厘米，横 33.8 厘米
故宫博物院藏

仇英（？—1552），字实父，一作实甫，号十洲，太仓（今属江苏）人，居吴县（今江苏苏州）。善绘人物、山水，工写俱佳，尤善临摹。该页选自仇英《人物故事图》册，画面描绘三位文士在园林中烹泉品茗、鉴赏古玩字画的场景。据画面内容可知，头戴软脚幞头，坐在藤墩上的男子应为园主人，身后有二仕女相伴，他在指挥仆人将所藏画轴、器物陆续呈上，列而玩之。画面上方竹林中的春笋说明此时正值春季。该页刻画精工，设色艳雅，人物形象生动、自然。本幅款署"实父"，钤"仇""英"朱文联珠印。

明中期后，致仕还乡的官员多热衷营造私人书斋、草堂或园林，并召集志趣相投者进行以诗画创作或古玩鉴赏为主题的文化雅集。明人认为，品古活动宜先治幽轩邃室，虽在城市，亦有山林之致。故此，园林景观与雅集博古活动的融合，体现了园主人的鉴藏品位与园林景观意趣的一致性。本幅画面运用了多样的园林装饰元素，如中景的屏风、朱栏，远景的湖石、石方桌、竹林。园林环境不仅体现了文人雅集所必需的客观物质条件，同时也蕴含着文人热衷园居、追求精致优雅的生活品味。（蒋彤）

77 *Appreciating Antiquities in a Bamboo Courtyard*

Ming Dynasty (1368-1644) , Qiu Ying
Album leaf, color on silk
L.:41.1cm, W.:33.8cm
Collection of the Palace Museum

Ming intellectuals believed that appreciating antiquities required a secluded and elegant space. Even in urban settings, these spaces evoked the tranquillity of mountains and forests. This painting employs various garden decorative elements, such as screens, red railings, lake rocks and bamboo groves. In the scene, scholars sip tea and admire antiques, while the garden owner directs servants to present more scrolls and artifacts. (Jiang Tong)

《竹院品古图》页所绘青铜器种类繁多，包括鼎、方鼎、簋等食器，尊、觚、提梁卣、罍、角等酒器，鉴、盉等水器，以及以钟为代表的乐器，基本涵盖了青铜器的主要品类。另外，可见其他材质仿铜器器形的仿古器，如斝、豆、樽等。画面结构清晰，以屏风为界，近景铜器之青绿与远景庭院翠竹交相辉映，更加凸显品古逸事之雅致，将承载历史之重器融入庭院的自然风光之中，不失和谐自然。本图后所列六件青铜器和瓷器，与图上器物相似。（哈少川）

78 父戊舟爵

西周（公元前 1046—前 771）
口纵 18 厘米，口横 8.9 厘米，高 23.2 厘米
故宫博物院藏

爵长流槽，曲口，尖尾上翘较甚。口沿有一对束伞形立柱，卵圆形腹上有三道扉棱，内侧有兽首鋬，三足较高。腹饰曲折角兽面纹，流下饰仰叶纹，颈饰三角纹。左柱铸铭文二字，流侧铸三字，共五字：作尊，父戊，舟。爵为最早出现的青铜礼器，早期单独出现，商代以后多与觚、斝等酒器成组出现。（哈少川）

78 Bronze *Jue* Vessel

Western Zhou Dynasty (1046-771 BCE)
L. (at mouth) :18cm, W.(at mouth):8.9cm, H.:23.2cm
Collection of the Palace Museum

79 冎父己鼎

商（公元前 1600—前 1046）
宽 17.5 厘米，高 21.8 厘米
故宫博物院藏

此件为清宫旧藏。鼎敛口，窄沿方唇，口沿上有一对立耳，外侈。鼓腹分裆，三柱足细高。腹部饰三组下卷角兽面纹浅浮雕，以云雷纹填地。腹内壁铸有铭文三字：冎父己。

鼎是青铜礼器中的主要食器，作为烹食器的一种，有烹煮肉食、实牲祭祀和燕飨等各种用途。在古代社会中，鼎被用作等级制度和权力的标志。青铜鼎由新石器时代的陶鼎发展而来，目前发现最早者出土于二里头遗址。商代铜鼎已有圆鼎、方鼎、高鼎和扁足鼎等多种形式。此鼎曾以“周社鼎”为名著录于《西清古鉴》第二卷，意指置于社内的祭鼎。（哈少川）

79 Bronze *Ding* Vessel

Shang Dynasty (1600-1046 BCE)
W.:17.5cm, H.:21.8cm
Collection of the Palace Museum

80 文王方鼎

明（1368—1644）
宽 18.1 厘米，高 25.5 厘米
故宫博物院藏

鼎直口，平折沿，上饰双立耳。浅腹，平底，四扁足。腹部四面均饰兽面纹，间饰夔纹，八条扉棱分布于四面正中及四角。扁足以夔龙为形，昂首卷尾擎起鼎身。腹部内侧铸有铭文七字：周公作文王尊彝。

文王鼎是仿古彝器中较为多见的一种，因其扁足形制少见，又以铭文中涉及周文王、周公旦二贤，深得后世喜爱。除青铜以外，还有铜胎掐丝珐琅、玉、陶等其他材质的文王鼎。除收藏以外，文王鼎还用作陈设或蹴香。乾隆时期出现将其充当礼器用在文庙祭祀活动中的情况。（哈少川）

80 Bronze Square *Ding* Vessel

Ming Dynasty (1368-1644)
W.:18.1cm, H.:25.5cm
Collection of the Palace Museum

81 涡纹盉

西周（公元前 1046—前 771）
宽 34 厘米，高 24 厘米
故宫博物院藏

81 Bronze *He* with Spiral Pattern

Western Zhou Dynasty (1046-771 BCE)
W.:34cm, H.:24cm
Collection of the Palace Museum

盉造型别致，作侧置鼓形，独具风格。上有椭方口，方唇直颈；下具四扁足，呈夔形对峙。管状流呈长直状，盉鋬呈顾首夔龙状，均饰以云雷纹。腹部纹饰为三层同心圆环状纹饰，由内到外分别饰以涡纹、重环纹与斜角雷纹。

盉是一种水器，与盘搭配，用于沃盥之礼，是用以倾倒液体的早期器形，为后世茶壶等器形的发展带来了深远影响。流与鋬为倾注器的普遍特征，早期器物的鋬装饰性高于实用性。因其器身大且较重，纤细的鋬不足以承重，在使用时常以双手捧持器腹。发展为酒具和茶具的倾注器，鋬部设计也随着材质的更迭而注重实用性。（哈少川）

82 仿哥釉碗

明（1368—1644）
口径 11.5 厘米，足径 3 厘米，高 4.9 厘米
故宫博物院藏

碗撇口，斜腹，小底，圈足。碗内、外施青釉，满布粗、细两种开片。口沿处镶铜扣，圈足端无釉，露出铁黑色胎体。此器模仿宋代斗笠盏，具有“金丝铁线”的开片特征和“紫口铁足”的做法，具哥窑瓷器神韵，显示出明代文人在器用方面对复古的追求。（单莹莹）

82 Imitation *Ge*-glazed Bowl

Ming Dynasty (1368-1644)
Diam.(at mouth):11.5cm, Diam.(at foot):3cm, H.:4.9cm
Collection of the Palace Museum

83 仿哥釉纸槌瓶

明（1368—1644）
口径 5.7 厘米，足径 8.3 厘米，高 14 厘米
故宫博物院藏

瓶平折沿，直颈，平肩，肩部下折，直腹，圈足。外壁通体施青釉，满布长而粗的黑色开片和短而浅的金色开片。圈足端施酱色护胎釉。

纸槌瓶是宋代经典瓶式，定窑、汝窑、龙泉窑均有烧造。此瓶仿哥窑瓷器风格，既有“金丝铁线”的开片特征，又有深色圈足的“铁足”特色，颇具古意。此类瓶一般用于陈设、插花。（单莹莹）

83 Imitation *Ge*-glazed Paper-Hammer-Shaped Bottle

Ming Dynasty (1368-1644)
Diam.(at mouth):5.7cm, Diam.(at foot):8.3cm, H.:14cm
Collection of the Palace Museum

84 造云石

元（1271—1368）
宽 70 厘米，高 63 厘米
故宫博物院藏

造云石系灵璧石，因其“如天风遏云抱崖绮布之势”，取郦道元《水经注》中“造云壁立”句而命名，“造云”二字刻于石额间。石上共八处题刻，五处元人，三处明人。玉山雅集是元代苏州玉山草堂主人顾瑛主持的文人雅集，造云石刻述了元至正十年（1350）秋以赏石为主题的雅集盛事。此石陈设于紫禁城建福宫花园东侧惠风亭内，出静怡轩或在静怡轩廊下，便可望见。（张淑娴）

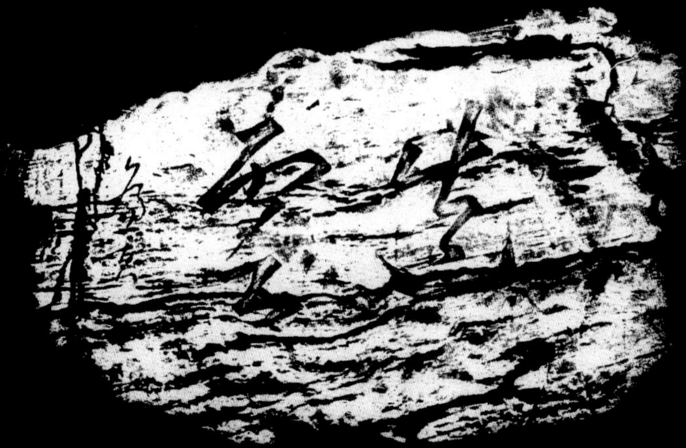

84 Cloud-embracing Stone

Yuan Dynasty (1271-1368)
W.:70cm, H.:63cm
Collection of the Palace Museum

This stone was originally placed in the Palace of Established Happiness (*Jianfu gong*) in the Forbidden City. Its inscription records one of the celebrated 'Yushan Elegant Gatherings', hosted by Gu Ying, the master of Yushan Cottage in Suzhou, in the autumn of the 10th year of Zhizheng reign (1350) during the Yuan Dynasty. This particular gathering was themed on stone appreciation. (Zhang Shuxian)

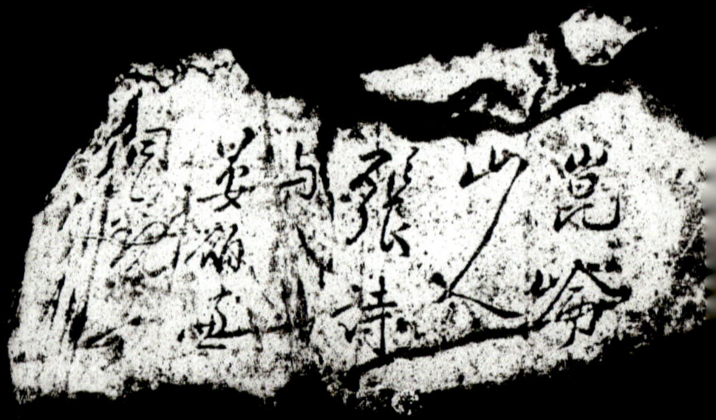

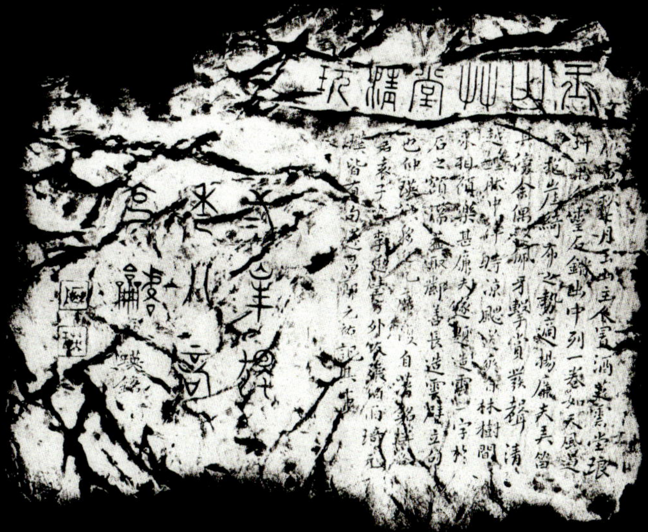

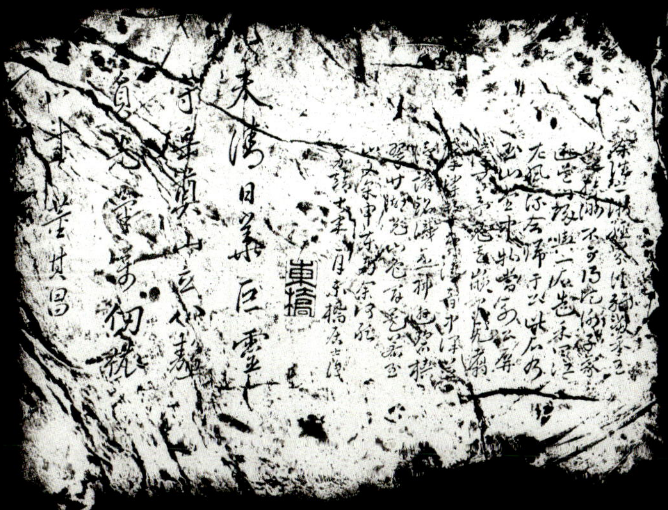

14 世纪，文艺复兴运动发端于意大利，人们推崇古希腊和古罗马的艺术成就。大量出土的古代艺术品被争相纳入教皇、贵族与富商们的收藏，用于装饰华美的宅邸和花园，以彰显权力与财富。约 1519 年，拉斐尔在致教皇利奥十世的信中写道："古城之荣光与壮丽，非由他人之述可知，乃由遗存之废墟而自显。盖此废墟，至圣教皇陛下，恍如亲见，彰显人类造物之极致与伟力。"（许愿）

In the 14th century, the Renaissance movement began in Italy, with people admiring the artistic achievements of ancient Greece and Rome. Numerous ancient artworks unearthed through archaeological discoveries were eagerly acquired by popes, nobles, and wealthy merchants to decorate their magnificent residences and gardens, thereby showcasing their power and wealth. In his letter to Pope Leo X around 1519, Raphael wrote: 'The glory and grandeur of the ancient city cannot be known through others' descriptions but are revealed through the ruins that remain. These ruins, Your Holiness, appear as if you are witnessing them in person, demonstrating the pinnacle and power of human creation.'(Xu Yuan)

85 古典风格的建筑随想

约 1755—1765，［意大利］乔瓦尼·保罗·帕尼尼、［法国］查尔斯·约瑟夫·纳托尔
黑色粉笔、钢笔，黑色和灰色墨水、灰色和棕色淡彩，点缀水彩和白色水粉
纵 36.2 厘米，横 24.5 厘米
大都会艺术博物馆藏

帕尼尼（Giovanni Paolo Panini）是 18 世纪罗马杰出的画家，其"建筑随想"风格的创作旨在唤起对罗马辉煌过往的挽歌式追忆。随想画作常借助废墟中可见的辉煌遗迹，或是与想象建筑的组合，营造出冲突与幻想并存的氛围。此画左下角，望景楼的残躯雕塑（The Belvedere Torso）及其他艺术品散落于废墟之中，便是画家想象中的图景。（雷浩）

85 Architectural Capriccio with Figures and Antiquities

Ca. 1755-1765, Giovanni Paolo Panini (Italian), Charles Joseph Natoire (French)
Black chalk, pen, black and gray ink, gray and brown wash, with touches of watercolor and white gouache
L.:36.2cm, W.:24.5cm
Lent by The Metropolitan Museum of Art, Harry G. Sperling Fund, 2018 (2018.863)

Panini's *Architectural Capriccio* evokes an elegiac vision of Rome in ruins. The theme of *capriccios* often leverages the grandeur of the past glimpsed through the ruins or the imagined juxtapositions of architectural elements to convey heightened emotions of conflict and fantasy. For example, in this painting, The Belvedere Torso in the lower left corner, along with other scattered statues and artworks amidst the once-majestic ruins, depicts a fantastical collection scene. (Lei Hao)

86 古代罗马

1757—1760，据［意大利］乔瓦尼·保罗·帕尼尼作品创作
纸本，石墨底稿，钢笔和墨水、水彩、淡彩，白色高光
纵 44.3 厘米，横 69.6 厘米
芝加哥艺术博物馆藏

帕尼尼的《古代罗马》与《现代罗马》系列启发了众多罗马风景画家。《古代罗马》中，他将罗马各地最著名的古迹，与多个庄园庭院收藏的古代雕塑，组合在一个想象的画廊场景中。18 世纪正值“大旅行（The Grand Tour）”风行时期，画作中的微型雕塑和遗址足以构成罗马经典庄园的浓缩行程。（郭欣悦）

86 Roma Antica

1757-1760, after Giovanni Paolo Panini (Italian)
Pen and ink, watercolor, wash, with white heightening, over graphite, on paper
L.:44.3cm, W.:69.6cm
Collection of the Art Institute of Chicago, Gift of Emily Crane Chadbourne (1955.745)

Giovanni Paolo Panini's *Ancient Rome* and *Modern Rome* series inspired numerous landscape painters. This work is a reproduction of his renowned painting *Ancient Rome*, in which he skillfully combined Rome's most famous ancient monuments with classical sculptures from various villa collections, arranging them within an imagined gallery setting. During the 18th century, when the 'Grand Tour' was at its peak, the sculptures and ruins depicted in his works offered viewers a condensed tour of Rome's iconic villas and landmarks. (Guo Xinyue)

87 美第奇宫的柱廊和花园

1870 后，[法国] 于贝尔 · 罗贝尔风格
木板，油画
纵 44.5 厘米，横 39.4 厘米
芝加哥艺术博物馆藏

罗马美第奇宫以石碑、石柱、檐饰、马赛克等古代艺术品装饰其建筑与花园。画家用花瓶取代了原本的墨丘利雕像，将这件藏品一同收入画中。美第奇宫被誉为“露天博物馆”，鲁本斯、普桑、安格尔等后世艺术家都曾驻留于此。画中亦刻画了摹写花园中收藏的艺术品的绘者形象。（许愿）

87 Colonnade and Gardens of the Medici Palace

After 1870, style of Hubert Robert (French)
Oil on panel
L.:44.5cm, W.:39.4cm
Collection of the Art Institute of Chicago (1968.616)

The Medici Palace in Rome is adorned with ancient artworks such as steles, columns, cornices, and mosaics. In this painting, the artist incorporates a vase into the scene, replacing the original Mercury statue. The palace is renowned as an open-air museum, and has hosted artists like Rubens, Poussin and Ingres. The painting also features an artist sketching these artworks. (Xu Yuan)

88 美第奇花瓶

选自《罗马和坎帕尼亚的六大美景》之一
1656 年，[意大利] 斯特凡诺 · 德拉 · 贝拉
纸本，蚀刻和雕刻版画
纵 30.9 厘米，横 27.5 厘米
芝加哥艺术博物馆藏

大理石瓶身上以精美的浮雕刻绘神话场景，曾陈列在罗马美第奇宫，现藏于佛罗伦萨乌菲兹美术馆。画中年轻的艺术家正在描摹花瓶，与花园中的柏树、方尖碑等共同营造出浓郁的文艺复兴尚古风情。(许愿)

88 The Medici Vase

Plate one from *The Six Large Views of Rome and the Campagna*
1656, Stefano della Bella (Italian)
Etching and engraving on ivory laid paper
L.:30.9cm, W.:27.5cm
Collection of the Art Institute of Chicago, John H. Wrenn Memorial Collection (1972.776)

The marble vase, carved with Greek mythological scenes, was made in 1st century Athens. During the Renaissance, the Medici family displayed it at their villa in Rome, hence the name 'Medici Vase'. In the image, a young artist is depicted sketching the vase. The scene, enriched by the presence of cypress trees and an obelisk in the garden, evokes a strong sense of Renaissance antiquarianism. (Xu Yuan)

89 沐浴的维纳斯

17 世纪初，据［尼德兰］詹博洛尼雕塑模型制作
青铜
高 33.7 厘米
大都会艺术博物馆藏

维纳斯，爱与美的女神，在文艺复兴艺术中广受青睐，亦是园林雕塑的经典主题。此作品采用文艺复兴后期杰出风格派雕塑家詹博洛尼（Giambologna）创立的“蛇形雕像”构图。文艺复兴时期对人体美学的深入探索及对古典主题的全新诠释在此作品中可见一二。（段颖）

89 Bathing Venus Statuette

Early 17th century, after a model by Giambologna (Netherlandish)
Bronze
H.:33.7cm
Lent by The Metropolitan Museum of Art, Gift of Ogden Mills, 1924 (24.212.16)

Venus, the goddess of love and beauty, was a popular figure in Renaissance art and a classic theme in garden sculptures. This piece uses the serpentine figure composition invented by Giambologna, an outstanding Mannerist sculptor of late Renaissance Italy. It reflects the Renaissance's exploration of human aesthetics. (DuanYing)

90 萨科马佐内游戏者

约 1621，［意大利］奥拉齐奥 · 莫奇
青铜
长 45.7 厘米，宽 30.4 厘米，高 36.3 厘米
芝加哥艺术博物馆藏

萨科马佐内（Saccomazzone）是一种盛行于意大利文艺复兴时期的游戏。两位玩家均蒙眼，一手触石，另一手持麻包或打结的布条尝试击打对方，受击少且坚持手不离石者为胜。奥拉齐奥 · 莫奇（Orazio Mochi）以该游戏为主题构思了这座雕塑并制作了多个小型青铜、蜡和石膏像。1620 年，科西莫二世 · 德 · 美第奇（Cosimo II de'Medici）委托罗摩洛 · 费鲁奇（Romolo Ferrucci）以莫奇的构思为原型制作了石质雕塑，用于装饰美第奇家族在佛罗伦萨的波波里花园（Boboli Gardens）。波波里花园是意大利文艺复兴时期最具代表性的花园之一，位于皮蒂宫（Pitti Palace）的正后方，园内藏有大量古罗马及文艺复兴时期的雕塑。（郭欣悦）

90 Saccomazzone Players

Ca. 1621, Orazio Mochi (Italian)
Bronze
L.:45.7cm, W.:30.4cm, H.:36.3cm
Collection of the Art Institute of Chicago, Gift of the Edward A. and Inge Maser Collection (2009.130)

Saccomazzone was a popular Renaissance festival game in which two blindfolded players kept one hand on a stone and used a sack or knotted cloth to strike each other while evading their opponent's blows. The original sculpture remains in the Boboli Garden to the rear of the Pitti Palace, part of the Medici family's estate in Florence. The garden, adorned with many ancient and Renaissance sculptures, is famed as an open-air museum. (Guo Xinyue)

91 尼普顿与海豚

17 世纪末或 18 世纪初，据［意大利］乔凡尼·洛伦佐·贝尔尼尼雕塑模型制作
青铜
长 39.1 厘米，宽 29.8 厘米，高 51.4 厘米
大都会艺术博物馆藏

此青铜雕像为大理石喷泉雕塑尼普顿及其子（Neptune and Triton）的缩小版，原作曾置于罗马教皇西克斯图斯五世（Sixtus V）建造的蒙塔尔托别墅（Villa Montalto）中喷泉的顶部（见后对页铜版画）。原作由 17 世纪雕塑家贝尔尼尼（Gian Lorenzo Bernini）创作，他开创了巴洛克雕塑风格，作品以动感与戏剧性著称。（施琪）

91 Neptune with a Dolphin Statuette

Late 17th/early 18th century, after a model by Gian Lorenzo Bernini (Italian)
Bronze
L.:39.1cm, W.:29.8cm, H.:51.4cm
Lent by The Metropolitan Museum of Art, Rogers Fund, 1946 (46.183)

This is a small-scale variant of the marble fountain sculpture Neptune and Triton by 17th century sculptor Gian Lorenzo Bernini. The original sculpture, was once placed atop the fountain in Villa Montalto in Rome. (Shi Qi)

92 男性坐像

16 世纪中期，意大利中部
青铜
长 56.5 厘米，宽 31.8 厘米，高 54.6 厘米
大都会艺术博物馆藏

男性坐像为意大利加尔佐尼庄园（Villa Garzoni）旧藏，曾被放置于壁炉台上方正中。庄园府邸入口设有雕像大道，另有环绕着雕塑装饰的天井庭院。几个世纪以来，这座庄园几经易手，以其室内外的珍贵藏品享有盛誉。此坐像风格呈现出典型的 16 世纪佛罗伦萨艺术特征。（郭欣悦）

92 Seated Male Figure Statuette

Mid-16th century, central Italian
Bronze
L.:56.5cm, W.:31.8cm, H.:54.6cm
Lent by The Metropolitan Museum of Art, Edith Perry Chapman Fund, 1966 (66.177)

This statue was once centered on the mantel of a fireplace at Villa Garzoni in Pontecasale, Italy. The villa features an avenue lined with statues leading to the entrance, and inside, a courtyard adorned with sculptures. Over the centuries, the villa has changed ownership several times, and it is renowned for its valuable collections. (Guo Xinyue)

FONTANA E PISCHIERA NEL GIARDINO MONTALTO

Architettura del Ca

Gio. Francesco Venturini del et inc.

LE TERME DIOCLETIANE SV L MONTE VIMINALE

Domenico Fontana. 17

G.Iac.Rossi le stampa in Roma alla pace con Priu. del S.Pont.

93 罗马望景楼花园（复制品）

16 世纪，[尼德兰] 亨德里克 · 凡 · 克利夫三世
版画
纵 17.5 厘米，横 24 厘米
哈佛艺术博物馆藏

93 The Belevedere, Rome (Belvedere Romanum) (Facsimile)

16th century, Hendrick van Cleve III (Netherlandish)
Print
L.:17.5cm, W.:24cm
Harvard Art Museums/Fogg Museum, Light-Outerbridge Collection, Richard Norton Memorial Fund, Photo© President and Fellows of Harvard College (M24480.2)

15 世纪末，教皇英诺森八世（Pope Innocent VIII）在梵蒂冈城墙北端的山丘上修建望景楼别墅（Villa Belvedere），作为避暑、接待与艺术收藏之所，其名称意为“美丽的景色”。16 世纪初，教皇尤利乌斯二世（Julius II）委托著名建筑师布拉曼特（Donato Bramante）扩建梵蒂冈建筑群，画中展现了这一时期的景象。画面中部，新建的连廊将左侧望景楼别墅与右侧教皇宫连成一体，形成宏伟的望景楼花园（Belvedere Garden）。尤利乌斯二世热衷艺术，其收藏的《拉奥孔群像》和《阿波罗 · 贝尔维第》被安置在望景楼花园上层台地的雕塑长廊（Gallery of the Statues），奠定了梵蒂冈古典雕塑收藏的基础。（许愿）

At the center of the painting is the Belvedere corridor, connecting the Villa Belvedere on the left with the Papal Palace on the right, forming the Belvedere Garden. The Gallery of Statues on the upper terrace once housed the famous *Laocoön* and *Apollo Belvedere*, laying the foundation for Vatican's collection of classical sculpture. (Xu Yuan)

94 梵蒂冈望景楼

1765—1766，[法国]让·雅克·德·博瓦修
纸本水洗，黑色粉笔打底，白色高光提亮
纵 24.7 厘米，横 36.2 厘米
芝加哥艺术博物馆藏

画中描绘了望景楼别墅的外墙及其周边景致，此处为梵蒂冈城墙北端的山丘。画面中最高建筑的穹顶下安置着著名的松果雕塑（The Pigna），这座高约 4 米的古罗马青铜雕塑也是教皇的收藏，寓意不朽与重生。（许愿）

94 View of an Italian Villa and Gardens (the Belvedere of the Vatican)

1765-1766, Jean Jacques de Boissieu (French)
Wash, with white heightening, over black chalk, on paper
L.:24.7cm, W.:36.2cm
Collection of the Art Institute of Chicago, Gift of Mr. John L. Strauss (1959.45)

This painting depicts the exterior walls of Villa Belvedere and its surrounding landscape, located on the hill at the northern end of the Vatican walls. Beneath the dome of the tallest building in the scene stands the famous *Pigna* (The Pinecone), a 4-meter-tall bronze sculpture from ancient Rome. As part of the Pope's collection, it symbolizes immortality and rebirth. (Xu Yuan)

从 18 世纪开始，英国的造园理念逐渐由传统规则式转向追求非对称和风景如画般的模仿自然，雕塑神庙、洞穴、亭台楼阁等装饰性建筑作为园中的收藏也为“鲜活画面”提供了氛围和趣味。与此同时，以中国园林为代表的东方艺术与异域风情引起了英国贵族和文人的广泛兴趣，造型奇异又具有中国风格的庭阁和宝塔开始在英国流行。这种带有东方元素的建筑收藏和装饰应用的英国园林风格传入法国后，催生出了影响欧洲园林世界风尚的“英中式园林（Jardin Anglo-Chinois）”。（雷浩）

Starting from the 18th century, British garden design gradually shifted from traditional formal layouts to an emphasis on asymmetry and picturesque imitations of nature. Decorative structures such as sculpted temples, grottoes, pavilions, and pagodas were incorporated into gardens as collectibles, adding atmosphere and charm to these ‘living paintings’. At the same time, Eastern art and exotic aesthetics, particularly those of Chinese gardens, sparked widespread interest among British aristocrats and scholars. Uniquely shaped garden pavilions and pagodas with Chinese influences became fashionable in Britain. When this style of garden design, incorporating Oriental elements, was introduced to France, it gave rise to the *Jardin Anglo-Chinois* (‘Anglo-Chinese Garden’), which went on to shape garden trends across Europe. (Lei Hao)

95 奇斯威克花园中的三岔道景观

1760—1767，据［英国］约翰 · 多诺威尔作品创作
版画，手工着色
纵 31.8 厘米，横 45.7 厘米
耶鲁大学英国艺术中心藏

18 世纪，英国著名园艺师肯特（William Kent）与伯灵顿伯爵共同设计建造了西伦敦奇斯威克宅邸。其花园宛如一座博物馆，汇集了意大利古典风格的装饰性建筑与古罗马风格的人像雕塑。画中的三岔道分别通向三座帕拉迪奥式的建筑小品。肯特在奇斯威克花园的创新尝试践行了其“自然”风格的园林设计理念，并率先推广以装饰性建筑引导自然式园林布局的造园手法，深刻影响了 18 世纪下半叶整个欧洲大陆的园林设计发展（见后对页油画）。（王雪涵）

95 **A View of the Three Walks Terminated by the Cassina, the Pavilion, and the Rustic House in the Garden of the Earl of Burlington, at Chiswick**

1760-1767, after John Donowell (British)
Hand-colored engraving
L.:31.8cm, W.:45.7cm
Collection of the Yale Center for British Art (B1995.13.163)

This garden resembled a museum, showcasing Italian classical-style ornamental buildings and ancient Roman statues. In the painting, a three-pronged path leads to three Palladian-style structures. The Earl of Burlington and his designer William Kent used ornamental buildings to guide the natural layout of the garden, profoundly influencing garden design across Europe in the second half of the 18th century. (Wang Xuehan)

96 邱园中的塔和桥

1762，［英国］理查德·威尔逊
布面，油画
纵 47.6 厘米，横 73 厘米
耶鲁大学英国艺术中心藏

邱园最初是奥古斯塔王妃庄园中一座占地 3.5 公顷的植物园，现已扩大为 120 公顷规模的英国皇家植物园，园内有大量异域风格的建筑小品。其中的中式宝塔据传参照南京大报恩寺琉璃塔所建，为此地平添了一分典雅如画的意境。宝塔建成后，成为诸多艺术作品的灵感主题。（王雪涵）

96 Kew Gardens: The Pagoda and Bridge

1762, Richard Wilson (British)
Oil on canvas
L.:47.6cm, W.:73cm
Collection of the Yale Center for British Art (B1976.7.172)

Kew Gardens, originally a 3.5-hectare botanical garden within the estate of Princess Augusta, is now the 120-hectare Royal Botanic Gardens. It features many exotic structures, including a Chinese pagoda modeled after the Porcelain Pagoda of Nanjing, adding a picturesque charm to the landscape. Since its completion, the Pagoda has been the inspiration for many artworks. (Wang Xuehan)

邱园中的中式宝塔，奥斯瓦尔德·喜仁龙（Osvald Siren）摄于 20 世纪 40 年代

97 马戛尔尼使团礼品图册之《邱园花园和建筑平面图、立面图、剖面图和透视图》（复制品）

18 世纪，[英国] 威廉 · 钱伯斯
纵 38 厘米，横 55 厘米
中国第一历史档案馆藏

威廉 · 钱伯斯爵士（Sir William Chambers）是 18 世纪英国著名建筑师之一，对中国建筑与园林十分推崇。乾隆五十八年（1793），英国马戛尔尼使团访华时携带的这本礼品图册为其所著，细致地呈现了邱园中异域建筑的风姿，包括中式宝塔、鸟舍、清真寺，以及古典风格建筑和中国亭，此为其中两页。（王雪涵）

97 *Plans, Elevations, Sections, and Perspective Views of The Gardens And Buildings At Kew in Surry, The Seat of Her Royal Highness The Princess Dowager of Wales* (Facsimile)

18th century, Sir William Chambers (British)
L.:38cm, W.:55cm
Collection of the First Historical Archives of China

Sir William Chambers, a renowned 18th century British architect, greatly admired Chinese architecture and gardens. This gift album, carried by the Macartney Embassy to China in 1793 during the Qianlong Emperor's reign, details the exotic architecture of Kew Gardens, including the mosque and classical-style buildings, the Great Pagoda, the Aviary, and the Chinese pavilion on this page. (Wang Xuehan)

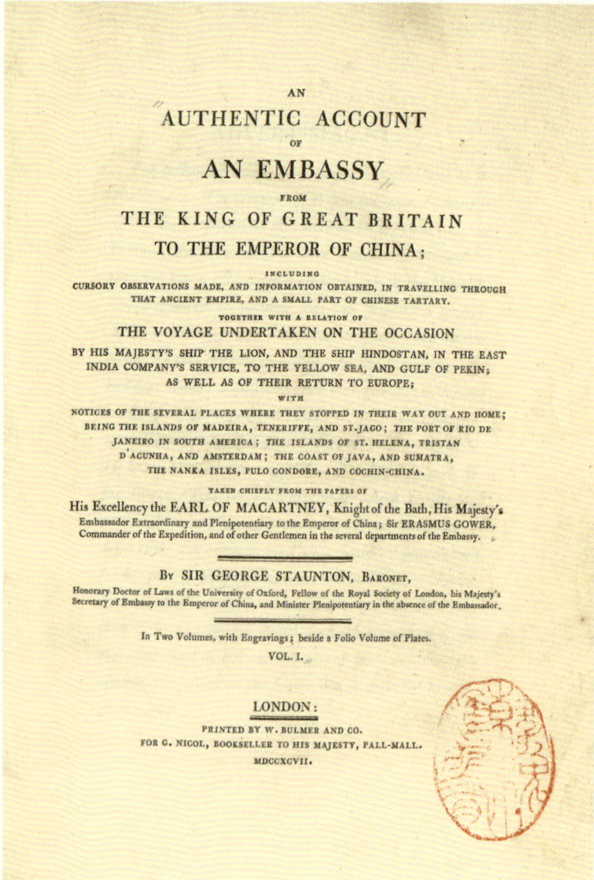

AN
AUTHENTIC ACCOUNT
OF
AN EMBASSY
FROM
THE KING OF GREAT BRITAIN
TO THE EMPEROR OF CHINA;
INCLUDING
CURSORY OBSERVATIONS MADE, AND INFORMATION OBTAINED, IN TRAVELLING THROUGH THAT ANCIENT EMPIRE, AND A SMALL PART OF CHINESE TARTARY.
TOGETHER WITH A RELATION OF
THE VOYAGE UNDERTAKEN ON THE OCCASION
BY HIS MAJESTY'S SHIP THE LION, AND THE SHIP HINDOSTAN, IN THE EAST INDIA COMPANY'S SERVICE, TO THE YELLOW SEA, AND GULF OF PEKIN; AS WELL AS OF THEIR RETURN TO EUROPE;
WITH
NOTICES OF THE SEVERAL PLACES WHERE THEY STOPPED IN THEIR WAY OUT AND HOME; BEING THE ISLANDS OF MADEIRA, TENERIFFE, AND ST. JAGO; THE PORT OF RIO DE JANEIRO IN SOUTH AMERICA; THE ISLANDS OF ST. HELENA, TRISTAN D'ACUNHA, AND AMSTERDAM; THE COAST OF JAVA, AND SUMATRA, THE NANKA ISLES, PULO CONDORE, AND COCHIN-CHINA.
TAKEN CHIEFLY FROM THE PAPERS OF
His Excellency the EARL OF MACARTNEY, Knight of the Bath, His Majesty's Embassador Extraordinary and Plenipotentiary to the Emperor of China; Sir ERASMUS GOWER, Commander of the Expedition, and of other Gentlemen in the several departments of the Embassy.

BY SIR GEORGE STAUNTON, BARONET,
Honorary Doctor of Laws of the University of Oxford, Fellow of the Royal Society of London, his Majesty's Secretary of Embassy to the Emperor of China, and Minister Plenipotentiary in the absence of the Embassador.

In Two Volumes, with Engravings; beside a Folio Volume of Plates.
VOL. I.

LONDON:
PRINTED BY W. BULMER AND CO.
FOR G. NICOL, BOOKSELLER TO HIS MAJESTY, PALL-MALL.
MDCCXCVII.

98 《英使谒见乾隆纪实》

1797，［英国］乔治·斯当东
纵 31.5 厘米，横 25.5 厘米
故宫博物院藏

本书出版于 1797 年，为马戛尔尼使团副使斯当东著。上册介绍使团从英国启程后的沿途见闻，下册介绍使团抵达中国后，所见的风土人情及觐见乾隆帝的过程。上下册扉页分别附有随团画师为马戛尔尼及乾隆帝所绘的肖像图。（王宁远）

98 *An authentic account of an embassy from the King of Great Britain to the Emperor of China*

1797, Sir George Staunton (British)
L.:31.5cm, W.:25.5cm
Collection of the Palace Museum

This book was published in 1797 and authored by George Staunton, deputy envoy of the Macartney Embassy. The first volume recounts their observations and experiences during their journey from Britain, while the second volume details their arrival in China, the local customs, and their audience with the Qianlong Emperor. The title pages of both volumes feature portraits of Macartney and the Qianlong Emperor, drawn by the embassy's accompanying artist. (Wang Ningyuan)

99 “中国风”花园入口设计图

约 1760—1770，[法国] 佚名
钢笔绘制（灰色墨水），画笔水洗（灰色和棕色水彩）
纵 19.4 厘米，横 43.8 厘米
大都会艺术博物馆藏

18 世纪下半叶，在欧洲尤其是法国，掀起了借鉴英国自然式造园手法的热潮。同时，受中国古典园林影响，花园装饰中出现大量“中国风”构筑物。此图展现了一处典型的欧洲中国风花园设计，包括放置古典风格雕塑的台阶与带有中国元素的大门。（雷浩）

99 Design for Stairs in a Garden, Ligned with Trelliswork and a Chinese-inspired Gate

Ca. 1760-1770, unknown painter (French)
Pen and gray ink, brush and gray and brown wash
L.:19.4cm, W.:43.8cm
Lent by The Metropolitan Museum of Art, The Elisha Whittelsey Collection, The Elisha Whittelsey Fund, 1954 (54.513.1)

In the second half of the 18th century, the English landscape garden style became popular in Western Europe, especially in France. Meanwhile, influenced by classical Chinese gardens, many Chinoiserie structures appeared in garden designs. This illustration depicts a typical European Chinoiserie garden design, featuring classical-style sculptures on steps and a gateway with Chinese elements. (Lei Hao)

遊山

中国古典园林大多是有“山”的。仙山、洞天的意象，虽居城市而咫尺林泉的追求，都寄托于山。

山水可“宅心”，可“养心”，可“修心”。

一峰则太华千寻，一勺则江湖万里。立一块湖石假山，仿佛见九天云烟；置一方楼台亭榭，便能引明月清风。中国古人游于园中假山，或游于真山真水，都在追求一片超脱于俗世的心灵空间。

在伊斯兰和欧洲的规则式花园中，自然被理想化为完美的几何结构。文艺复兴时期的意大利台地园依山层叠而建，游台地园也成为一种游山之旅。模仿山地洞穴的人工洞窟、隐喻自然之神的独特巨人雕像，使园林成为艺术入主自然的产物。文艺复兴的浪潮让时人在自然与神性之间，于游观中重新觅得人本的光辉。

Gardens and Nature

Many classical Chinese gardens included ‘mountain’. These evoke the imagery of sacred peaks and celestial grottoes, embodying the desire to stay close to nature even within the city. Mountains and waters were believed to ‘settle the heart’, ‘nourish the spirit’, and ‘cultivate the mind’. As the saying goes, ‘a single peak reflects the grandeur of the highest mountains; a scoop of water captures the vastness of rivers and lakes’. In Chinese gardens, Taihu rockeries evoke heavenly mists, while pavilions and terraces invite the breeze and moonlight. Whether wandering through artificial mountains or exploring real ones, Chinese people of the past sought a realm beyond the ordinary.

Formal gardens of the Islamic and European traditions shaped nature into perfect geometric forms. The terraced gardens of Renaissance Italy were built upon slopes, making a stroll in the gardens akin to a mountain hike. Artificial grottoes mimicking natural caves and monumental statues representing gods of nature symbolized humanity’s artistic mastery over the natural world. Within these landscapes, the Renaissance revived humanism through the interplay of nature and the divine.

100 乾隆花园第三进院烫样

现代
综合材质
长 77.8 厘米，宽 99 厘米，高 21 厘米
故宫博物院藏

这件烫样表现的是乾隆花园的第三进院落。三进院的格局布置极为大胆，在存世园林中堪称罕例。三座建筑围合成的院落，被一组体量巨大的假山完全填满，在有限空间内营造出深山大壑的气象。山体之中辟有蜿蜒曲折的山洞，人既可以从洞中穿行，也可以沿山石踏跺攀上山顶，再从另一面的石阶下山。山顶有平台可供休憩，建有耸秀亭，登临可以眺望整座院落。假山西端原有小桥，与西面的延趣楼二层相通，嘉庆年间拆除。

萃赏楼是第三进院的主体建筑，也是连接三四进院的过渡空间。楼上明间前后开门，向南可以凭栏眺望，向北则有一座小石桥，通往后院的另一组假山和山顶的碧螺亭。

苏州园林中，狮子林素以崎岖奇崛的洞壑式假山闻名，乾隆皇帝曾多次到访，对这座园林欣赏有加，甚至专门命人制作了整座园林的烫样，运送回京，在圆明园和避暑山庄依样仿建。研究者推测，乾隆花园假山的设计很可能也曾受到狮子林的启发。（刘梦雨）

100 Model of the Third Courtyard of the Qianlong Garden

Contemporary
Mixed media
L.:77.8cm, W.:99cm, H.:21cm
Collection of the Palace Museum

This model represents the third courtyard of the Qianlong Garden, which is known for its unique and bold layout, and is a rarity among classical gardens. The space enclosed by three buildings is nearly filled with an enormous artificial rock formation, creating the illusion of a vast mountain landscape. Visitors can wander through cave passages or ascend stone steps to the peak. The Qianlong Emperor greatly admired the Lion Grove Garden in Suzhou and once ordered a complete model of it to be made and sent to Beijing. He then commissioned replicas in the Old Summer Palace and the Chengde Mountain Resort. The rockery design in the Qianlong Garden may also have been inspired by the Lion Grove Garden. (Liu Mengyu)

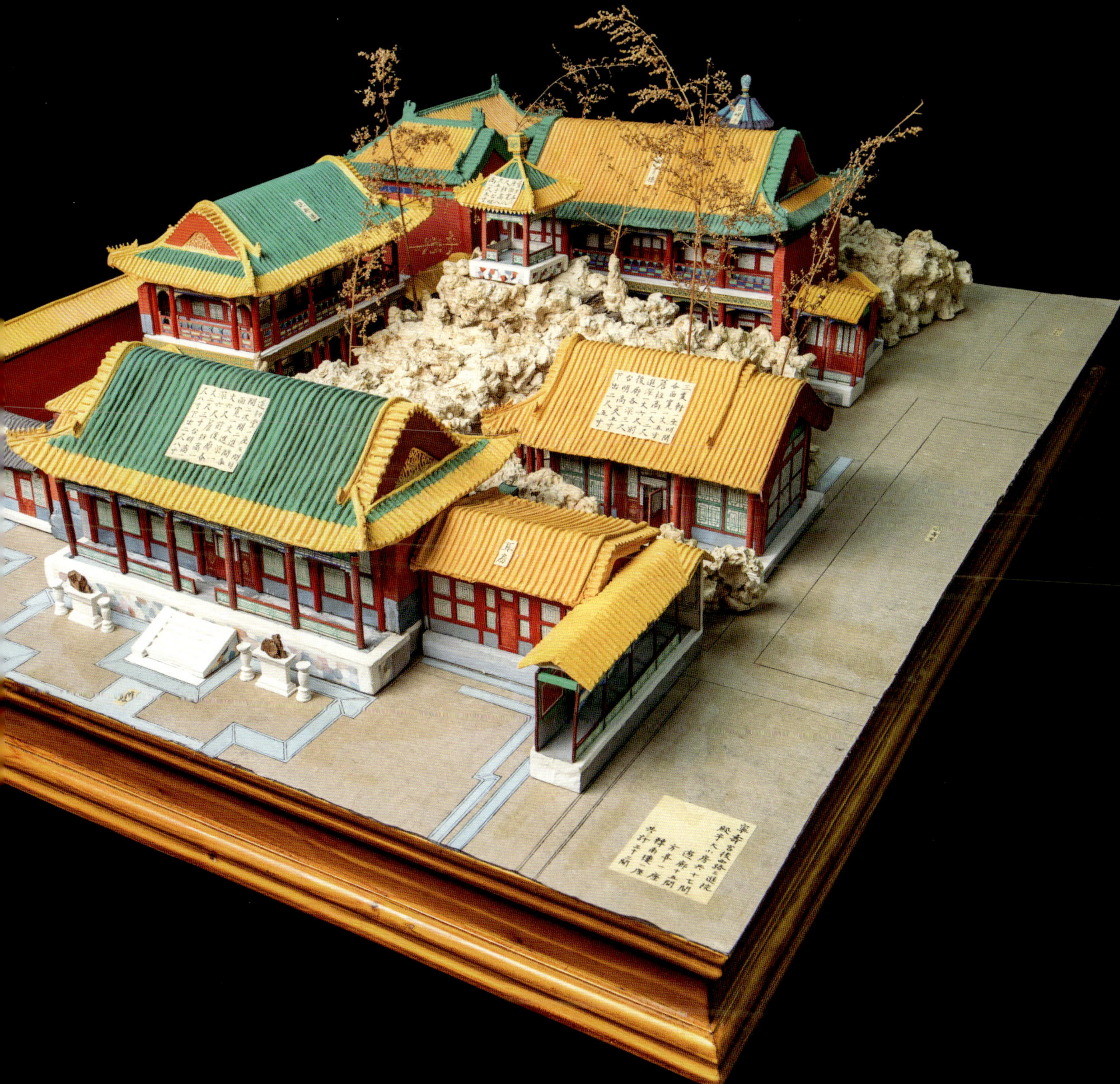

101 御花园古柏图轴

清乾隆（1736—1795），钱维城
纸本，设色
纵 118.8 厘米，横 180.7 厘米
故宫博物院藏

此作以横轴的方式，呈现了紫禁城御花园的东北隅。画面左侧为堆秀山，堆秀山上为御景亭。堆秀山始建于明代，初名堆绣山，是重阳节皇帝与后妃登高之处。堆秀山正面有岩洞，内为砖砌穹隆式石雕蟠龙藻井，石洞门两侧设有石蟠龙喷泉，蟠龙口中可喷水，为宫中仅存之水法。山右侧（东侧）为摛藻堂，为宫中藏书之所。摛藻堂再右侧（东侧）为凝香亭，摛藻堂前方（南侧）为浮碧亭，这些建筑现存完好。

画面中的建筑以界画方式呈现，设色典雅，形象精准。堆秀山、树木和远山皆用水墨为主，呈现出文人画的样貌。画面左下方署款："臣钱维城恭绘"，下钤"臣钱维城""笔沾春雨"二印。

钱维城，号稼轩，武进人，乾隆年间状元，官翰林院修撰、工部右侍郎等，是乾隆朝典型的词臣画家。画面右上方有乾隆帝御题诗《御花园古柏行》，描述并歌颂了御花园的古柏。下钤"得象外意""乾隆宸翰"二印，诗文右侧另有乾隆帝"水月两澄明""欢喜园"二印。（许彤）

101 *Ancient Cypress Trees in the Imperial Garden*

Qing Dynasty, Qianlong Reign (1736-1795), Qian Weicheng
Hanging scroll, color on paper
L.:118.8cm, W.:180.7cm
Collection of the Palace Museum

This painting shows the northeastern corner of the Imperial Garden in the Forbidden City. On the left is the Mountain of Accumulated Elegance, topped by the Imperial Prospect Pavilion. Built in the Ming Dynasty, it served as a spot for the emperor to ascend during the Double Ninth Festival. At its front is a grotto with a brick dome and a stone-carved dragon caisson. Flanking the entrance are stone dragon fountains, the palace's only surviving waterworks. All these structures remain well preserved. (Xu Tong)

御花園古柏行近作
乾隆御筆

102 庄园南立面与大草坪

选自《普拉特里诺庄园图景集》
1658，［意大利］斯特凡诺·德拉·贝拉
纸本，蚀刻版画
纵 32.5 厘米，横 45 厘米
芝加哥艺术博物馆藏

16 世纪中期，美第奇家族在佛罗伦萨建造了普拉特里诺庄园（the Villa Pratolino），巧妙利用山地环境建造了精美的洞窟与先进的水力装置，营造出与平地不同的独特游园体验。（许愿）

102 South Facade of the Villa and the Great Meadow

From *Views of the Villa Pratolino*
1658, Stefano della Bella (Italian)
Etching on cream laid paper
L.:32.5cm, W.:45cm
Collection of the Art Institute of Chicago, Celia and David Hilliard Endowment (1990.501.3)

In the mid-16th century, the Medici family built the Villa Pratolino in Florence. They skillfully used the mountainous terrain to create elaborate grottoes and advanced hydraulic systems, offering a unique garden experience distinct from that of flat landscapes. (Xu Yuan)

103 亚平宁山巨像（上）

选自《普拉特里诺庄园图景集》
约 1653，［意大利］斯特凡诺·德拉·贝拉
纸本，蚀刻版画
纵 32.5 厘米，横 45 厘米
芝加哥艺术博物馆藏

图中巨像是普拉特里诺庄园中至今还留存的醒目景观，半人半山的巨人形象是亚平宁山脉的化身，由著名雕塑家詹博洛尼（Giambologna）于 1579 年至 1580 年间雕刻而成。巨像内设有多个洞窟和房间，足可容纳小型乐团。庄园的建造者弗朗切斯科一世·德·美第奇（Francesco I de'Medici）曾在巨人头部的洞窟中享受垂钓之乐，鱼线从眼缝穿出。（许愿）

103 The Colossal Statue of the Apennines (up)

From *Views of the Villa Pratolino*
Ca. 1653, Stefano della Bella (Italian)
Etching in black on ivory laid paper
L.:32.5cm,W.:45cm
Collection of the Art Institute of Chicago, Mr. and Mrs. Gaylord Donnelley Fund (1990.501.6)

The colossal statue in the picture remains a striking feature of the Pratolino estate today. The half-man, half-mountain Colossus symbolizes the Apennine Mountains and was sculpted by the renowned sculptor Giambologna between 1579 and 1580. The statue represents the integration of artistic creativity and natural landscapes during the Renaissance period. (Xu Yuan)

104 潘神与菲墨女神窟（下）

1653 或 1655，［意大利］斯特凡诺·德拉·贝拉
纸本，蚀刻版画
纵 32.9 厘米，横 45 厘米
芝加哥艺术博物馆藏

这幅画以对称的双联构图同时展现了洞窟两端的潘神（左）和菲墨女神（右）形象，利用自然光线和流水，创造出一种神秘而迷人的氛围。洞窟作为古典神话、自然景观与人文理念的融合体，是文艺复兴时期一种独特的艺术表现形式。（罗越）

104 The Grotto of Pan and Fame (down)

1653/1655, Stefano della Bella (Italian)
Etching on ivory laid paper
L.:32.9cm, W.:45cm
Collection of the Art Institute of Chicago, Helen Davis Baily and Mr. and Mrs. Henry C. Woods endowments (1990.501.4)

This double-perspective print depicts the interior of the grotto in the Villa Pratolino, as shown in the left-hand work. On either side of the statue symbolizing the River Mugrone, are Pan (left) and the nymph Fame (right). As a fusion of classical mythology, natural landscapes, and humanistic ideals, the grotto exemplifies the artistic expression of the Renaissance period. (Luo Yue)

105 万松金阙图卷

宋（960—1279），赵伯骕

绢本，设色

纵 27.7 厘米，横 135.2 厘米

故宫博物院藏

此卷以并不浓重的青绿色表现了一处理想的山水景致。这一景致并非完全出于艺术想象，而是有具体描绘对象的，即临安（今浙江杭州）的凤凰山万松岭一带。

此卷并没有作者款印，之所以定为赵伯骕所画，是根据尾纸上元人赵孟頫跋文的判断。赵伯骕，字希远，宋太祖七世孙，官至和州防御使。南渡后居临安，入内廷，以丹青知名，受到高宗眷爱。

万松岭在西湖南，其上皇家殿宇众多。画面起始处水面开阔，山坡上松林郁郁葱葱，金顶的宫殿富丽堂皇，若隐若现。笔法工中带拙，形成了一种富有文人气息、笔墨蕴藉的青绿山水新风格，是典型宋代贵族文人笔下青绿设色与水墨意趣的结合与探索。（许彤）

105 ***Ten Thousand Pines and Golden Palace***

Song Dynasty (960-1279), Zhao Bosu
Handscroll, color on silk
L.:27.7cm, W.:135.2cm
Collection of the Palace Museum

The painting bears no artist's signature or seal but includes a colophon by Zhao Mengfu. It depicts the scenery of the Wansong Ridge area of Mount Phoenix in Lin'an (present-day Hangzhou). Wansong Ridge lies beside the West Lake and is dotted with numerous imperial buildings. The painting begins with an expansive water surface, leading up to lush pine forests on the hillside, where golden-roofed palaces are faintly visible. The brushwork blends precision and simplicity, creating a refined blue-and-green landscape style. (Xu Tong)

107 友松图卷

明（1368—1644），杜琼
纸本，设色
纵 28.8 厘米，横 92.5 厘米
故宫博物院藏

本幅无款题，钤“东原斋”“一丘一壑”“杜氏用嘉”。卷后文伯仁题跋，称为杜琼之作。晚明张丑《清河书画舫》、清初高士奇《江村书画目》著录，《清河书画舫》著录本尚有刘珏、张祯及张丑三跋，今皆佚。据张丑题，此卷系杜琼为其姐丈魏友松所作别号图。

魏茂，别号友松，家居吴中闹市，其子魏昌开辟屋后之地，艺树凿池，奇石间列，名之“成趣轩”。其时吴门名士如徐有贞、刘珏等，皆往来其间，为雅集之乐。画卷描绘园林中文士书写对谈，室内榻上两人品评画卷，屋外遒松三株盘旋而上，寓意“友松”之清高。用笔细雅苍秀，风格融合王蒙、吴镇。

杜琼（1396—1474），字用嘉，号东原，吴县（今江苏苏州）人，其山水宗法董源及元人，为沈周之师。（马顺平）

107 *Friend of Pines*

Ming Dynasty (1368-1644), Du Qiong
Handscroll, color on paper
L.:28.8cm, W.:92.5cm
Collection of the Palace Museum

This scroll was created by Du Qiong as a sobriquet painting for his brother-in-law Wei Yousong. The scroll depicts scholars in the garden engaging in writing and conversation, with two figures inside a room appreciating a scroll on a couch. Outside, three vigorous pines twist upward, and at the end of the scroll, several artificial mountains are arranged in a balanced yet dense composition. (Ma Shunping)

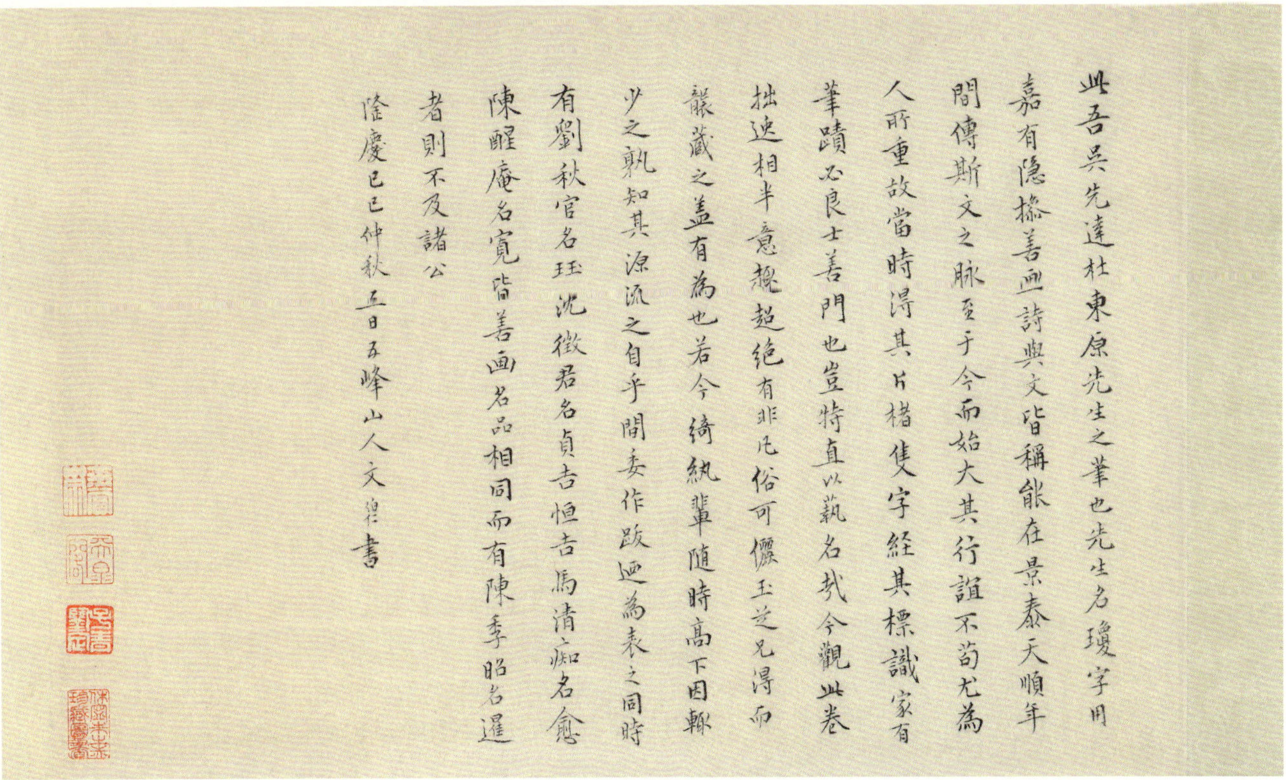

此吾吳先達杜東原先生之筆也先生名瓊字用
嘉有隱操善畫詩與文皆稱能在景泰天順年
間傳斯文之脈至于今而始大其行誼不苟尤為
人所重故當時得其片楮隻字經其標識家有
筆蹟必良士善門也豈特直以藝名哉今觀此卷
拙逸相半意趣超絶有非凡俗可儗玉芝兄得而
襲藏之蓋有為也若今綺紈輩隨時高下因鄙
少之孰知其源流之自乎聞妄作跋迺為表之同時
有劉秋官名珏沈徵君名貞吉恒吉馬清痴名愈
陳醒庵名寬皆善畫名品相同而有陳季昭名暹
者則不及諸公
隆慶己巳仲秋五日五峰山人文伯仁書

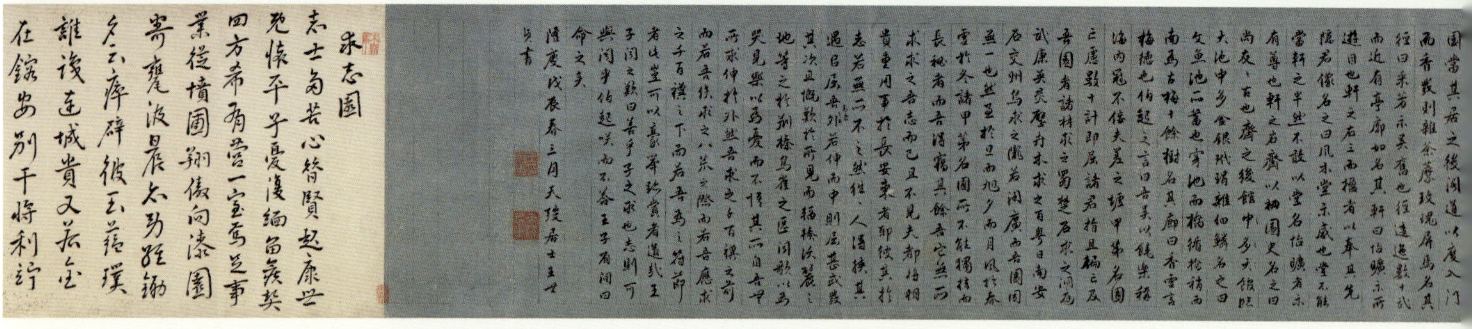

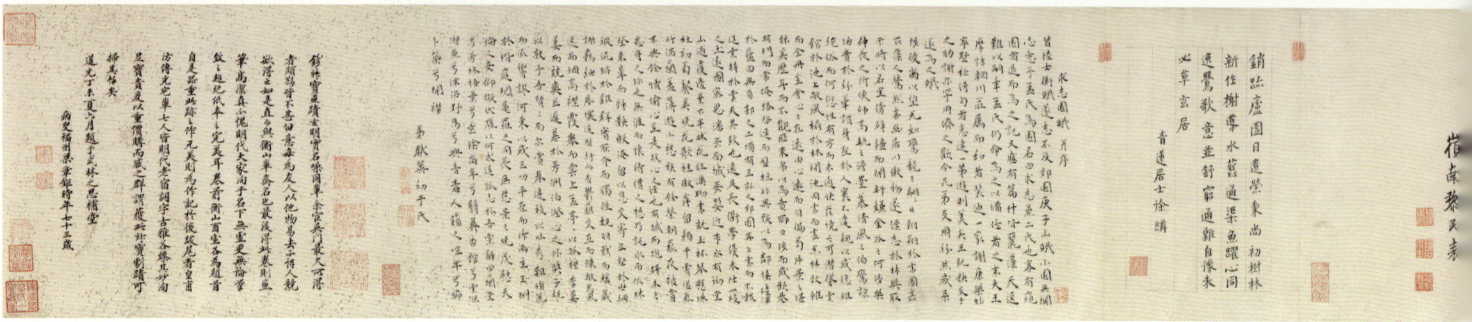

108 求志园图卷

明嘉靖四十三年（1564），钱榖
纸本，设色
纵 29.8 厘米，横 190.2 厘米
故宫博物院藏

本幅款署“嘉靖甲子夏四月，钱榖作求志园图。”下钤“悬磬室画印”“叔宝”白文印，可知作于嘉靖四十三年（1564），钱榖时年五十五岁。此图描绘的求志园，是明代著名文人张凤翼的园居。张凤翼（1527—1613）字伯起，号灵虚，长洲（今江苏苏州）人。与其弟燕翼、献翼俱有才名，时人称为“三张”。图前两段引首分别为文徵明所书“文鱼馆”和王穉登所书“求志园”，可见张凤翼与吴门画家有十分密切的交往。

图中以“怡旷轩”及其两侧的“风木堂”“尚友斋”为画面中心，并在其前后绘制“采芳径”“文鱼池”“香雪廊”等园中景致。与图后明代著名文人王世贞于隆庆二年（1568）所作《求志园记》中的记述大致相符，可知此图当是参考实景绘制而成的。此类描绘园林的绘画，是明代中后期吴门地区的一种流行题材，这与当时文人士大夫热衷造园的风尚有直接关系。由于求志园今已不存，因而此图对于我们了解明代造园艺术有着十分重要的参考价值。

卷后另有皇甫汸、傅光宅、李攀龙、黄姬水、黎民表、徐緌等人诗题及张献翼《求志园赋》。他们赞赏了求志园景色的优美，并阐发出张凤翼营造园林的旨趣，既反映了当时吴门地区文人雅集的盛况，也是

108 *Qiuzhi Garden*

1564, Qian Gu
Handscroll, color on paper
L.:29.8cm, W.:190.2cm
Collection of the Palace Museum

研究吴门文人群体交游的重要资料。

此图用笔轻柔，设色雅澹，风格秀逸，营造出一派轻松闲适的氛围，既展现了钱榖高超的画艺，也很好地反映出园林主人的审美追求，为钱榖传世的重要代表作。此作曾经清陈淮、戴植、梁章钜等人鉴藏，《过云楼书画记》著录。

钱榖（1508—1579）字叔宝，号罄室，苏州人。善山水，师从文徵明，为吴门画派后期重要画家之一。（郁文韬）

This scroll was created in the 43rd year of the Jiajing reign (1564) in the Ming Dynasty. Qiuzhi Garden was the residence of the Ming scholar Zhang Fengyi. The painting generally aligns with Wang Shizhen's *Qiuzhi Garden Record*, written at the end of the scroll, confirming that the painting was based on actual scenery. (Yu Wentao)

109 青花山水人物图长方花盆

清康熙（1662—1722）
口径 41.7 厘米 ×29.1 厘米，底径 31.5 厘米 ×19 厘米，
高 21.2 厘米
故宫博物院藏

花盆长方形，通体以青花为饰，口沿一周为花卉纹，器身四壁绘山水人物纹。内外底均露胎无釉，底有两圆孔，以便渗水。外口沿书“大清康熙年制”青花六字楷书款。

外壁所绘山水人物风景，四面构图独立，又可贯穿连续。将四壁画幅展开，可以看作一幅气韵连贯的通景画。画中有远山、瀑布，又有亭台、小桥、柳树，画中人物有的泛舟湖上，有的憩于阁中，还有的相携观瀑，表达了文人高士寄情山水的主题。（孙悦）

109 Blue and White Rectangular Flowerpot with Landscape and Figures Pattern

Qing Dynasty, Kangxi Reign (1662-1722)
D.(at mouth):41.7cm×29.1cm, D.(at base):31.5cm×19cm,
H.:21.2cm
Collection of the Palace Museum

The exterior of this flowerpot is decorated with landscape and figure paintings. The artwork features distant mountains, waterfalls, pavilions, small bridges, and willow trees. Figures are shown boating on a lake, resting in a pavilion, or admiring a waterfall, embodying the literati ideal of finding solace in nature. The base has two round drainage holes, indicating that it was originally crafted as a

110 西苑千尺雪图卷

清乾隆十八年（1753），董邦达
纸本，墨笔
纵 28 厘米，横 142.5 厘米
故宫博物院藏

“千尺雪”本为明代苏州名士赵宧光（1559—1625）于其隐居的寒山别业中建造的一处景观。此景因泉水沿峭壁倾泻，形似千尺飞雪而得名。乾隆帝对此景极为喜爱，每次南巡必游此地，并在西苑、热河避暑山庄、盘山静寄山庄三地仿造其景构舍，作为赏景品茗之所。这三处仿建虽结合本地特色，但都保留了苏州“千尺雪”的名称和意境。

本卷为董邦达奉乾隆帝之命所绘西苑“千尺雪”。此卷引首有乾隆帝御书“液池清派”四字，卷首御笔“西苑千尺雪”五字。全卷水波荡漾，叠石起伏，石桥、水榭等依水而建。笔墨苍润细密，具有文人趣味。

卷尾款署：“乾隆癸酉孟春，臣董邦达奉敕敬绘。”癸酉为乾隆十八年（1753），董邦达时年五十八岁。拖尾处有乾隆帝跋文：“西苑千尺雪并寒山、田盘及山庄而四适，披斯卷，清秋新霁，珠玉喷溅。皑如其色，铿尔其声。觉苑中人巧，固当逊此天真耳。乾隆丁丑秋八月，御识于山庄千尺雪。”

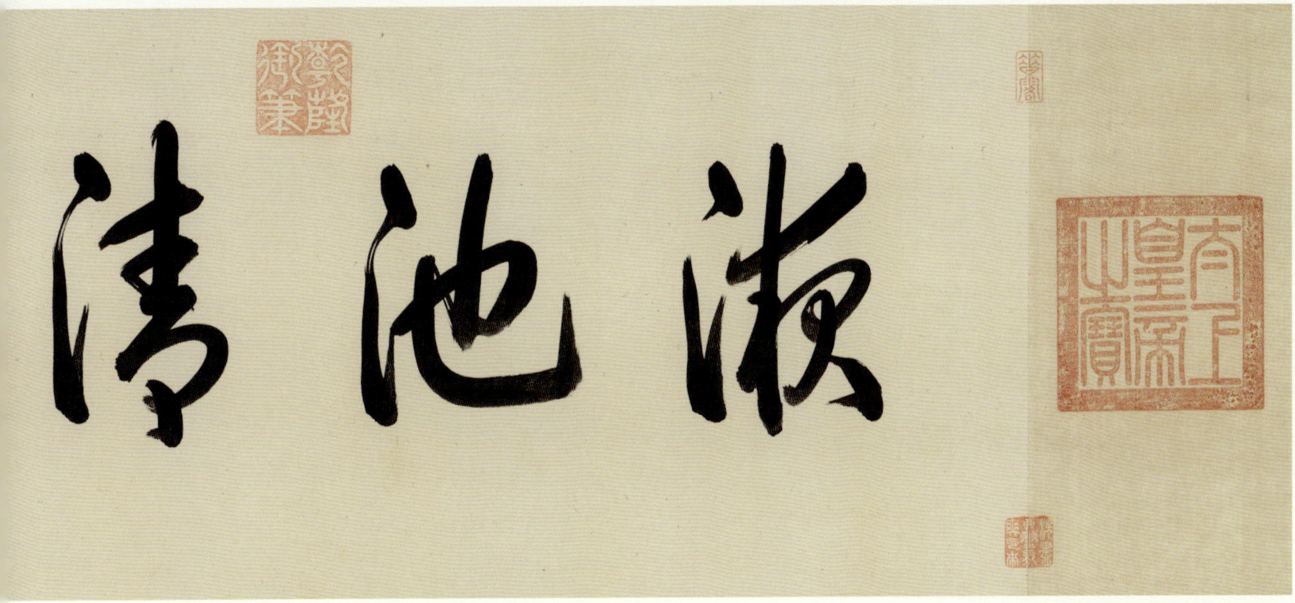

110 ***Thousand-feet Snow in the Western Garden***

1753, Dong Bangda
Handscroll, ink on paper
L.:28cm, W.:142.5cm
Collection of the Palace Museum

董邦达（1696—1769），字孚存，一字非闻，号东山，浙江富阳人。董邦达为雍正十一年（1733）进士，累官工部尚书、礼部尚书，卒后谥文恪。董邦达工山水，画风苍逸古厚，其山水画远师董源、巨然及黄公望，近学董其昌、王原祁。（胥瑞頔）

The Thousand-feet Snow was originally a scenic spot at the Han Mountain retreat of Zhao Yiguang, a renowned Ming Dynasty (1368-1644) scholar from Suzhou. The site was named for its cascading waters resembling flying snow. Deeply fond of this scene, the Qianlong Emperor had it replicated at several imperial complexes, preserving both the name and ambiance of Thousand-feet Snow as a place to enjoy the view and drink tea. (Xu Ruidi)

111 蒂沃利大瀑布

选自《罗马美景集》
1766，[意大利]乔凡尼·巴蒂斯塔·皮拉内西
纸本，蚀刻版画
纵 50.5 厘米，横 72 厘米
芝加哥艺术博物馆藏

112 蒂沃利的小瀑布与湍流

选自《罗马美景集》
1769，[意大利]乔凡尼·巴蒂斯塔·皮拉内西
纸本，蚀刻版画
纵 55.7 厘米，横 78.2 厘米
芝加哥艺术博物馆藏

111 View of the Grand Cascade at Tivoli

From *Views of Rome*
1766, Giovanni Battista Piranesi (Italian)
Etching on ivory laid paper
L.:50.5cm, W.:72cm
Collection of the Art Institute of Chicago, The Charles Deering Collection (1927.4780)

112 View of the Small Waterfall and Rapids Tivoli

From *Views of Rome*
1769, Giovanni Battista Piranesi (Italian)
Etching on ivory laid paper
L.:55.7cm, W.:78.2cm
Collection of the Art Institute of Chicago, Clarence Buckingham Collection (1910.324)

此二图绘意大利蒂沃利古镇的山地风光，自古罗马时期，这里便是贵族们喜爱的度假胜地。阿尼耶内河（Aniene）流经蒂沃利，顺高差形成壮美瀑布。埃斯特庄园正是因为有充沛的水源，才建成壮观的水景，并保有沿用至今的复杂水利工程。作者皮拉内西是 18 世纪杰出的版画家与建筑师，热衷于绘制罗马的古迹与城市景观。（许愿）

The two works depicts the mountainous landscape of the ancient town of Tivoli, Italy. Starting from the Roman era, this was a favored retreat for nobility. The Aniene River flows through Tivoli, cascading down to form magnificent waterfalls. The Villa d'Este's renowned fountains and intricate hydraulic system, still in use today, owe their existence to this abundant water source. (Xu Yuan)

113 蒂沃利宫和花园

选自《罗马壮丽之镜》
1581，[意大利]乔凡尼·安布罗吉奥·布兰比拉（传）
蚀刻版画
纵 36.7 厘米，横 52 厘米
大都会艺术博物馆藏

意大利台地园是文艺复兴时期兴起的一种园林风格，以其对称的几何布局、层次分明的台地设计、精妙绝伦的水景艺术以及古典艺术的融合而闻名于世。作为意大利台地园的经典代表作之一，16 世纪建造的埃斯特庄园（蒂沃利宫及其花园）在图中一览无余。花园以几何图形划分六层台地，精心布置着喷泉、雕塑和整形绿篱，展现了意大利文艺复兴时期手法主义园林的独特魅力。（雷浩）

113 Tivoli Palace and Gardens

From *Speculum Romanae Magnificentiae*
1581, Attributed to Giovanni Ambrogio Brambilla (Italian)
Etching
L.:36.7cm, W.:52cm
Collection of The Metropolitan Museum of Art, Harris Brisbane Dick Fund, 1941 (41.72 :3.65)

The Italian Terraced Garden emerged during the Renaissance. It was renowned for its symmetrical geometric layout, clearly defined terraced designs, exquisite water features, and fusion of classical art. A classic example is the 16th century Villa d'Este (Tivoli Palace and Gardens), where six geometrically divided terraces are adorned with fountains, sculptures, and trimmed hedges, showcasing the charm of Italian Renaissance gardens. (Lei Hao)

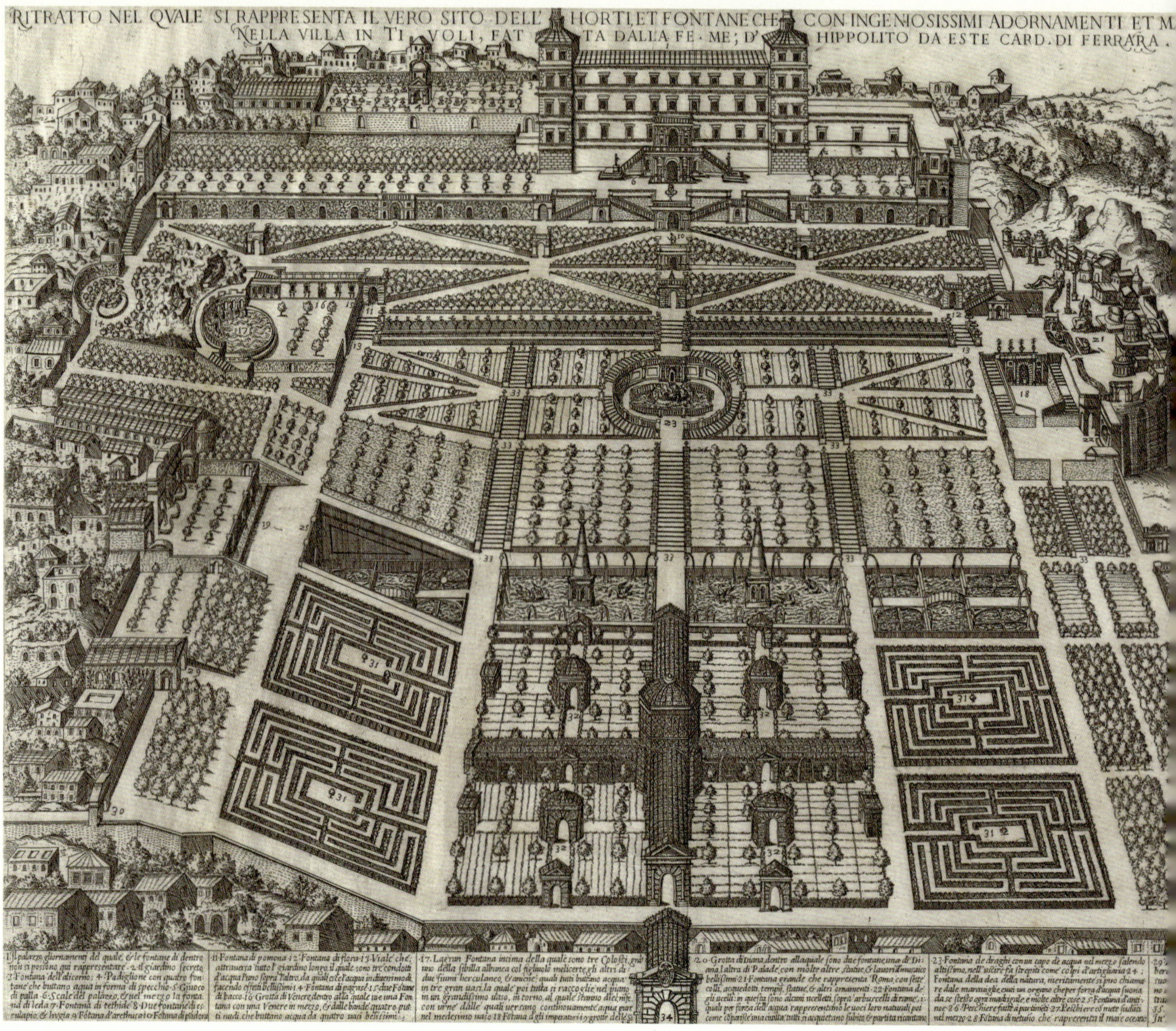

114 埃斯特庄园图景

1817，［法国］克劳德·蒂耶农
纸本，版画
纵 48.5 厘米，横 33.2 厘米
芝加哥艺术博物馆藏

站在埃斯特庄园入口，花园的台地特征清晰可见，沿中轴线拾级而上，经龙喷泉、百泉台，最终抵达府邸大门。建成两百年后，花园草木葱郁，弥漫着如绘的风景气质。（罗越）

114 View of the Villa d'Este

1817, Claude Thiénon (French)
Lithograph in black on ivory wove paper
L.:48.5cm, W.:33.2cm
Collection of the Art Institute of Chicago, Harold Joachim Purchase Fund (1972.717)

From the entrance of the Villa d'Este, the garden's terraced layout is clearly visible. Walking along the central axis, visitors pass the Dragon Fountain and the Hundred Fountains Terrace before finally reaching the mansion's grand entrance. Two hundred years after its completion, the garden remains lush, exuding a picturesque charm. (Luo Yue)

115 埃斯特庄园

选自《罗马美景集》
1773，［意大利］乔凡尼·巴蒂斯塔·皮拉内西
纸本，蚀刻版画
纵 55 厘米，横 77.7 厘米
芝加哥艺术博物馆藏

皮拉内西在这幅版画中将埃斯特庄园描绘成了乌托邦式的天堂，远景可辨真实的府邸建筑和花园。近景中巨大的雕塑是艺术家的想象，充满强烈的浪漫主义色彩。（许愿）

115 View of the Villa d'Este

From *Views of Rome*
1773 , Giovanni Battista Piranesi (Italian)
Etching on heavy ivory laid paper
L.:55cm, W.:77.7cm
Collection of the Art Institute of Chicago, Clarence Buckingham Collection (1910.330)

In this etching, Piranesi depicts the Villa d'Este as a utopian paradise. The distant view reveals the real mansion and garden, while the imagined foreground conveys a strong sense of Romanticism. (Xu Yuan)

116 埃斯特庄园的椭圆形喷泉

选自《蒂沃利埃斯特花园的喷泉集》
约 1676—1700，[意大利] 乔凡尼 · 弗朗西斯科 · 文图里尼
蚀刻和雕刻版画
纵 24.7 厘米，横 34 厘米
大都会艺术博物馆藏

作为埃斯特庄园中最主要的喷泉之一，椭圆形喷泉引阿尼耶内河水，利用重力建成壮观的水景。椭圆形壁龛内，八位仙女手持流水瓶，中心喷泉之上是古罗马神话中的预言女神西比尔（Sibyl）及其子。人们游憩其中，感受山水与神话的奇妙交汇。（罗越）

116 The fountain of the Tiburtine sibyl in the garden at Villa d'Este

From the series *The fountains of the Este garden in Tivoli*
Ca. 1676-1700, Giovanni Francesco Venturini (Italian)
Etching and engraving
L.:24.7cm, W.:34cm
Lent by The Metropolitan Museum of Art, Bequest of Phyllis Massar, 2011 (2012.136.406)

In the niche of the Oval Fountain (*la Fontana dell' Ovato*), eight nymphs hold water vessels. Centered above the oval is the figure of Sibyl, the prophetic goddess from ancient Roman mythology, along with her son. Figures stroll and relax within, experiencing the enchanting fusion of nature and mythology. (Luo Yue)

117 蒂沃利的梅塞纳斯庄园与瀑布

1794—1797，［英国］约瑟夫·马洛德·威廉·透纳、［英国］托马斯·吉尔丁，传据［英国］约翰·罗伯特·科岑作品创作
石墨底稿，蓝灰色淡彩
纵 22.5 厘米，横 40.6 厘米
大都会艺术博物馆藏

悬于意大利蒂沃利山巅的复层拱廊曾被认为是罗马奥古斯都皇帝的顾问盖乌斯·梅塞纳斯（Gaius Maecenas）的庄园遗址，现代研究基本将其确定为公元前 2 世纪的赫拉克勒斯·维克托神庙（Santuario di Ercole Vincitore）。青年时代的透纳（Joseph Mallord William Turner）与吉尔丁（Thomas Girtin）据风景画大师科岑（John Robert Cozens）的素描，合作完成了此图。图中遗址外观与周围地形与今天十分相似。有意思的是，画面所呈现出的光影与水汽交融般的气韵折射出些许东方山水意境。（郭欣悦）

117 Tivoli, Villa of Maecenas and Part of the Cascatelle

1794-1797, Joseph Mallord William Turner (British), Thomas Girtin (British), probably after John Robert Cozens (British)
Brush and blue-gray wash, over graphite
L.:22.5cm, W.:40.6cm
Lent by The Metropolitan Museum of Art, Rogers Fund, 1906 (06.1042.18)

The Superimposed Arcades, perched atop the hills of Tivoli, Italy, were once believed to be the villa of Gaius Maecenas, an Augustan-era imperial advisor. However, they have now been largely identified as the Sanctuary of Hercules Victor (*Santuario di Ercole Vincitore*). In their youth, J. M. W. Turner and Thomas Girtin collaborated on this work based on a sketch by the landscape watercolorist John Robert Cozens, imparting to the scene a subtle evocation of Eastern landscape aesthetics. (Guo Xinyue)

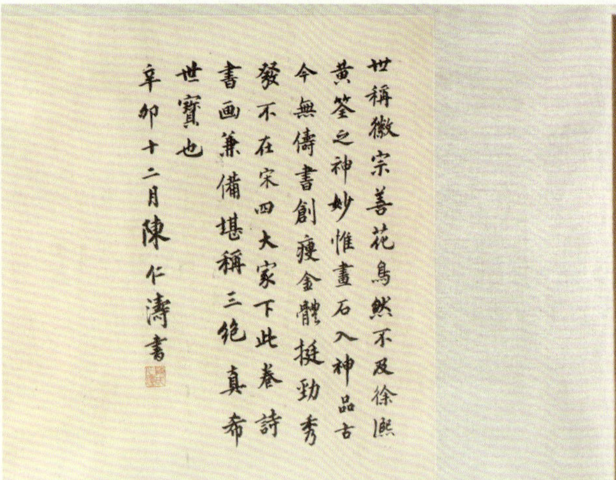

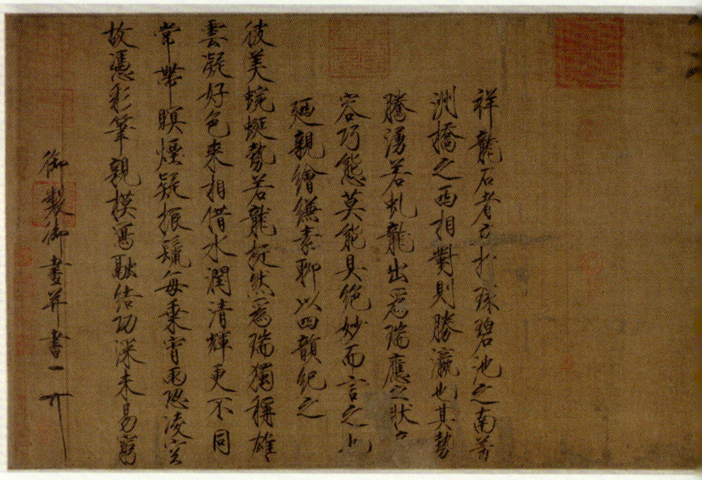

118 祥龙石图卷

北宋（960—1127），赵佶
绢本，设色
纵 53.8 厘米，横 127.5 厘米
故宫博物院藏

本卷原为册页，后裱为手卷，系徽宗真迹，因石上有徽宗楷书“祥龙”二字，故得名。卷左有徽宗瘦金体御书：“祥龙石者，立于环碧池之南，芳洲桥之西，相对则胜瀛也。其势胜涌，若虬龙出，为瑞应之状，奇容巧态，莫能具绝妙而言之也。乃亲绘缣素，聊以四韵纪之。”并御制七律一首赞美祥龙石。末款署“御制御画并书”“天下一人”押字。钤“御书”“宣和殿宝”（真伪待考）。拖尾有清代陈仁涛、吴荣光题跋。又“天历之宝”印，知曾为元文宗内府所有，藏于杨桂山方伯家，后流散香港为文物小组救购，重归宫藏。

本幅中祥龙石以顶天立地之姿独立入画，墨线勾勒、细劲挺锐，峰角峥嵘，瘦皱透漏毕现，又设色青黑苍润。鱼鳞纹遍布石身，植物与奇石映衬共生，用色浓淡薄厚亦随形而不同，近极写生之所能。目为瑞应祝福之形，绘入睿览册内。

宣和殿乃徽宗起居殿，亦是其法帖、绘画、图书、金石收藏之要地。《宋史·地理志》云：“宣和殿，在睿思殿后，绍圣二年四月殿成。其东侧别有小殿曰凝芳，其西曰琼芳，前曰重熙，后曰环碧。”可知环碧池与宣和殿相近，同属宫城西北角的后苑区域。又据蔡京《太清楼侍宴记》可考，宣和殿后则有名曰“环碧”的池沼，东西建有“临漪亭”和“华渚亭”，而祥龙石推测其静立于殿宇与环碧池沼之间，是构成花园景致的要点。徽宗痴石，除皇城北隅修筑艮岳外，为建立祥瑞表征，在东京皇宫花园中营造具有瑞应之状的微缩景观，纵横不满百步，可见“太湖之石”“泗滨之磐”林立。祥龙石置于皇帝生活和宴游的花园空间之境，从自然之物衍化为可游、可赏的美学意象。

赵佶（1082—1135），即宋徽宗，宋朝第八位皇帝，号宣和主人、教主道君皇帝等，画家、书法家，擅长山水、人物、花鸟、墨竹，创“瘦金体”。邓椿在《画继》中评价他“天纵将圣，艺极于神”。（吴凡）

118 *Auspicious Dragon Stone*

Northern Song Dynasty (960-1127), Zhao Ji
Handscroll, color on silk
L.:53.8cm, W.:127.5cm
Collection of the Palace Museum

This scroll is widely recognized as an authentic work by Emperor Huizong of the Song Dynasty. It depicts a Taihu rock inscribed with the characters *Xiang Long* (Auspicious Dragon) in regular script in the inner garden of the Song imperial city. (Wu Fan)

119 遂初堂前对置石峰（其一）

清（1644—1911）
长约 80 厘米，高约 60 厘米
故宫博物院藏

古人爱石，故常于园林中置石峰。宋人孔傅《云林石谱序》称，石乃“天地至精之气”凝结而成，“物象宛然，得于仿佛”。坐园赏石，“则三山五岳、百洞千壑，䙰缕簇缩，尽在其中”——正是所谓卧而游天下。

乾隆花园中有大小石峰近二十处，其中遂初堂前四块为对置灵璧石。所谓“对置”，就是将体量、姿态相似的山石沿轴线左右对称布置，构成一幅规矩、完整的画面。遂初堂前原本没有对置石峰。乾隆四十五年（1780），两淮盐政伊灵阿送到灵璧石八十块。乾隆帝随即调整遂初堂前布置，下旨“（遂初堂）两稍间阶前石座分位并对南边游廊柱中及对垂花门柱中，交工程处各添安石座一对，交造办处安设灵璧石”。

灵璧石，产自安徽宿州灵璧县。《云林石谱》称其：“或成物状，或成峰峦，巉岩透空，其眼少状妙有宛转之势。或多空塞，或质偏朴，或成云气、日月、佛像，或状四时之景。”展览中选用的石峰位于遂初堂西南角，单摆浮搁于汉白玉基座之上。石峰基座两侧又对称摆放有莲花状石座，上设铜瓶。石峰远观宛如仙山，下有云气，中有洞府；近看则沟壑纵横，线纹明显。孙绰《遂初赋》云：“处幽潜德，含圣神兮。”乾隆帝居庙堂之高，无法放旷林泉，便寄情于卷山之中，用石峰抒发“遂初”之愿。（刘仁皓）

119 Rock Formation in Front of the Hall of Fulfilling Original Wishes (*Suichu tang*) (Piece I)

Qing Dynasty (1644-1911)
L.:80cm, H.:60cm
Collection of the Palace Museum

This rock formation, made of Lingbi stone, is positioned in the southwest corner of the Hall of Fulfilling Original Wishes. It is placed symmetrically alongside three other rock formations. The stone sits independently atop a white marble pedestal, with lotus-shaped stone bases on either side, each supporting a bronze vase. From a distance, the rock resembles a celestial mountain, with mist swirling below and a grotto within. Up close, its deep ravines and intricate textures evoke the spirit of natural landscapes, recalling woods and flowing springs. (Liu Renhao)

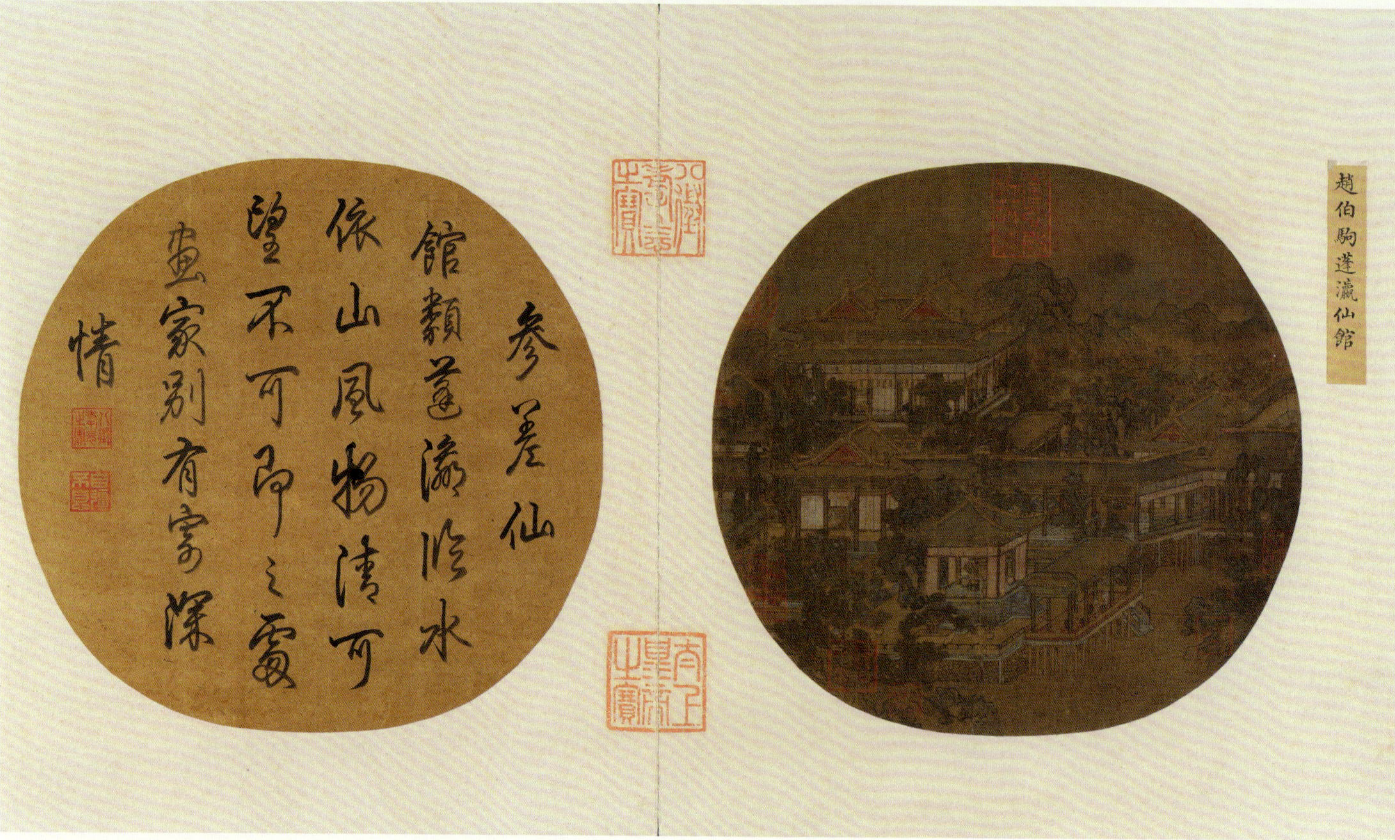

120 蓬瀛仙馆图页

宋（960—1279），佚名
绢本，设色
纵 26.4 厘米，横 27.9 厘米
故宫博物院藏

本幅无款，钤乾隆内府诸收藏印。蓬瀛，泛指蓬莱、瀛洲等传说中的仙山。画中楼台参差，亭榭栉比，曲槛回廊错落有致，奇石异木、各式陈设精致非凡。山水环绕楼阁，云气缥缈宛似仙境。乾隆帝在对幅题绝句："参差仙馆类蓬瀛，临水依山风物清。可望不可即之处，画家别有寄深情。"与画面相得益彰。

此幅描绘了想象中的仙山楼阁，是以建筑为主体之园林仙境的理想缩影。可观、可居、可游的人间仙境构建于园林之中，要有崇楼峻宇四面环水，更要有奇石异卉伴祥禽瑞兽，不仅可远观，还可经飞阁亲临，如乘虚往来于人间与仙境。（尚倩宇）

120 *Peng-Ying Immortal Pavilion*

Song Dynasty (960-1279), unknown painter
Album leaf, color on silk
L.:26.4cm, W.:27.9cm
Collection of the Palace Museum

'Peng-Ying' refers to Penglai and Yingzhou, mythical mountains of the immortals in Chinese legends. In this painting, the pavilions and towers are staggered in arrangement, surrounded by unique stones and rare trees, and accompanied by various exquisite furnishings. On the opposite leaf, the Qianlong Emperor inscribed a poem, 'The tiered halls of the immortals recall Peng-Ying; by water and mountain, the scene is pure and serene. That which can be seen but never reached; the artist captures with profound feeling.' (Shang Qianyu)

121 方壶图轴

明（1368—1644），吴彬
纸本，设色
纵 320 厘米，横 101 厘米
故宫博物院藏

本幅作者自识："方壶图。天启丙寅春暮，枝隐吴彬写。"钤"吴彬之印""文中"印两方。"天启丙寅"为明天启六年（1626）。

中国古典园林自秦汉发端，秦始皇用一池一山来象征海上蓬莱，汉武帝开凿太液池，以"一池三山"创造了此后历代宫苑营造的典型。居游于"宛若仙境"的园林，中国古人追求着内心的超脱之境，而中国园林中的"山"则成为这一追求的依托物象。

画家笔下的方壶仙山外形奇诡，山峰之间洞天别具，琼楼玉宇隐现其中，有仙人凭栏远眺。山脚下海浪拍岸，惊涛翻涌，山间云气缥缈，充满着神话中仙山的奇幻之感。画中夸张变形的山体，似水，似云，远观如仙境，近看又如园林中的假山赏石，山与石在观者眼中相互变幻，人间与仙境似乎也仅一步之遥。这也许就是造园者追求的亦真亦幻之感。

吴彬（1550—1643），字文中，又字文仲，号枝隐、枝隐庵头陀，福建莆田人。与同时代画家相比，吴彬中晚年作品多尺幅巨大、变形的奇山异水，此幅即为其晚年之作。枝隐为吴彬于南京栖霞寺的隐居室名。（尚倩宇）

121 *Fanghu Immortal Mountain*

Ming Dynasty (1368-1644) , Wu Bin
Hanging scroll, color on paper
L.:320cm, W.:101cm
Collection of the Palace Museum

The Fanghu Immortal Mountain in this painting features an extraordinary and fantastical shape. The mountain peaks form unique landscapes, with jade-like palaces and towers vaguely visible amidst them. An immortal stands by the railing, gazing into the distance. At the base of the mountain, waves crash against the shore while misty clouds swirl around the peaks, evoking the mythical aura of an immortal mountain. The exaggerated, distorted mountain forms in the painting appear, from a distance, like a fairyland, yet upon closer inspection they resemble the artificial rocks and scholar's stones in a traditional Chinese garden. (Shang Qianyu)

122 碧玉刻乾隆帝御制诗仙山楼阁图山子

清乾隆（1736—1795）

长 23.4 厘米，宽 6 厘米，高 18.7 厘米

故宫博物院藏

山子碧玉质，细腻光洁。两面随形分层镂雕山水楼阁人物图。正面镌刻楷书填金乾隆帝御制诗《孙琰仙山楼阁》："树有清风溪有澜，楼台高迥石林端。笏峰丛倚白云表，辋水乱飞碧落寒。法是山樵传派正，地惟王子得居安。个中妙处称生动，楮叶精工有底难。"背面雕山水楼阁图，远山、近水、楼阁、人物有深有浅，层次分明。此山子旧藏遂初堂。

此诗作于乾隆十六年（1751），收录于乾隆帝《御制诗二集》卷二十四。孙琰于乾隆时期供奉内廷，善山水，常绘仙山楼阁图。此件玉山子将孙琰的《仙山楼阁图》与乾隆帝题画诗相结合，充满诗情画意。（黄英）

122 Jasper Carving with Poem by the Qianlong Emperor

Qing Dynasty, Qianlong Reign (1736-1795)

L.:23.4cm, W.:6cm, H.:18.7cm

Collection of the Palace Museum

This jasper carving (*Shanzi*) is intricately pierced and layered on both sides, depicting a landscape with pavilions and figures in accordance with the natural shape of the stone. The front side is inscribed in gold-filled regular script with a poem by the Qianlong Emperor. The reverse side is carved with complementary landscape and architectural motifs in the same intricate layered style. It was formerly housed in the Hall of Fulfilling Original Wishes (*Suichu tang*). (Huang Ying)

静修

“人道我居城市里，我疑身在万山中。”

这是元代高僧、苏州狮子林的第一位主人天如禅师在园中修行所感。中国古典园林中四时变换的自然景观、壶中天地的造园意象、出则入世进则出世的喻示，都是习静修心、恰到好处的观照对象。

御苑之内，假山环绕的佛楼中，乾隆帝曾拈香礼佛；依山傍水的寺观园林中，日耕夜禅的修行传统已延续千年。欧洲中世纪的修道院，修士们在回廊院中祈祷冥想，在十字划分的中央方庭中种植草药花卉，以双手的园艺劳作慰藉心灵。

园林，是理想的栖居地，也是理想的静修涤心之所。

Gardens and Retreats

‘Others say I live in the city, but I find peace as though I dwell among ten thousand mountains.’

This was the sentiment of Tianru, a monk of the Yuan Dynasty and the first owner of Lion Grove Garden in Suzhou, as he practiced his spiritual devotion within the garden. The ever-changing natural scenery across the seasons, the concept of creating a miniature universe, and the symbolic elements reflecting a balance between worldly engagement and spiritual transcendence–these defining features of classical Chinese gardens create an ideal setting for inner cultivation.

Imperial gardens could also serve as places of devotion; in the Qing Dynasty, the Qianlong Emperor lit incense in a Buddhist hall nestled among artificial mountains. In temple gardens, set against real mountains and flowing waters, generations of monks have farmed by day and meditated by night. In European medieval monasteries, monks prayed and meditated in cloisters, tending herbs and flowers in cross-shaped enclosed gardens, finding solace for their souls through the work of their hands. Gardens are timeless retreats, offering both a place to dwell and a sanctuary to soothe the soul.

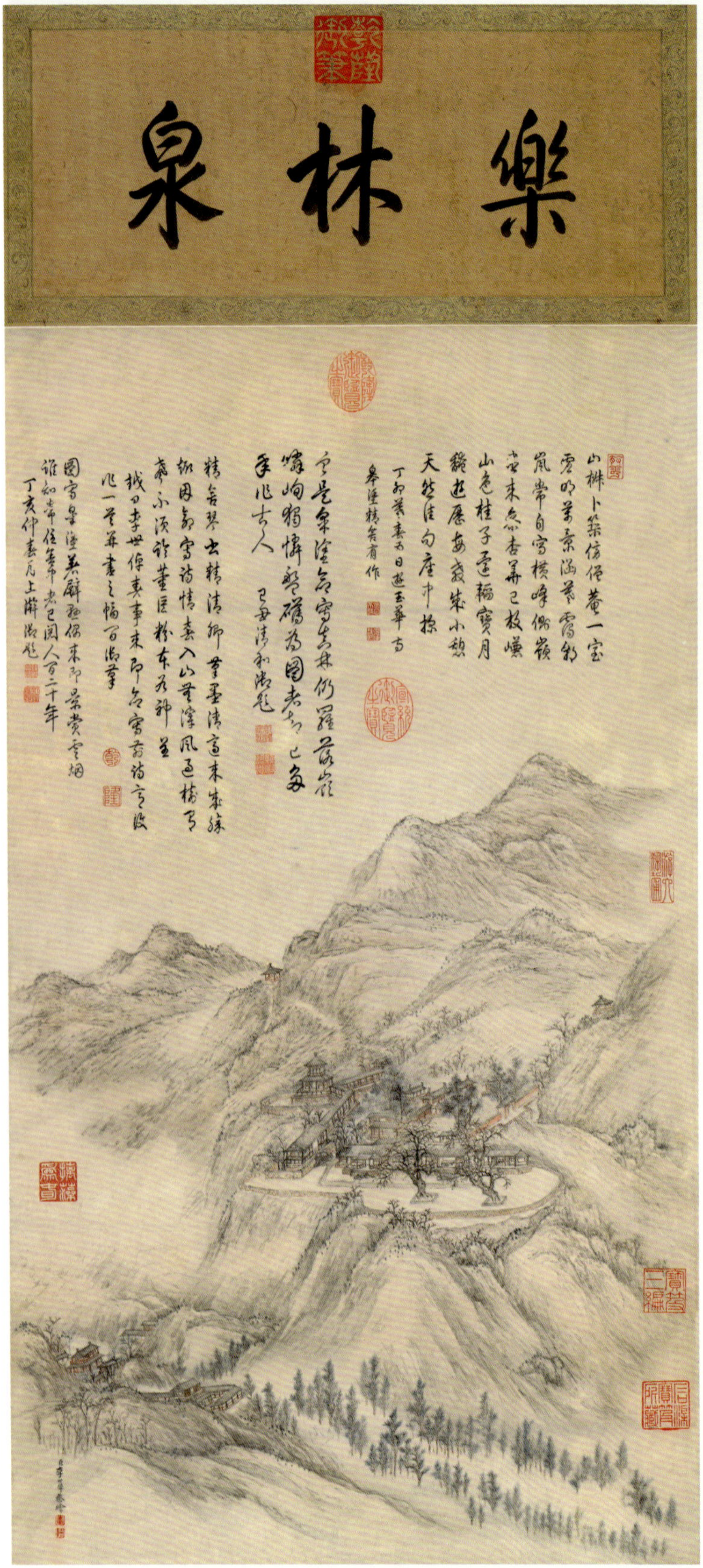

123 皋涂精舍图轴

清乾隆（1736—1795），李世倬
纸本，淡设色
纵 84 厘米，横 48.5 厘米
故宫博物院藏

本幅款署"臣李世倬恭绘"，钤"李""世倬"连珠印。另有"乾隆御览之宝""石渠宝笈所藏""宝笈三编""宣统御览之宝"等鉴藏印。

静宜园二十八景中有一处名为玉华岫，中有玉华寺，寺里乾隆帝的书屋取名"皋涂精舍"。此图以俯瞰角度作画，建筑地处半山，四周群山环伺，气势非凡。建筑被安排在画面的视觉中心，描绘得尤为精心。院落廊台，一丝不乱，古木高大，修竹挺秀，足见是写实之笔。

画家用墨简淡而不干涩，皴笔短促而细密，恰到好处地表现出北方山水园林的荒寒之意。

李世倬（1687—1770），字天章，一字汉章，号谷斋，别号十石居士、清在居士等，三韩（今辽宁铁岭）人。正蓝旗汉军，历官知州、副都御史、太常寺卿等职。擅作画，山水、人物、花卉俱臻其妙。（聂卉）

123 *Gaotu Vihara Pavilion*

Qing Dynasty, Qianlong Reign (1736-1795), Li Shizhuo
Hanging scroll, ink and light color on paper
L.:84cm, W.:48.5cm
Collection of the Palace Museum

Among the Twenty-Eight Scenic Spots of Jingyi Garden, there is a site called *Yuhuaxiu* (Jade Essence Peak), which houses Yuhua Temple. Within the temple, the Qianlong Emperor's study was named *Gaotu Vihara*. This painting captures the scene from a bird's-eye view, depicting the building situated on a mountainside and surrounded by imposing, towering peaks. The artist employed light yet rich ink, with short and dense brushstrokes, skillfully conveying the desolate beauty of northern landscapes and gardens. Qianlong Emperor's inscription, *Le Lin Quan* (rejoicing in woods and springs), expresses the delight found in nature's landscapes. (Nie Hui)

124 荒野中的忏悔者圣哲罗姆

约 1497，［德国］阿尔布雷特・丢勒
纸本，雕刻版画
纵 32 厘米，横 22.4 厘米
芝加哥艺术博物馆藏

125 意大利风景中读书的圣哲罗姆

约 1653，［荷兰］伦勃朗・凡・莱因
纸本，蚀刻、雕刻、干刻版画
纵 27 厘米，横 21.9 厘米
芝加哥艺术博物馆藏

124 **Saint Jerome Penitent in the Wilderness**

Ca. 1497, Albrecht Dürer (German)
Engraving in black on cream laid paper
L.:32cm, W.:22.4cm
Collection of the Art Institute of Chicago, Clarence Buckingham Collection (1938.1459)

125 **Saint Jerome Reading in an Italian Landscape**

Ca. 1653, Rembrandt van Rijn (Dutch)
Etching, burin, and drypoint on ivory laid paper
L.:27cm, W.:21.9cm
Collection of the Art Institute of Chicago (1941.1048)

126 荒漠中的圣哲罗姆

约 1553，[佛兰德] 海欧纳莫斯·考克据 [佛兰德] 老彼得·勃鲁盖尔作品创作
纸本，蚀刻和雕刻版画
纵 31.1 厘米，横 42.6 厘米
芝加哥艺术博物馆藏

126 Saint Jerome in the Desert

Ca. 1553, Hieronymus Cock (Flemish), after Pieter Bruegel the elder (Flemish)
Etching and engraving on ivory laid paper
L.:31.1cm, W.:42.6cm
Collection of the Art Institute of Chicago, Joseph Brooks Fair Fund (1981.225)

圣哲罗姆（Saint Jerome，约 342—420）是基督教早期的重要神学家，也是《圣经》拉丁文译本（Vulgate）的译者。其生平与形象是西方艺术创作的重要主题之一，丢勒与伦勃朗曾以圣哲罗姆为题材创作了许多经典作品。在艺术作品中，圣哲罗姆或以荒野中苦修的隐士形象出现，或以内省与探索的学者形象出现。其于自然中的修行意象映射出基督教世界的自然观念，对欧洲中世纪和文艺复兴之后的造园艺术产生了一定影响。（雷浩）

Saint Jerome (Ca. 342-420) was an important theologian of early Christianity and translator of the Vulgate (a late-4th century Latin translation of the Bible). Saint Jerome is often portrayed in art either as a hermit engaged in ascetic practices in the wilderness, or as a scholar engaged in self-reflection and intellectual exploration. His image of cultivating himself through practice in nature reflects Christian views of nature and influenced the development of European garden art from the Middle Ages to the Renaissance and beyond. (Lei Hao)

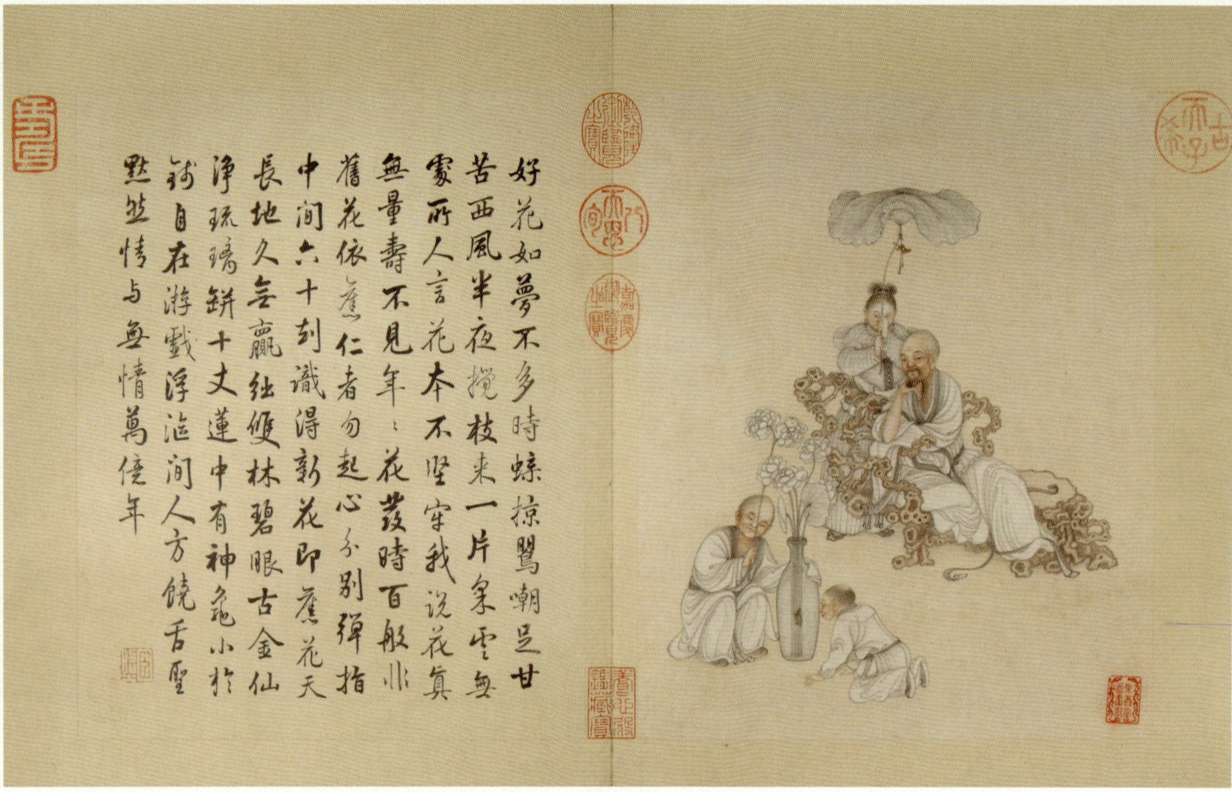

127 罗汉图册

清（1644—1911），冷枚

纸本，设色

纵 35.5 厘米，横 29.1 厘米

故宫博物院藏

该册共二十开，绘菩萨、罗汉、高僧等佛教形象和故事，今选其中四开。

此四开画，分别绘华苹献寿、达摩面壁、指上蜃楼、顽石点头故事，画家用图绘表现了深奥的佛理。图中人物淡设色，衣纹近于白描，笔法飘逸，人物生动，面目精细，画家运用了凹凸晕染表现人物的立体形象。对幅有张照题诗，阐述了画中佛理。

第一幅华苹献寿，绘僧人坐于藤椅中，支颐谛观瓶中荷花。华苹一说即并蒂莲。佛教中观想荷花也是一种修行的方式。自然界中鲜花开谢，周而复始，强行分别，则有旧花今花的区别。但年年开花，对比周围不断变换的景色，反而衬托出花的永恒。

第二幅面壁示禅，绘禅宗祖师菩提达摩面壁趺坐，身后二祖慧可大师跪侍求法，这是禅宗传法的典故。面壁虽然历经九年，但对于悟道的人来说，心中了无分别，和屈伸手臂的时间并无二致。其实修道者真正用功处，是不断检查自己的心中过错。

第三幅指上蜃楼，绘僧人与两人谈论，并用手指示空中宫阙，可能化用佛教中乾闼婆城或者海市蜃楼的典故。佛教中常用乾闼婆城比喻虚幻不实之物。张照对题中所用“指月”典故，出自六祖惠能，寓意文字经典只是手指，而真正需要关注的是经典中所讲述的佛理。

第四幅顽石点头，绘僧人坐于拳石之中，此典出于东晋高僧竺道生顽石点头的故事。道生认为众生皆有佛性，此论在当时还不被人接受，生公于是在苏州虎丘山聚石说法，说完群石纷纷点头，以此表明道生说法符合佛意。

冷枚（约 1669—1742），字吉臣，号金门画史，山东胶州人，清雍正、乾隆时宫廷画家，善画人物、界画，尤精于仕女。冷枚师承焦秉贞，画法上吸收西画特点，注重画面透视、明暗对比等，笔墨洁净，设色雅丽，人物造型典丽。（施含牧）

127 ***Meditation Facing the Wall***

Qing Dynasty (1644-1911), Leng Mei
Album leaf, color on paper
L.:35.5cm, W.:29.1cm
Collection of the Palace Museum

This four-leaf album combines paintings by the artist with accompanying poems written by Zhang Zhao, elaborating on Buddhist principles. The artist employed shading techniques for a three-dimensionl effect.

In the first painting, a monk sits in a rattan chair, resting his chin on his hand as he contemplates a vase of flowers. In nature, flowers bloom and wither in an endless cycle. One may insist on distinguishing them into old and new, yet flowers bloom every year, and the surrounding scenery constantly changes. Could this not also be a form of eternity?

The second painting portrays Bodhidharma seated in meditation, with his disciple Huike behind him seeking the *Dharma*. Bodhidharma meditated for nine years, a span not unlike a snap of fingers for those who attain enlightenment.

The third painting illustrates monks discussing whether, when one sees a dharma banner move, the movement is of the banner itself or of the mind, which leads to a debate on whether celestial palaces on the moon truly exist. In Zhang Zhao's accompanying poem, he makes a reference to the Buddhist parable of 'Pointing to the Moon, not the Finger'.

In the forth painting, a monk is seated among rugged stones, a reference to the story of monk Zhu Daosheng in the Eastern Jin Dynasty (317-420 CE). Zhu asserted that all sentient beings possess Buddha-nature, a theory that was not widely accepted at the time. So he preached among rocks in Suzhou. When he finished, the rocks nodded in unison, signifying that his teachings were in harmony with the true essence of Buddhism. (Shi Hanmu)

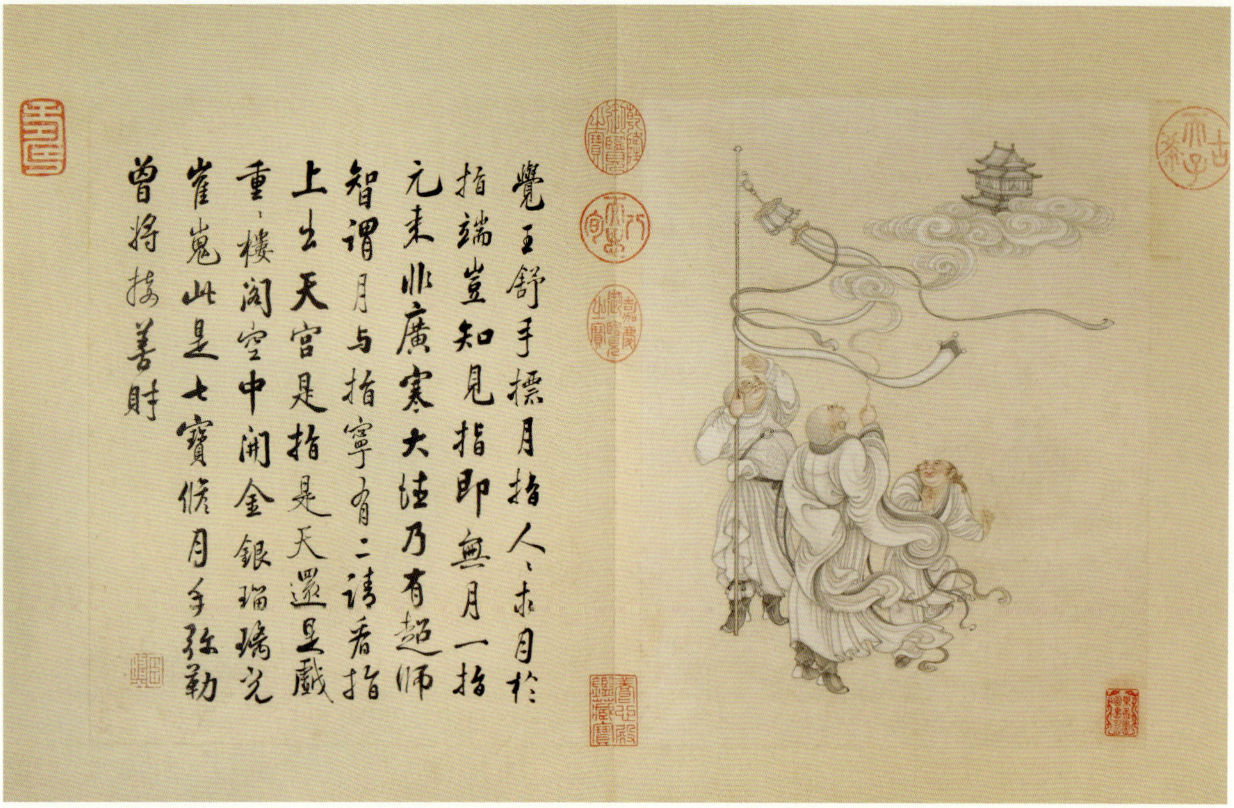
覺王舒手標月指人人未見月於
指端豈知見指即無月一指
元來非廣寒大士乃有超師
智謂月與指寧有二請看指
上出天宮是指是天還是戲
重重樓閣空中開金銀瑠璃光
崔嵬此是七寶俟月手弥勒
曾將接善財

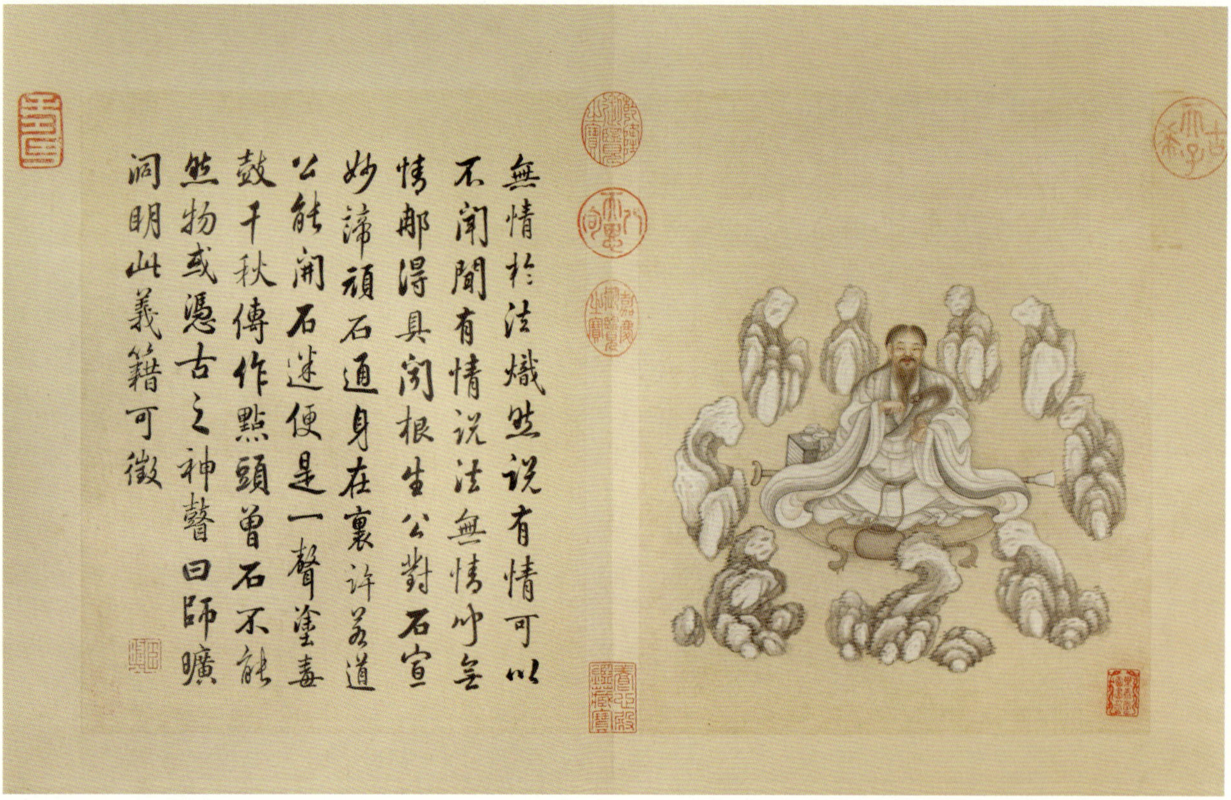
無情於法識然說有情可以
不聞聞有情說法無情[illegible]會
情那得具聞根生公對石宣
妙諦頑石通身在裏許吾道
公能聞石迷便是一聲塗毒
鼓千秋傳作點頭曾石不能
然物或憑古之神瞽曰師曠
洞明此義籍可徵

128 白玉炉瓶盒三事

清（1644—1911）
炉：口径 8.4 厘米，底径 6 厘米，高 9.8 厘米
瓶：口径 2.6 厘米 ×1.9 厘米，底径 2.6 厘米 ×1.9 厘米，通高 11.8 厘米
盒：口径 5.65 厘米，底径 3.65 厘米，高 3.5 厘米
故宫博物院藏

炉、瓶、盒三事均为白玉质，温润细腻。炉圆形，覆碗式盖，圆形盖钮，兽形耳，圈足底，三兽形足，器身雕兽面纹。瓶扁圆形，双贯耳，圈足，腹部浮雕兽面纹，颈部浮雕蕉叶纹。盒圆形，薄壁，矮圈足，盖面圆形开光内浮雕勾连云纹。

炉、瓶、盒是文人雅士几案上的常用陈设，三件成组搭配，被称为“三事”，是焚烧香料的器具，具有祛除异味、散发香气的功用。炉用于燃香；瓶放置铲、箸，用于铲除香灰和夹取香料；盒存储香料。“三事”既有实用功能，又兼具陈设功能，是清代宫廷常见器物，亦常见于寺观园林的绘画作品中。（黄英）

128 White-Jade Incense Burner, Vase, and Box (Three Essentials)

Qing Dynasty (1644-1911)
Burner: Diam.(at mouth):8.4cm, Diam.(at base):6cm, H.:9.8cm
Vase: Diam.(at mouth):2.6cm×1.9cm, Diam.(at base): 2.6cm×1.9cm, H. (in total):11.8cm
Box: Diam.(at mouth):5.65cm, Diam.(at base):3.65cm, H.:3.5cm
Collection of the Palace Museum

The incense burner, vase, and box were common items on the desks of scholars, and collectively known as the 'Three Essentials'. The burner held incense, the vase stored tools like a spatula and tongs for handling ashes and incense, and the box contained fragrant materials. Serving both practical purposes, such as spreading pleasant aromas, and decorative functions, they were often featured in paintings of Buddhist and Daoist temples and gardens. (Huang Ying)

129 玉洞仙源图轴

明（1368—1644），仇英
绢本，设色
纵 169 厘米，横 65.4 厘米
故宫博物院藏

图绘青山石洞，白云缭绕，琼楼水阁点缀其间，苍松掩映下一白衣高士在洞侧临水盘膝，抚琴而思，随侍童仆们则汲水、生火、烹茶、备食。该图以细线勾勒山石，少皴，青绿设色明丽雅致，是仇英继承宋代院体青绿山水而又融入士人趣味的佳作。本幅右下有"仇英实父制"款，钤"仇实父氏"朱文方印，有安岐、卞永誉等钤鉴藏印。

山洞在道教文化中有"洞天福地"的意涵，即山中有洞室可通达天上。明清两代，在营造形式的影响下，"洞天"进入到许多园林造景中，尤其是标榜隐居理想的文人士大夫园林中，如苏州拙政园中就有"别有洞天"景观。明万历年间造园家计成在《园冶》一书中提出的"掇山"要领便有"岩、峦、洞、穴之莫穷，涧、壑、坡、矶之俨是"，还专门论述了园林中的"理洞法"，以及洞上应堆土植树，或作台，或置亭屋等景观建筑。园林中的山洞、楼阁等亦成为文人修身养性的特殊空间。（蒋彤）

129 *Jade Grotto Immortal Residence*

Ming Dynasty (1368-1644) , Qiu Ying
Hanging scroll, color on silk
L.:169cm, W.:65.4cm
Collection of the Palace Museum

In Daoist culture, mountain grottoes symbolize 'paradise realms', the sacred mountains and paradisiacal realms where Daoist immortals are believed to reside. During the Ming and Qing Dynasties, garden design trends incorporated this concept, particularly in the private gardens of scholar-officials who idealized reclusive living. The painting depicts verdant mountains and a rocky grotto, where a scholar in white robes sits by the water, playing the *qin*. (Jiang Tong)

130 塞辛科特林园中的神庙、喷泉和洞穴

1819，[英国] 托马斯 · 丹尼尔
布面，油画
纵 120.3 厘米，横 145.4 厘米
耶鲁大学英国艺术中心藏

这座位于英格兰格洛斯特郡的塞辛科特庄园（Sezincote House）融合了印度莫卧儿的设计风格，其建筑与园林深受印度教影响，充满了异国情调和浪漫气息（见后对页照片）。画作展示了园林内的神庙水池，中央有一个长满青苔的凸出喷泉，喷泉后的神庙中心供奉着印度教太阳神苏里耶的浮雕形象。(雷浩)

130 Temple, Fountain and Cave in Sezincote Park

1819, Thomas Daniell (British)
Oil on canvas
L.:120.3cm, W.:145.4cm
Collection of the Yale Center for British Art (B1976.7.106)

Sezincote House in Gloucestershire, England is heavily influenced by Mughal-inspired Hindu aesthetics, giving it an exotic and romantic atmosphere. This scene depicts the temple pool in its garden. At the center stands a moss-covered protruding fountain. Behind it, a relief of *Surya*, the Hindu sun god, is enshrined at the heart of the sanctuary. (Lei Hao)

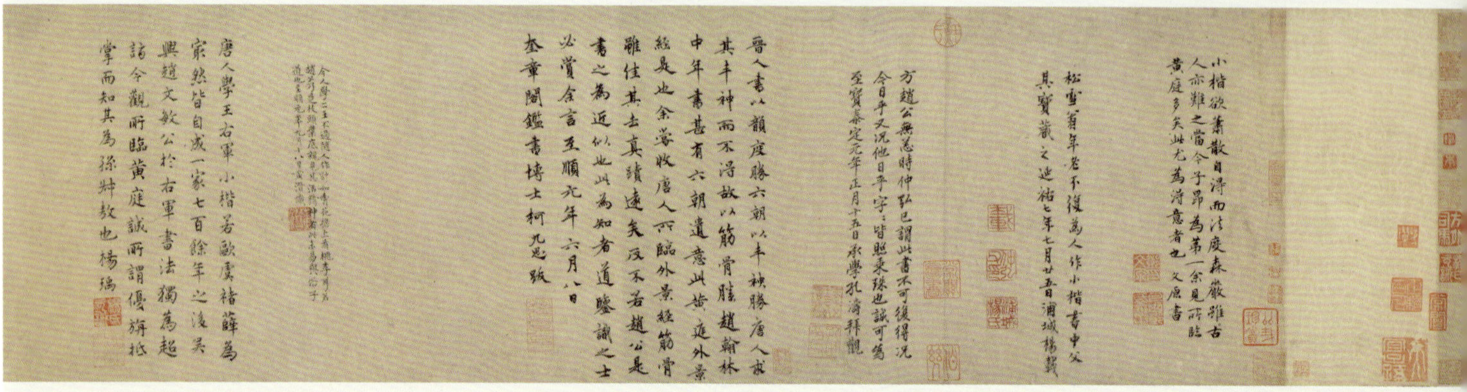

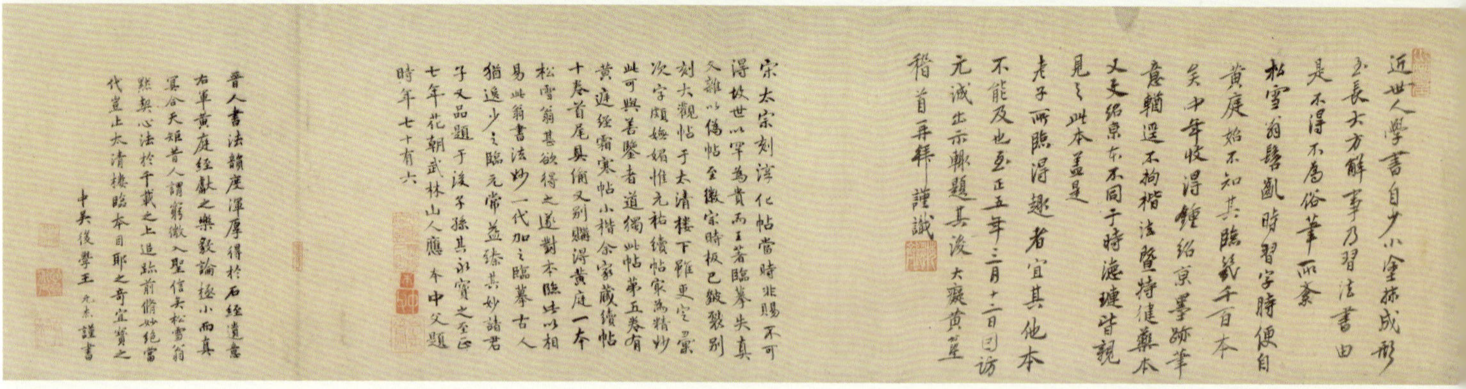

131 临黄庭经卷

元（1271—1368），赵孟頫
纸本，楷书
纵 26.5 厘米，横 114.8 厘米
故宫博物院藏

本幅款署“子昂”，钤“赵氏子昂”朱文印一方，无年款。《黄庭经》是道教上清派的主要经典，成书于东汉晚期到晋朝以前，被称为“学仙之玉律，修道之金科”。说的是道教存神养气的修持之术，只要存思诸神，凝神静虑，使人的意念与天神相汇，便可达到天人合一的境界。王羲之生活的时代有着浓厚的仙道观念，他以所书《黄庭经》换鹅的故事广为流传。此卷为赵孟頫临《太清楼帖》本王羲之小楷《黄庭经》，体势极近右军。卷后有邓文原、杨载、孔涛、柯九思等十六人题跋。杨载说赵孟頫“年老不复为人作小楷书”，柯九思也说是其中年所作，“有六朝遗意”。后又有邓文原称当今小楷“子昂为第一”，此卷“尤为得意”。据说，世祖皇帝见赵孟頫“神采秀异，珠明玉润，照耀殿庭”，称其为“神仙中人”。当然，这种赞赏很大程度来源于他的艺术修养。他主张以复古为旗帜，改变南宋书画的颓势，同时也体现着他追寻魏晋风度的审美理想，所表现出来的个人修养和精神气质与此不无关系。卷中有“仙人道士非可神，积精所致和专仁”等语，多为丹家之要旨，精神之园林。

赵孟頫（1254—1322），字子昂，号松雪道人、鸥波、水精宫道人等，是宋太祖之子秦王德芳的后裔。世居湖州，故赵孟頫为吴兴人。官至兵部郎中、翰林院学士承旨、荣禄大夫。为官三十余年，荣际五朝。延祐六年（1319）辞官归湖州，至治二年（1322）卒，年六十九岁。有《松雪斋集》行世。工诗文、善书画。《元史》称：“篆籀分隶真行草无不冠绝古今，遂以书名天下。”小传见《元史》卷一七二。（焦东华）

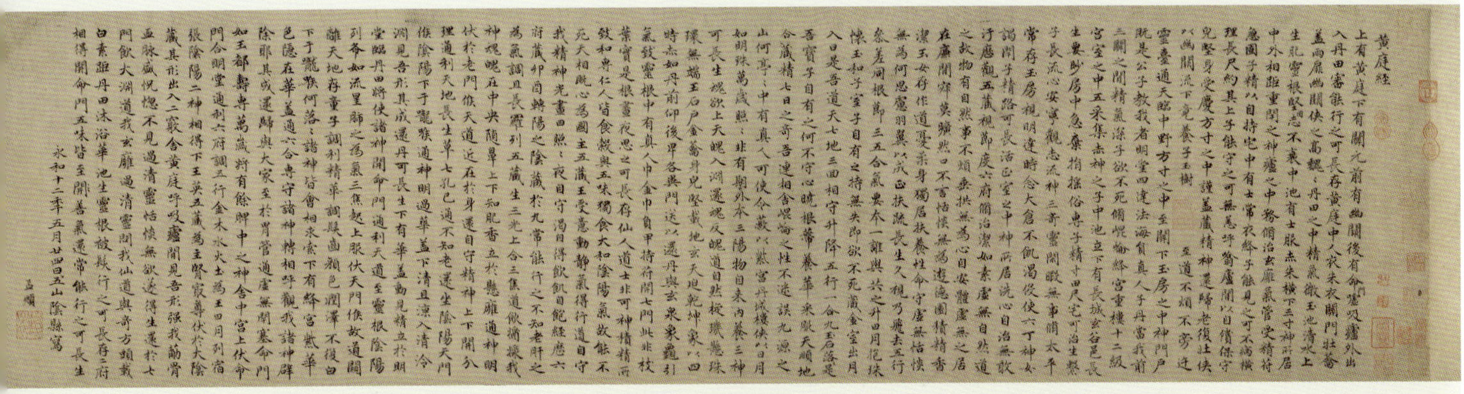

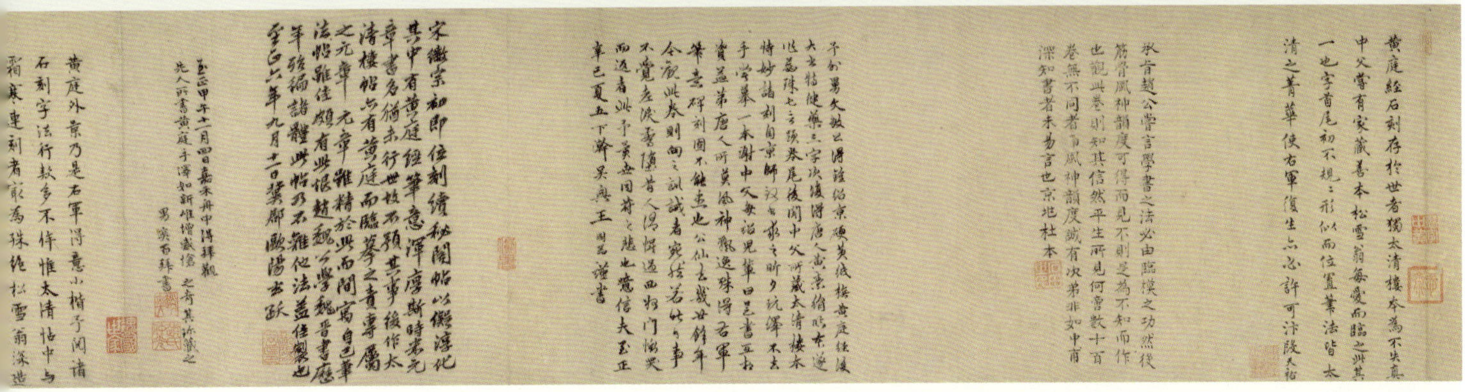

131 **After *Yellow Court Scripture***

Yuan Dynasty (1271-1368), Zhao Mengfu
Handscroll, paper, regular script
L.:26.5cm, W.:114.8cm
Collection of the Palace Museum

The *Yellow Court Scripture* is a key classic of Daoism. This scroll is a copy by Zhao Mengfu of Wang Xizhi's version of the scripture. Zhao advocated for a revivalist approach, using the aesthetics of earlier artists to counter the decline he saw in the calligraphy of his time.

The scroll contains phrases such as 'Immortals and Daoists are achieved through focused cultivation rather than divinity', encapsulating the core principles of Daoists' gardens of the spirit. (Jiao Donghua)

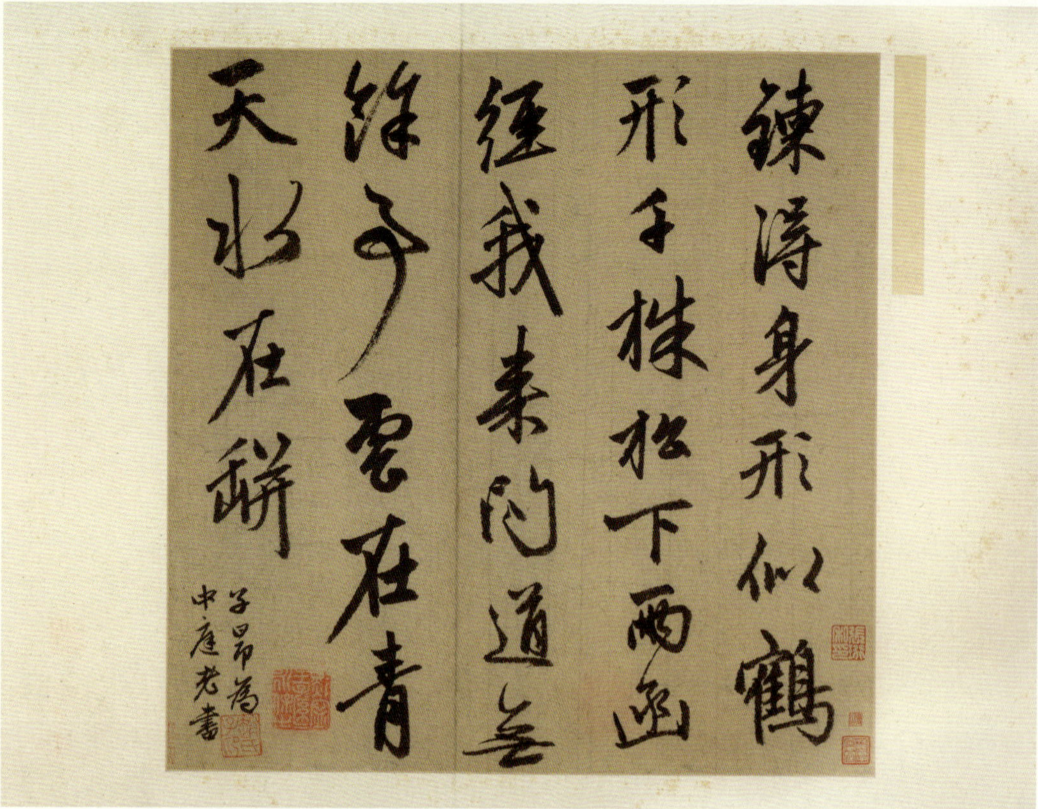

132 行书诗册页

元（1271—1368），赵孟頫
纸本，行书
纵 34.7 厘米，横 35.3 厘米
故宫博物院藏

此册页为赵孟頫书唐李翱的《赠药山高僧惟俨二首》之一："练得身形似鹤形，千株松下两函经。我来问道无余说，云在青天水在瓶。"末识"子昂为中庭老书"，钤"赵氏子昂"印。另有鉴藏印"张珩私印""博山""潘厚审定""黔宁王子子孙孙永保之"四方。

李翱在任朗州（今湖南常德）刺史时，拜访药山道场高僧惟俨禅师后作此诗。药山寺是一座寺观园林，唐代蔚为壮观。此诗不仅刻画了高僧惟俨清雅出尘的形象，也描写了药山道场的修行环境。最后一句"云在青天水在瓶"更是将自然景观与哲学思想相结合，既表现了高僧修行的不凡境界，也体现了诗人对天地自然与禅宗思想的感悟，深得禅趣。

此册页笔力雄健，富于变化，既有雍容之姿，也有苍劲厚重之笔，为赵孟頫晚年大字行书佳作。
（魏婧洁）

132 Calligraphy Album of Poems

Yuan Dynasty (1271-1368), Zhao Mengfu
Album leaf, paper, running script
L.:34.7cm, W.:35.3cm
Collection of the Palace Museum

This page features Zhao Mengfu's transcription of Tang Dynasty (618-907) poet Li Ao's *Presented to the Eminent Monk Weiyan of Yaoshan*. Yaoshan Monastery, known for its combination of temple and garden landscapes, was an awe-inspiring sight during the Tang Dynasty. The final line, 'Clouds in the sky, water in the jar', blends philosophical thought and natural scenery. (Wei Jingjie)

133 莲舟仙渡图页

宋（960—1279），佚名
绢本，设色
纵 21.4 厘米，横 23.5 厘米
故宫博物院藏

画中仙人着树叶状的羽衣，手持帛书，赤脚坐于莲舟之上，逍遥游于云气之间。莲舟之侧，峰峦在变幻莫测的云气间若隐若现。苏轼有诗："纵一苇之所如，凌万顷之茫然。"在莲舟不盈一握的尺寸之外，是浩渺天地，而仙人泰然独坐，自成清静气象。画作旨在

表达仙人修行的自在境界，乘一莲舟即可泰然遨游天宇，这与中国园林“壶中天地”的意趣殊途同归，一园虽小，却可小中见大，于拳山勺水中静观天地。

《苕溪渔隐丛话》记载了北宋画家李公麟绘有这样一幅作品，诗人韩驹诗云：“太一真人莲叶舟，脱巾露发寒飕飗。轻风为帆浪为楫，卧看玉宇浮中流。”其所指传为李公麟《太乙莲舟图》，与此件题材相同。后世多有对太乙莲舟的吟咏之作，如元好问有诗：“我与太虚同一体，也无莲叶也无波。”而李公麟图式又经历代传续演变，见于瓷器、玉器，亦有仙人乘槎酒具等。（尚倩宇）

134 朱碧山款银槎

元（1271—1368）
长 20 厘米，宽 9 厘米，高 18 厘米
故宫博物院藏

银槎通身银制，酒器，为元代名匠朱碧山根据仙人乘槎上天河的传说而作。槎尾有铸款“龙槎”“至正乙酉，渭塘朱碧山造于东吴长春堂中，子孙保之”“华玉”篆书印。铸诗于腹下：“百杯狂李白，一醉老刘伶。知得酒中趣，方留世上名。”槎口下铸：“贮玉液而自畅，泛银汉以凌虚。杜本题。”（邢娜）

133 *Immortal Aboard Lotus Boat*

Song Dynasty (960-1279), unknown painter
Album leaf, color on silk
L.:21.4cm, W.:23.5cm
Collection of the Palace Museum

This painting depicts an immortal leisurely traversing clouds in a tiny lotus boat and simple clothing. The carefree state the artwork aims to convey aligns with the serene state of an immortal's cultivation–drifting effortlessly through the vast heavens aboard a lotus boat. (Shang Qianyu)

134 Silver Raft Cup with Zhu Bishan's Mark

Yuan Dynasty (1271-1368)
L.:20cm, W.:9cm, H.:18cm
Collection of the Palace Museum

This wine vessel is made entirely of silver and was crafted by the renowned Yuan Dynasty artisan Zhu Bishan. Inspired by the legend of an immortal traveling up the Milky Way on a wooden raft, the vessel takes the form of a silver raft. (Xing Na)

135 萃赏楼圣容漆泥子佛挂龛

清乾隆（1736—1795）
纵 93 厘米，横 166 厘米，厚 7.3 厘米
故宫博物院藏

萃赏楼位于乾隆花园第三进院落最北端，为二层楼阁式建筑。其室内二层西稍间为佛堂，内设高低床，床上有两槽带玻璃方窗的板墙。板墙与佛堂墙壁之间，密密麻麻挂满了佛画像。按乾隆朝造办处《活计档》记载："门南边应供挂像佛一轴，门斗应供挂像佛一轴，西进深应供挂像一轴，前面罩上应挂像佛一轴，东进深挂像佛二轴，罩上里外供挂像佛十轴，后层罩里中间挂像佛一轴，两傍（旁）边挂像佛二轴，东进深挂像佛一轴，西进深漆泥子佛挂龛一座，中间挂像佛一轴，两边挂像佛二轴。"诸多挂像中唯一不同的，便是此面圣容漆泥子佛挂龛。

漆泥子佛，更常见的名称是"擦擦佛"，指的是用陶土或泥，用脱模的方式制作而成的佛造像，表面通常施有彩绘。而将这些擦擦佛陈列在如佛龛般的挂屏中，便是所谓漆泥子佛挂龛。目前故宫博物院存有漆泥子佛挂龛共 9 件，而其中 7 件的主尊为文殊扮相的乾隆帝，萃赏楼这件挂龛即其中之一。

本件龛内共 95 尊擦擦像，分别为格鲁派本尊神、藏传佛教祖师，与结跏趺坐正中壶门内右手施说法印、左手托法轮的乾隆帝。乾隆帝坐垫上有藏文偈颂，译作：睿智文殊人之主，游戏圣主法之王。金刚座上安奉足，意愿天成善福缘。对比乾隆帝御制《萃赏楼》诗："层屋假山巅，翼然复洒然。不教陈御座，可以供金仙。亦却庄严丽，惟欣清净便。题名曰萃赏，却弗涉因缘"，乾隆帝似要通过萃赏楼室内的佛堂设计，传达自己在退位后不愿再为政务烦恼，而想要专注修行，于山水园林内怡情养性之意。正如他在另一首御制《萃赏楼》诗中所云："洗心得静怡，真赏欣荟萃。"（刘仁皓）

135 Hanging Shrine with Lacquered Clay Figurines

Qing Dynasty, Qianlong Reign (1736-1795)
L.:93cm, W.:166cm, Th.:7.3cm
Collection of the Palace Museum

This shrine is in the Buddhist chapel on the second floor of the Building for Enjoying Lush Scenery (*Cuishang lou*) in the Qianlong Garden. The shrine contains 95 *tsa-tsa* figurines, representing deities of the Gelug school, masters of Tibetan Buddhism, and, at the center of the arch-shaped opening, the Qianlong Emperor himself. (Liu Renhao)

136 乾隆帝御笔佛日楼楹联

清乾隆四十三年（1778）
纸本，墨笔
纵 156 厘米，横 35.5 厘米
故宫博物院藏

此联在宁寿宫佛堂区域佛日楼二层正门内两侧悬挂，上联：鸟雀听经皆宿慧，下联：风幡说偈自高标。上联右上钤椭圆形双龙篆书“奉三无私”印，下联左下钤正方形篆书“惟精惟一”印和正方形篆书“乾隆宸翰”印。纸质为黄笺蜡纸，笺上绘有墨色花卉纹，此联主体结构为木框，上部有挂环，边缘有明黄绫包边。

据清宫造办处《活计档》，乾隆四十三年（1778）正月：“十三日，太监张进喜来说，首领董五经交御笔黄笺纸字对二副（佛日楼），御笔绿笺纸字横披一张（重华宫），传旨俱做一玉玉璧子托贴横披，随托钉挺钩，其字对厢一寸蓝绫边，随托钉倒环，钦此。”

上联中的“宿慧”即先天的智慧或慧根。禅宗六祖慧能在《金刚般若波罗蜜经》“序”中认为：“宿植上根者，一闻便了。若无宿慧，读诵虽多，不悟佛意。”下联中的“风幡”即佛幡，清宫在佛堂装饰和佛事活动中常用佛幡。乾隆帝曾亲自下旨选定咸若馆的佛幡样式，宝华殿前的地面上至今遗存有系留佛幡的专用构件。“偈”即偈颂，又称偈子，是佛经中用以唱颂的诗作，多以四字为一偈，蕴含佛教义理。“高标”即（竖起的）高木杆。结合佛日楼二层的陈设与宫廷唐卡内容推测，此处的“高标”可能暗指宾头卢尊者妄显神通的典故。据《四分律》卷五十一所载，王舍城中有长者制作了华丽的旃檀钵置于“高标”上，宣称城中的阿罗汉可施展神通自取，宾头卢尊者在与目犍连尊者推让一番之后，飞到空中施展神通，取走了旃檀钵。佛陀知晓此事后，训斥宾头卢尊者不应在居士面前显示神通，认为其行为是出于私欲，不符合佛教的威仪和戒律，更不是利益众生的行为。

此对联为乾隆帝老年所作，用笔从容洗练。由于纸质的原因，个别笔画有墨迹剥落的情况。对联内容不仅切合佛日楼所处的幽静环境，显现了乾隆帝的佛教学养，而且体现了他善于利用佛教教化人心，维护统治的特点。（尹昌杰）

136 ‘Birds and Banners’ Couplet Inscribed by the Qianlong Emperor

1778
Ink on paper
L.:156cm, W.:35.5cm
Collection of the Palace Museum

This couplet was hung on both sides of the main entrance inside the second floor of the Sanctuary of Buddhist Light (*Fori lou*), located in the Buddhist Hall of the Palace of Tranquil Longevity (*Ningshou gong*). It poetically reflects the pavilion’s religious atmosphere. (Yin Changjie)

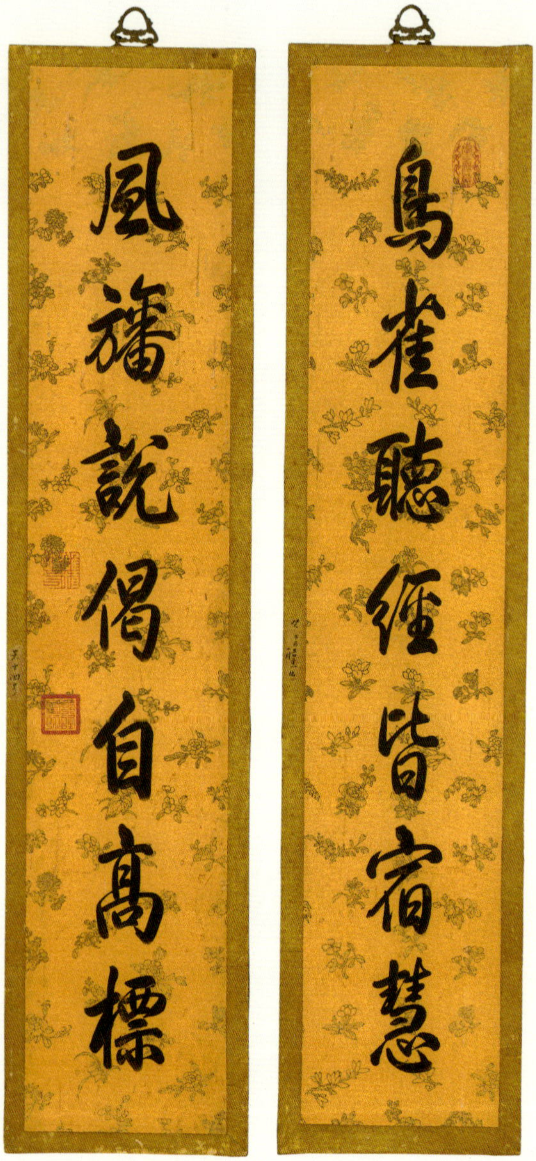

137 玉海晏河清图鼎式提炉（一对）

清（1644—1911）
鼎：口径 8.5 厘米，底径 8.3 厘米，高 11 厘米，带链高 68 厘米
架：高 88 厘米
几：长 51 厘米，宽 51 厘米，高 76 厘米
故宫博物院藏

此对提炉供于铜丝罩五龙捧圣海灯侧，玉质，圆口束颈，口沿饰回纹，腹为圆柱形。其上雕刻三道竖棱线，上刻寿字，三道竖棱线把腹部分割成三个单元，每单元雕刻一种图案，其中一种是祝寿主题的海屋添筹。双耳形如蟠螭，耳中穿环环相套的玉链，相当于提炉挂绳，玉链挂钩为铜质，上饰玛瑙珠。珐琅鼎盖，狮钮。圈足底，饰莲瓣纹。鼎口沿外刻“蓝田瑞玉海晏河清鼎”。

海晏河清，出自五代欧阳熙《龙寿院光化大师碑铭》：“旋闻海晏河清，远播民舒物泰。”指沧海波平，黄河水清，形容国内安定，天下太平。

描金漆木架立于红漆描金二龙戏珠纹香几上。底盘雕刻成海水江崖形，从海中涌出须弥山以承托海晏河清鼎式炉，木架的三根支柱雕刻成从海水中涌现的三条升龙，十分巧妙地构成了用于悬挂的支架。

配红漆描金二龙戏珠纹香几，几面方形，束腰，腰上角柱之间开光雕刻碗花结带纹。鼓腿足上丰下敛，雕刻二龙戏珠纹，其上描金，下承方形托泥。

高功举办大斋时所持法器炉，法师一揖放炉执简，这时道众奏乐引至坛内，法师至炉案上香，知磬举步虚词。

清宫造办处《活计档》记载：“（乾隆二十年二月二十八日）员外郎白世秀来说，太监胡世杰传旨：钦安殿现安玉提炉香几糙，看照中间安海灯香几样式一样成做香几一对。钦此。于本年四月十三日催总六达子将做得香几一对持进安讫。”（王子林）

137 Jade Censer in the Shape of a *Ding* Vessel (a pair)

Qing Dynasty (1644-1911)
Vessel: Diam.(at mouth):8.5cm, Diam.(at base):8.3cm, H.:11cm, H.(with chains):68cm
Frame: H.:88cm
Table: L.:51cm,W.:51cm, H.:76cm
Collection of the Palace Museum

This censer was intended for worship of a major Daoist deity, *Zhenwu Dadi* (Great Perfect Warrior Emperor) at the Hall of Imperial Peace (*Qin'an dian*). The lacquered wooden stand, decorated with gold, is carved to resemble waves and river cliffs. The stand is supported by three intricately carved dragons emerging from the waves, skillfully forming a suspension frame to hang the censer. (Wang Zilin)

138 嵌玉十六尊者像屏

清乾隆（1736—1795）
单扇纵 213 厘米，横 58.7 厘米，厚 6 厘米
故宫博物院藏

此屏风原作为五代时前蜀画家、诗人、书法家贯休所画《十六应真像》，绢本已不复存在，现存多为后人摹本，故宫博物院藏此组屏风即为根据此画原件制作进献入宫。乾隆帝见后甚是喜爱，将其置于乾隆花园内的云光楼供奉。十六罗汉（或称十六尊者）是释迦牟尼的得道弟子。十六罗汉的名字早有佛经所载，传入中国后，约于唐末至五代十国时期演变为十八罗汉。

屏风一组十六扇，居中十扇，两翼各三扇，正面为罗汉形象，顶部一侧刻御制像赞辞与钤印，另一侧则是乾隆帝请格鲁派高僧、三世章嘉根据藏传佛教传统考定的罗汉名号。屏面嵌玉，屏背漆地描金绘花果植物，富贵雍容。（彭希贤）

138 Sixteen Arhats Screen in Jade Inlay

Qing Dynasty, Qianlong Reign (1736-1795)
L. (single screen):213cm, W.:58.7cm, Th.:6cm
Collection of the Palace Museum

This screen was originally based on *The Sixteen Arhats* by Guanxiu, a renowned painter, poet, and calligrapher of the Five Dynasties period (907-960). The Qianlong Emperor grew deeply fond of the piece and placed it in the Qianlong Garden for worship. One side of its top contains an ode to the Arhat portraits composed by the Qianlong Emperor, along with his seal, and the other side lists the names of the Arhats, authenticated by the Third Changkya Khutukhtu upon Qianlong's request. (Peng Xixian)

139 养和精舍木雕莲花纹壶门

清乾隆（1736—1795）
纵 234.5 厘米，横 164 厘米
故宫博物院藏

“精舍”一词最早出现于《管子·内业》。该文认为，虽然世上万物总在变化，但人应当保持自身的安定，使心成为“精之所舍”。在此之后，随着儒释二家的普及，“精舍”一词逐渐被用来指代学舍、僧房（Vihāra，也称毗诃罗）一类与内在修养相关的建筑。

乾隆花园中的养和精舍就是取毗诃罗之意所建的一座静修之所，是乾隆帝的养“和”之处。其一层东稍间木雕莲花纹壶门内的空间，即乾隆花园八所佛堂之一，整体似仿毗诃罗窟形制而建。观者想要进入室内，须越过精舍西面两组仕女、婴戏主题通景画；见过嘉庆御笔七言联“此心时欲归于是，进步方超最上乘”；跨过象征清净的莲花纹壶门。经此“三过”后，站在佛堂回首，便可看到壶门内高悬的乾隆帝御笔“西方极乐世界安养道场”横条和杨大章绘制的腊梅天竺积雪贴落。在整组空间内的体验，就像是修行本身，是超脱世俗的。踏过莲花，最终能够抵达净土。再看繁花，或许就如乾隆帝《养和精舍有会》所言：“试看芽纽蕾含始，却胜嫣红姹紫罗。方寸有为及无欲，斯之未信敢言他。”（刘仁皓）

139 Wooden Carved Lotus-patterned *Kunmen* from the Building of Luminous Clouds

Qing Dynasty, Qianlong Reign (1736-1795)
L.:234.5cm, W.:164cm
Collection of the Palace Museum

This *kunmen* gate is located inside the Buddhist hall in the Qianlong Garden. The hall itself is likely to have been modeled on a *vihara*, or Buddhist monastery. Standing at the entrance, the main frame of the *kunmen* was carved from nanmu wood in the shape of a lotus flower. The interior is fitted with a replica of pale blue gauze featuring circular dragon motifs intertwined with treasures. (Liu Renhao)

修道院回廊

修道院园林在中世纪欧洲园林发展中至关重要，既是宗教修行的场所，也是农业与园艺活动的中心。起初，修道院花园与绿地主要用于种植粮食和草药，后来逐渐具备装饰功能。在传统布局中，修道院的围合区域设有回廊，其柱廊庭院的形式源于古罗马的周柱中庭。回廊中的草地最初称为庭院，后因植物的引入演变为花园。四周的连拱廊既是有顶的通道，也是修士们阅读、修行和工作的庇护所。（雷浩）

Cloisters of the Monastery

Monastic gardens were pivotal in the development of medieval European landscapes, serving as both spiritual retreats and hubs for agriculture and horticulture. Initially, these gardens and green spaces were primarily used for cultivating food and medicinal herbs, later evolving to include decorative purposes. Traditionally, the enclosed area of a monastery featured a 'cloister', a columned courtyard inspired by the Roman peristyle. The grassy area within the cloister, originally called the 'courtyard', transformed into a garden as plants were introduced. The surrounding arcades served as covered walkways and provided sheltered spaces for monks to read, meditate, and work. (Lei Hao)

140 修女院的回廊

约 1835，[德国] 西蒙・奎格利奥
水彩
纵 13.7 厘米，横 18.2 厘米
大都会艺术博物馆藏

画家巧妙地运用光影，展现出修道院内部典雅的拱门与精美的石柱，加之画中祈祷的修女，共同构建出庄重、静谧的空间。画面左侧，花园一角绿意盎然，与回廊的沉静肃穆相映成趣。（曹旭卿）

140 Cloisters in a Nunnery

Ca. 1835, Simon Quaglio (German)
Watercolor
L.:13.7cm, W.:18.2cm
Lent by The Metropolitan Museum of Art, Harry G. Sperling Fund, 2004 (2004.378)

The artist skillfully uses light and shadow to highlight the elegant arches and exquisite stone columns within the nunnery. Along with the praying nun depicted in the scene, these elements create a solemn and tranquil atmosphere. On the left, a lush garden corner contrasts beautifully with the flowerpot, adding depth and harmony to the composition. (Cao Xuqing)

141 锡拉库萨岩壁上的回廊院

1835，［德国］安东尼·霍尔曼
石墨
纵 38.9 厘米，横 55.3 厘米
大都会艺术博物馆藏

意大利西西里锡拉库萨（Syracuse）古城中的修道院孤寂地坐落在高地之上，仿佛与陡峭的岩壁融为一体，自然与人工的界限在此消弭。早期修道院远离尘世的修行追求在作品细腻清冷的描绘中流露出来。（曹旭卿）

141 Cloister in Syracuse

1835, Anton Hallmann (German)
Graphite
L.:38.9cm, W.:55.3cm
Lent by The Metropolitan Museum of Art, Gift of Thomas and Gianna Le Claire, 2010 (2010.494.1)

Located in Syracuse, an ancient city in Sicily, Italy, this monastery stands in solitude, firmly perched atop a height and seemingly merging with the steep cliffs, blurring the boundary between nature and human craftsmanship. The delicate and cool-toned depiction in this artwork reflects early monasteries' pursuit of seclusion and spiritual retreat from the worldly realm. (Cao Xuqing)

142 水淹的修道院废墟

约 1824，[德国] 卡尔 · 布勒钦
钢笔，墨水、水彩淡彩，刮刻
纵 26 厘米，横 33.3 厘米
大都会艺术博物馆藏

画作描绘了一座中世纪修道院的废墟，掩映在高耸的植被之中。拱形穹顶支撑的高墙上清晰可见被水侵蚀的痕迹，透出岁月的沧桑。德国浪漫主义风景画大师卡尔 · 布勒钦（Karl Blechen）将自然幽古与修道院融为一体，带动观者的思绪，不知作者曾否发问："人的修行与自然、时间是怎样的关系？"（曹旭卿）

142 Inundated Ruins of a Monastery

Ca. 1824, Karl Blechen (German)
Pen and black ink, watercolor washes, sgraffito
L.:26cm, W.:33.3cm
Lent by The Metropolitan Museum of Art, Acquisitions Fund and 2005 Benefit Fund, 2008 (2008.109)

This image depicts the ruins of a medieval monastery, hidden amidst towering vegetation. The high walls, with faint reliefs of crosses visible on them, clearly show the marks of erosion by water. The German Romantic landscape painter Karl Blechen masterfully integrates the timelessness of nature with the monastery, evoking deep reflection in the viewer. (Cao Xuqing)

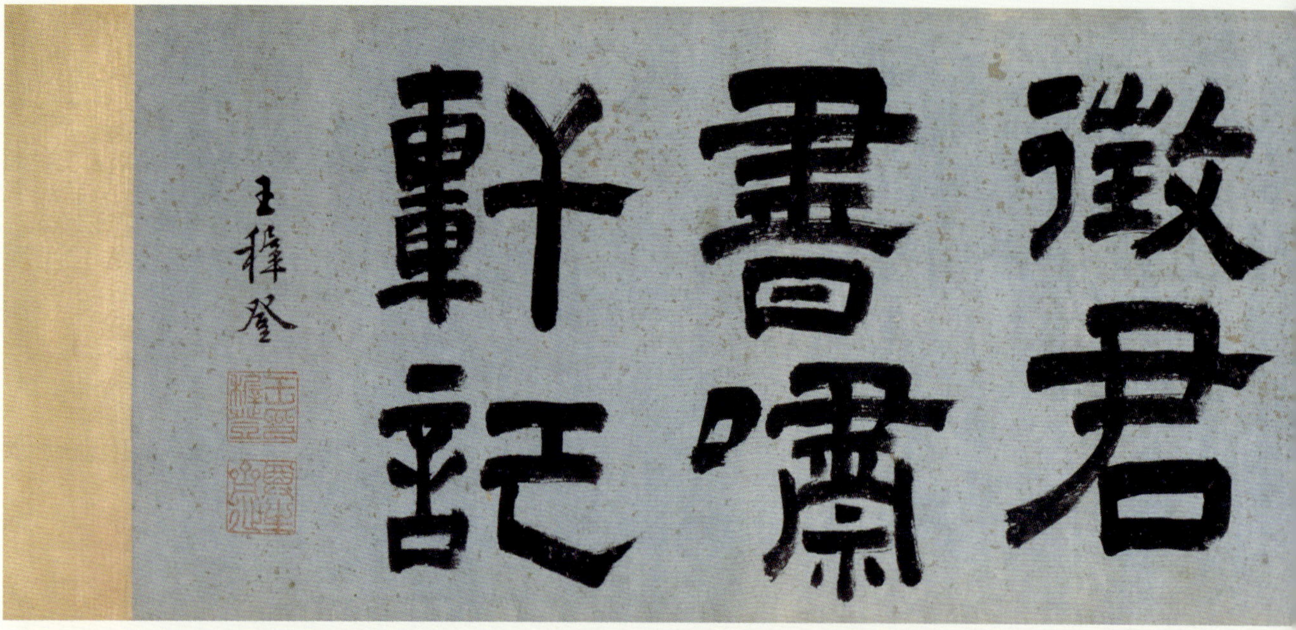

143 定慧禅院图卷

明嘉靖四十年（1561），钱穀
纸本，设色
纵 31.5 厘米，横 129 厘米
故宫博物院藏

此图作于 1561 年立春日，是钱穀为玄照和尚所画的定慧寺图。定慧寺原址在苏州城东，今在凤凰街，是一座千年古刹。寺院原为苏州双塔寺的子院，北宋大中祥符（一说天禧）年间改名为定慧禅院，后又称定慧禅寺、定慧寺等。据《东坡志林》《东坡七集》记载，定慧寺僧人守钦，倾心苏轼的才学和人品，但两人初不相识。苏轼被贬谪惠州，守钦特命弟子卓契看望苏轼，并寄给他十首自己仿唐代诗僧寒山子的诗作。苏轼读后大为感慨，称赞守钦为人“清逸绝俗”，其诗有佛家深刻的禅理，但没有诗僧贾岛、可名诗中荒寒的诗病。苏轼为此和诗八首，收录在其诗集中。

此图为实景之作，设色雅致，虽用笔不求工细，但定慧寺的形制、布局、佛像、天王像等十分明了。画中山门上方有两座高塔耸立，疑即苏州双塔寺。全寺围绕在长松巨柏之中，环境清幽，禅意深远。后院堂里，童仆奉茶，一僧一俗坐谈其中，颇具文人情趣。

143 ***Dinghui Monastery***

1561, Qian Gu
Handscroll, color on paper
L.:31.5cm, W.:129cm
Collection of the Palace Museum

钱穀（1508—？），字叔宝，号磬室，苏州人，文徵明弟子。善画山水、兰竹，继承文氏衣钵，并能结合沈周特点，用笔稍显粗重。引首有王穉登题："钱山人定慧寺图，黄徵君书啸轩记。"后有黄姬水书其父省曾《吴郡定慧禅寺苏文忠公啸轩碑记》一文。（施含牧）

This artwork depicts Dinghui Monastery, a thousand-year-old Buddhist temple in Suzhou. During the Northern Song Dynasty (960-1127), monk Shouqin of Dinghui Monastery admired Su Shi's literary talents, and they exchanged poems and writings during Su's exile. Based on actual scenery, the painting provides a detailed portrayal of the monastery's design, architectural layout, and interior elements, capturing the refined and leisurely spirit of the literati. (Shi Hanmu)

144 圆明园慈云普护图页

清乾隆九年（1744），唐岱、沈源
绢本，设色
纵 63.1 厘米，横 64.2 厘米
法国国家图书馆藏

慈云普护，圆明园四十景之一。其址在圆明园九州，建于雍正帝皇子时的潜居期间，旧称“涧阁”，雍正九年（1731）后，称“慈云普护”。这里是圆明园中一组释道兼容的园中园，以欢喜佛场、如祈应祷、慈云普护三殿为主体，供奉观音大士、关圣帝君、圆明园昭福龙王。清帝常于园居期间的初一、十五等日到此拈香行礼，曾有太监僧人于殿中念经。

作为园林中的宗教道场，慈云普护一区看似因供奉而散点布局，实则主次分明，错落有致。由临水被藤架、花圃攒簇的欢喜佛场出发，沿半廊向北，过如祈应祷龙王殿，半廊渐升为二层，折而西即到半入水中的慈云普护殿。二层供奉的观音大士居高抚水，乾隆帝称这里有“宛然天台”之象。本景的正北，一座三层钟楼孑然耸立，二层设自鸣钟，装“鱼子金漆大表盘”，称“时时如意时刻钟”，楼顶立金凤试风旗一架。旧时，殿中神像在缭绕供香的烘托下显得宁静神秘，霎时传来一段自鸣钟声，试风旗的修长凤尾随风摇曳，这一刻，庄严妙相，动静等观。

本幅为宫廷画师唐岱、沈源的承旨之作，除远山为写意渲染外，建筑、植被均写慈云普护实景。1860 年 10 月，圆明园被英法联军劫掠并焚毁后，本页所属的《圆明园四十景图咏》被杜潘上校（Pierre Charles François Dupin）裹挟至法国。（王志伟）

144 *Ciyunpuhu, The Old Summer Palace*

1744, Tang Dai and Shen Yuan
Album leaf, color on silk
L.:63.1cm, W.:64.2cm
Collection of the National Library of France

Ciyunpuhu, one of the Forty Sceneries of the Old Summer Palace, was built when Yongzheng Emperor was still a crown prince. It was formerly known as '*Jian*' Pavilion. After the 9th year of Yongzheng Emperor's reign (1731), it was named *Ciyunpuhu*. This is a garden within the garden in the Old Summer Palace that integrates Confucianism and Buddhism. (Wang Zhiwei)

145 圆明园月地云居图页

清乾隆九年（1744），唐岱、沈源
绢本，设色
纵 63.1 厘米，横 64.2 厘米
法国国家图书馆藏

月地云居，圆明园四十景之一。其址在圆明园西路，雍正时称“乐志山村”，乾隆初年被改造为一区佛宇，清宫称为“清净地”。乾隆九年（1744），乾隆帝御书“月地云居”额于此。作为一组宗教园林，月地云居地处圆明园西偏，与中路诸景以障山相隔，仅有一山口、一水道与外通连，颇为幽静。该区主体建筑在西侧，呈轴线分布，由南而北依次为清净地山门、妙证无声都罡殿、呀吗达嘎及吣密二坛、月地云居殿、旃檀佛殿、莲花法藏楼等。总体上看，虽然主体建筑保持了传统汉地佛寺“伽蓝七堂”的特征，但因其供奉杂糅显密，所以这里是一座汉藏合璧式寺院。

乾隆四十五年（1780），为迎接六世班禅入觐，乾隆帝将东侧院落改造为法源楼，并在东首开辟“静室”一座。当年九月，班禅大师驻跸北京期间，曾于法源楼留宿二日。

本幅为宫廷画师唐岱、沈源的承旨之作，图中建筑、植物均写月地云居实景，唯东侧法源楼处尚为乾隆九年状态。（王志伟）

145 *Yuediyunju, The Old Summer Palace*

1744, Tang Dai and Shen Yuan
Album leaf, color on silk
L.:63.1cm, W.:64.2cm
Collection of the National Library of France

Yuediyunju, one of the Forty Sceneries of the Old Summer Palace, is located on the West Road of the Old Summer Palace. During the reign of Yongzheng Emperor, it was called ‘*Lezhi*’ Mountain Village. In the early years of Qianlong Emperor’s reign, it was renovated into a Buddhist building complex, and was known as the ‘Pure and Quiet Place’ in the imperial palace of the Qing Dynasty. In the 9th year of Qianlong Emperor’s reign (1744), Qianlong Emperor personally inscribed the plaque with the characters *Yuediyunju* here. (Wang Zhiwei)

觀花

观花、画花、咏花是文人居于园林的日常。

“不要人夸好颜色，只留清气满乾坤。”梅花的清芬因王冕的凛然高洁而盈满宇内。“无可奈何花落去，似曾相识燕归来。小园香径独徘徊。”晏殊怅然地望向园中的落花，回想逝去的往昔……中国文人把对品格的追求赋予花，把对生命的理解倾注于花园，也把这些意象写于诗中、绘于卷上，留与今日的我们相见。

而在巴洛克时期的园林中，几何对称、色彩繁复的花坛，壮丽震撼；18 世纪的英国，园艺繁荣，缤纷各异的时令花束盛放在精致雕刻的石花钵中，绚烂浪漫；19 世纪的欧洲，在古典园林之外，我们看到了莫奈花园，如穿越时空的瑰宝，彰显东西气韵的融通之美。

Gardens and Inspiration

Appreciating flowers, painting them, and composing poetry about them were part of daily life for literati in gardens.

Plum blossoms care not for praise of their vibrant hues,
Only that their fragrance lingers throughout the land.

Echoing the poet Wang Mian, plum blossoms–blooming in the harshness of winter–embodied resilience and moral virtue, seeking no worldly praise.

Helplessly, the flowers fade,
Yet the swallows return, familiar as ever.
Alone, I wander the fragrant garden paths, lost in thought.

Thus the poet Yan Shu gazed wistfully at the fading blossoms in the garden, pondering the passage of time.

Chinese literati infused flowers with their ideals of moral integrity, poured their understanding of life into gardens, and captured these visions in poetry and paintings, leaving them for us to encounter today.

In the gardens of the European Baroque period, elaborate flowerbeds with bold geometry and vivid colors created a striking spectacle. In 18th century England, a flourishing horticultural tradition brought forth seasonal bouquets arranged in intricately carved stone urns. In the 19th century, beyond the classical gardens, Monet's garden at Giverny emerged–a timeless masterpiece reflecting the harmonious fusion of Eastern and Western aesthetics.

146 弘历观荷抚琴图轴

清乾隆（1736—1795），佚名
绢本，设色
纵 299.5 厘米，横 156.5 厘米
故宫博物院藏

本幅无作者款印，画心上部钤盖乾隆帝御玺三方，分别是“八徵耄念之宝”“五福五代堂古稀天子宝”“太上皇帝之宝”。

图中画乾隆帝身着汉装，在一临水亭中抚琴，水中荷花亭亭玉立，远景有高山流水，寓示着画中人物高逸超脱的心境。此图与这一时期的很多宫廷绘画一样，由西洋画家完成人物肖像部分，再由中国画家补画山水背景，画风和谐，体现了宫廷绘画典雅富丽的审美品位。（聂卉）

146 *Hongli Viewing Lotus and Playing the Qin*

Qing Dynasty, Qianlong Reign (1736-1795), unknown painter
Hanging scroll, color on silk
L.:299.5cm, W.:156.5cm
Collection of the Palace Museum

This image symbolizes the Qianlong Emperor's elevated and otherworldly state of mind. Like many other court paintings of the same period, it was a collaboration between a Western painter (the human figures) and a Chinese painter (the landscape), forming a coherent whole. It reflects the elegant and magnificent aesthetics of court paintings. (Nie Hui)

147 墨荷图轴

清康熙（1662—1722），石涛
纸本，墨笔
纵 90.6 厘米，横 50.6 厘米
故宫博物院藏

本幅作者自识："韩园虽好梵宫荒，歌妓游归恨杳茫。堤外莲花千万朵，不知谁是旧人香。大涤子邗上秋日写。"钤"靖江后人"白文方印、"大涤堂"白文方印。

此画为石涛定居扬州后所绘，诗中所指"韩园"应为今徐园所在，清初时为韩醉白别墅，建于长堤之上。诗歌中使用反衬的修辞手法，用园林之美好对比梵宫之荒芜，用莲花千万对比故人之思，反映出画家心中的盘郁与孤闷。画幅右下钤"大涤堂"印，说明此图作于大涤堂建成后，即康熙三十五年（1696）之后。此时画家回返扬州，虽豪商新贵应接不暇，但故友相继魂归道山，只留画家孑立于世。画家通过表现荷叶、茨菰、莲蓬、蒲草间的错综复杂，来隐喻扬州城繁荣聒噪的气氛，并以此来表达内心的孤独落寞。

在本幅中，石涛运用独特技法，即以深浅不一的墨点填充花叶的空白间隙，意在衬托荷花的白皙娇嫩，烘托出月下观荷的诗意氛围。

石涛（1642–1707），姓朱，名若极，后更名原济、元济、超济，小字阿长，字石涛，号大涤子，又号清湘老人、清湘陈人、清湘遗人、粤山人、湘源济山僧、零丁老人、一枝叟，晚号瞎尊者，自称苦瓜和尚。靖江王朱赞仪十世孙，行迹颇广，晚年居扬州，筑大涤草堂。他擅画山水、花鸟、兰竹，兼工人物，提出"一画论""搜尽奇峰打草稿"等观点，影响深远。（喻理）

147 *Lotus in Ink and Wash*

Qing Dynasty, Kangxi Reign (1662-1722), Shi Tao
Hanging scroll, ink on paper
L.:90.6cm, W.:50.6cm
Collection of the Palace Museum

Shi Tao finished this painting after he settled in Yangzhou. With the overlapping and staggered lotus leaves, Chinese arrowheads, lotus pods, and cattails in the pond, and the blank space among them filled with ink dots of varying shades, the artist sets off the tenderness and purity of the lotus flowers and portrays the poetic atmosphere of viewing lotus flowers in the moonlight. (Yu Li)

148 红莲图轴

清康熙十年（1671），唐苂、恽寿平
纸本，设色
纵 135.7 厘米，横 59 厘米
故宫博物院藏

恽寿平（1633—1690），初名格，字寿平，后以字行，又字正叔，号南田，别号云溪外史，江苏常州人。幼年时随伯父恽向学画，初师法黄公望山水，后改习花鸟，以没骨法创立新风，“清初六家”之一。

唐苂（1626—1690）字子晋，号匹士，江苏常州人，常州收藏大家唐宇昭长子，王翚、恽寿平早年都曾受唐宇昭资助，在其家临摹大量古代作品，唐苂与恽寿平同辈，交情更深，两人都擅长没骨技法，常探讨画理、画技，开创了常州画派没骨画法。

图绘三朵莲花，一为含苞，一为初放，一为盛开；荷叶有五，展现了含苞、初展、尽放的不同形态。花、叶因其形态，晕染也有所不同。画风清新淡雅，将荷花、荷叶之风中摇曳之姿描绘得别具一格。荷花因其“出淤泥而不染”“香远益清”之特性，备受历代文人画家的喜爱，凝聚着独特的精神内涵和文化气息。

根据题款可知，此图是康熙十年（1671）唐苂为王翚四十岁寿辰所画，恽寿平补荇藻。恽寿平在题跋中表述了自己和唐苂“研思写生”的创作过程，对于黄筌、赵昌的技法流露出不满。但是，徐熙的没骨法只有文献记载，没有画作可循，他与唐苂所用之没骨法“损益古法，参之造化”，是一种基于古人的新创。从王时敏、王鉴的评价中也可以看出，两位师古派领袖人物对于唐苂、恽寿平的没骨创新是非常认可的。（鲁颖）

148 *Red Lotuses*

1671, Tang Guang and Yun Shouping
Hanging scroll, color on paper
L.:135.7cm, W.:59cm
Collection of the Palace Museum

Tang Guang painted this work for Wang Hui's 40th birthday in 1671. The algae were later added by Yun Shouping. With its graceful style, the painting vividly depicts the charm of the lotus flowers swaying in the breeze. The lotus flower has been loved by literati painters of all generations for its special fragrance that is 'even more refreshing from afar', and it has a unique spiritual and cultural significance. (Lu Ying)

149 红荷图轴

明（1368—1644），陈洪绶
绢本，设色
纵 177.4 厘米，横 52.5 厘米
故宫博物院藏

图绘池塘中三株红荷，或含苞，或初绽，或怒放，各具姿态。荷叶环绕花朵，俯仰倾斜，衬托出花的丰腴娇媚。一只蝴蝶在上方翩翩起舞。画面前方的两块湖石，或雄奇锐利，或厚重沉凝，与塘中的水纹涟漪形成动静对比。画家的笔触细腻灵动，用色清新脱俗。花瓣以淡墨勾勒，点缀胭脂，叶片则采用没骨法绘出形状，再通过墨线勾勒和色墨晕染表现结构。荷花的柔红与荷叶的碧绿相互映衬，不仅象征着纯洁与高雅，也传达出一种超脱尘世的意境。荷花在古代园林中不仅是重要的观赏植物，更象征着“出淤泥而不染”的高洁品质。

本幅作者自识：“云溪老渔陈洪绶仿王若水画于永柏书堂。”钤“陈洪绶印”。

陈洪绶（1598—1652），字章侯，号老莲，浙江诸暨人，明末曾被召为舍人，内廷供奉，明亡后在绍兴云门寺出家为僧，改号“僧悔”“悔迟”等，晚年鬻画为生。擅用夸张变形的手法。（杨博然）

149 ***Red Lotuses***

Ming Dynasty (1368-1644), Chen Hongshou
Hanging scroll, color on silk
L.:177.4cm, W.:52.5cm
Collection of the Palace Museum

In traditional Chinese courtyards, lotus flowers are not only ornamental, but moreover a symbol of the noble quality of ‘remaining pure and untainted despite the muddy surroundings’. (Yang Boran)

150 符望阁紫檀嵌瓷玉夹双面绣花窗

清乾隆（1736—1795）
长 117 厘米，高 100 厘米
故宫博物院藏

此面花窗嵌于符望阁西面南次间炕床东墙上，与紫檀嵌珐琅夹臣工书画炕罩、松竹梅大漆地迎风板、缠枝莲纹炕沿板等内檐装修构件一起，装点了符望阁西面的空间。花窗主体由紫檀制成，使用深、浅两种青色瓷片和白玉作为嵌件，拼成花朵，堆簇于花窗边角。花窗内夹缠枝莲四合如意纹双面绣，为故宫博物院与匠人合作，按照符望阁他处残存绣片复制的。

类似这样的花窗符望阁内还有四面，两面位于北面，上嵌雕如意云头纹的青玉，两面位于东面，上嵌缠枝莲蝙蝠纹珐琅。通过不同材料、不同工艺的组合，乾隆帝于威武严肃的禁中创造出一组富丽堂皇、花团锦簇的空间。观者走入符望阁，常常有“乱花渐欲迷人眼”之感。（刘仁皓）

150 Red Sandalwood Window Frame from the Belvedere of Viewing Achievements (*Fuwang ge*) with Double-sided Embroidery

Qing Dynasty, Qianlong Reign (1736-1795)
L.:117cm, H.:100cm
Collection of the Palace Museum

Embedded in the Belvedere of Viewing Achievements, each corner of the frame is decorated with patterns of clustered flowers formed by inlaid porcelain pieces and jade. In the otherwise stately and solemn Forbidden City, the Qianlong Emperor created a space splendidly decorated with flowers. (Liu Renhao)

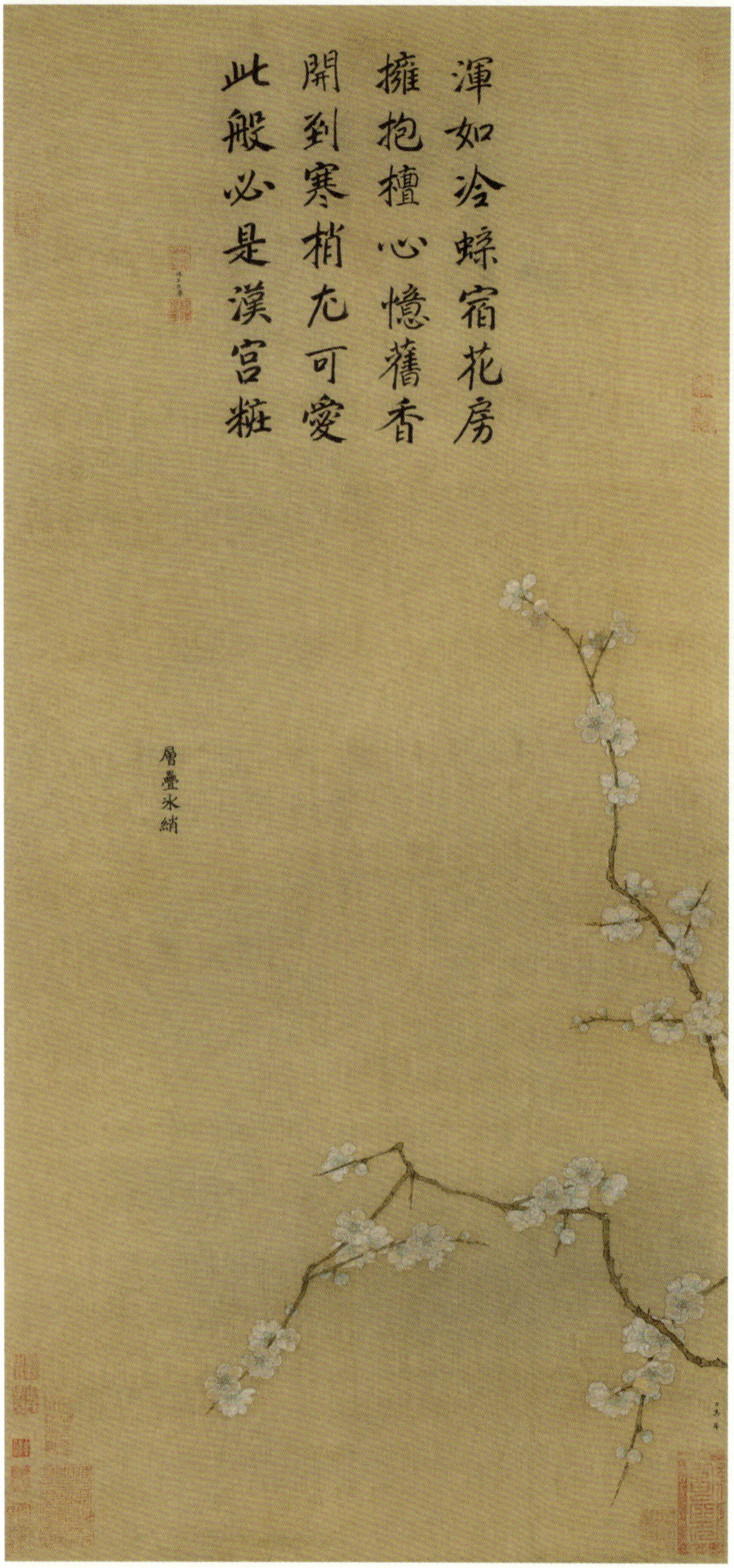

151 层叠冰绡图轴

南宋（1127—1279），马麟
绢本，设色
纵 101.7 厘米，横 49.6 厘米
故宫博物院藏

本幅作者自识："臣马麟。"上有宋宁宗皇后杨氏题诗曰："浑如冷蝶宿花房，拥抱檀心忆旧香。开到寒梢尤可爱，此般必是汉宫妆。赐王提举。"钤"丙子坤宁翰墨""杨姓之章"二印。另钤有"纪察司印""项元汴印""墨林秘玩""商丘宋荦审定真迹""顾氏芸阁珍藏"等鉴藏印。

图绘三枝绿萼梅自画右斜出，梅花繁密多姿，与清瘦的枝干形成鲜明对比，充分展现其清艳之美。画面采用双钩填色法，用笔精致细腻，枝干的转折和花朵的阴阳向背得到了细致描绘。花瓣以白粉和石绿层层渲染，凸显梅花冰清玉洁的特质，与宋宁宗皇后杨氏所题的"层叠冰绡"相契合。这幅作品不仅生动地描绘了花卉之美，还传递出一种超凡脱俗的意境，仿佛引领观者进入宁静而幽雅的花园。

马麟，钱塘（今浙江杭州）人，马远之子，传家学，工画人物、山水、花鸟。（杨博然）

151 *Ice-like Purity and Silk-like Delicacy*

Southern Song Dynasty (1127-1279), Ma Lin
Hanging scroll, color on silk
L.:101.7cm, W.:49.6cm
Collection of the Palace Museum

This painting depicts three twigs of Chinese plum blossoms densely clustered, entering the frame at a slanting angle. The artist outlined the objects before coloring them with delicate and exquisite brushwork. The petals are rendered in multiple layers with white and malachite powder to highlight the blossoms' crystal purity. This painting not only vividly depicts the beauty of flowers, but also conveys a transcendental artistic conception, as if bringing the viewer into a quiet and elegant garden. (Yang Boran)

152 符望阁紫檀镌螺钿框漆地嵌木字御笔中堂、对联挂屏

清乾隆（1736—1795）
中堂：宽 86 厘米，高 188 厘米
对联：宽 40 厘米，高 188 厘米
故宫博物院藏

此套挂屏位于符望阁西面南次间南墙，包含一面乾隆帝御笔中堂和两面御笔七言对。中堂出自乾隆帝《御制题明人影宋钞清波杂志》，诗中借《清波杂志》之名评宋高宗赵构。赵构于绍兴三十二年（1162）内禅其位于赵昚，移居德寿宫。乾隆帝对此颇有微词，常在宁寿宫相关诗文中提及，可见归政一事对其重要意义。七言对则出自《登宝月楼》："居中揽外襟怀畅，击毂摩肩职植殷。"诗句虽不是为乾隆花园所写，但意境与符望阁高阁耸立、俯瞰禁中的情景契合。乾隆帝也曾题诗称："登临仰宸翰，奎耀焕皇都。"

在诗句意境之外，这组御笔中堂、对联挂屏最值得称赞的，就是其材料与工艺。挂屏边框为紫檀雕绳壁纹，但通常用玉璧装饰的部分别出心裁地使用了雕刻好的厚螺钿，镶嵌于紫檀之上。通过调整背后的衬色，螺钿片呈现出金银两色，晶莹透彻。挂屏地髹大漆，饰以万字不到头纹。其上镶嵌文字为桄榔木雕刻染色而成，细看有密布的天然斑点。远看典雅大方，近看精致细腻，结合了珍贵的材料和高超的工艺，这是乾隆花园内檐装修的典型特征。（刘仁皓）

152 Wall Screens from the Belvedere of Viewing Achievements (*Fuwang ge*)

Qing Dynasty, Qianlong Reign (1736 1795)
Hanging screen: W.:86cm, H.:188cm
Couplet: W.:40cm, H.:188cm
Collection of the Palace Museum

This set of wall screens, located inside the Belvedere of Viewing Achievements in the Qianlong Garden, features excerpts from writings by the Qianlong Emperor whose artistic conception aligns with that of the Belvedere. The red sandalwood frames of the wall screen are carved with decorative knitted-rope patterns and inlaid with extravagantly carved mother-of-pearl pieces. The background is lacquered in high gloss with inscriptions carved from palm wood. (Liu Renhao)

153 克劳德·莫奈肖像

1922，[法国] 阿尔伯特·安德烈
布面，油画
纵 158.3 厘米，横 125.8 厘米
芝加哥艺术博物馆藏

印象派画家克劳德·莫奈不仅在画布上捕捉光影交错的美景，也在吉维尼的花园中营造出自然与艺术交融的诗意空间。画中莫奈悠然站立在自己的花园前，身后盛开的繁花与茂盛绿植是他精心配植的自然景致。他倾心于自由生长、不受约束的园艺风格，这片土地亦成为他晚年创作的灵感之源。（王萌萌）

153 Claude Monet

1922, Albert André (French)
Oil on canvas
L.:158.3cm, W.:125.8cm
Collection of the Art Institute of Chicago, The Stickney Fund (1923.149)

Monet not only captured the interplay of light and shadow on canvas, but also created a poetic space where nature and art merge in his garden at Giverny. In this painting, Monet stands peacefully in front of his garden, with its carefully curated natural scenery designed by himself. Monet was devoted to a gardening style allowing plants to grow freely and unconstrained, and this space became the source of inspiration for his later works. (Wang Mengmeng)

154 睡莲

1906，[法国] 克劳德 · 莫奈
布面，油画
纵 103.2 厘米，横 107 厘米
芝加哥艺术博物馆藏

"自然的瞬息之间，蕴藏着无限。"克劳德 · 莫奈在谈到他晚年在吉维尼家中创作的水景杰作时这样说道。在这些模糊空间界限的画作中，画家以俯视视角聚焦水面倒影与光线变化，将池水、天空与植物的界限消融，充盈着浓郁的东方意境。（芝加哥艺术博物馆）

154 Water Lilies

1906, Claude Monet (French)
Oil on canvas
L.:103.2cm, W.:107cm
Collection of the Art Institute of Chicago, Mr. and Mrs. Martin A. Ryerson Collection (1933.1157)

'One instant, one aspect of nature contains it all,' said Claude Monet, referring to his late masterpieces, the water landscapes that he produced at his home in Giverny. In this spatially ambiguous canvas, the artist looked down, focusing solely on the surface of the pond, with its cluster of plants floating amidst the reflected sky and trees. The artist's brushwork dissolves boundaries through a poetic meditation on liquid reflections and luminous transformations, featuring Eastern contemplative harmony. (AIC)

155 风铃草与蜻蜓

选自《花鸟系列》
约 1833 或 1834，［日本］葛饰北斋
彩色木版画
纵 26 厘米，横 37.5 厘米
芝加哥艺术博物馆藏

蜻蜓经常出现在描绘自然的版画中。此幅画中，黄色的背景下一只蜻蜓飞向一丛中国风铃草，微风则从右边徐徐吹来。（芝加哥艺术博物馆）

155 Bell-Flower and Dragonfly

From an *Untitled Series of Large Flowers*
Ca. 1833/1834, Katsushika Hokusai (Japanese)
Color woodblock print, oban
L.:26cm, W.:37.5cm
Collection of the Art Institute of Chicago, Clarence Buckingham Collection (1925.3377)

Dragonflies quite often appear in prints of nature. In this work, against a yellow ground, a dragonfly flies down towards some Chinese bell-flowers. At the same time a slight breeze is blowing from the right. (AIC)

167 蒲田梅园

选自《名所江户百景》
1857，［日本］歌川广重
彩色木版画
纵 36.6 厘米，横 24.4 厘米
芝加哥艺术博物馆藏

早春时节，蒲田地区的梅树繁花盛开，蔚为壮观。画中描绘的被称为“梅园”（Umeyashiki）的花园虽是私人财产，却对公众开放，身着传统服饰的行人三两成群享受盛开的梅花，茶馆和酒馆也为游人提供了休憩的场所。（雷浩）

167 The Plum Orchard at Kamata (Kamata no umezono)

From the Series *One Hundred Famous Views of Edo*
1857, Utagawa Hiroshige (Japanese)
Color woodblock print, oban
L:36.6cm, W.:24.4cm
Collection of the Art Institute of Chicago, Frederick W. Gookin Collection (1939.1461)

In the early spring season, the plum trees in Kamata are in full bloom, creating a magnificent sight. The ‘Plum Mansion’ (Umeyashiki) depicted in the painting, though privately owned, is open to the public. Passersby dressed in traditional attire stroll in small groups, enjoying the blossoming plum trees, while teahouses and taverns provide visitors with places to rest. (Lei Hao)

168 从真崎边看水神之森、内川和关屋之里之图

选自《名所江户百景》
1857，［日本］歌川广重
彩色木版画
纵 36 厘米，横 24.1 厘米
芝加哥艺术博物馆藏

此景或取自位于真崎稻荷神社旁的茶屋“甲子屋”。画面意境雅致，左侧一隅点缀着一簇白色山茶花。圆窗的障子向右拉开，展现出一幅完美的河川景致。右侧林木间可见水神社的鸟居和石灯笼，窗外盛放的梅花标志着时值早春，令人想起每年二月在稻荷神社举办的初午祭典。（芝加哥艺术博物馆）

168 **View from Massaki of Suijin Shrine, Uchigawa Inlet, and Sekiya (Massaki hen yori Suijin no mori Uchigawa Sekiya no sato o miru zu)**

From the Series *One Hundred Famous Views of Edo*
1857, Utagawa Hiroshige (Japanese)
Color woodblock print, oban
L.:36cm, W.:24.1cm
Collection of the Art Institute of Chicago, Clarence Buckingham Collection (1925.3747)

This particular view may be from Kinoeneya, the most famous of the dengaku establishments. It is at an elegant setting, with a fresh arrangement of white camellias discreetly shown in section to the left, and the shoji screen of the round window is slid open rightward to reveal a perfectly framed view over the river. In the grove of trees to the right, we can make out the torii and lanterns of Suijin Shrine.The view of plums blooming just outside the window would trigger an association with the tradition of visiting Inari shrines on the first day of the Horse in the Second Month.(AIC)

169 **Precincts of Kameido Tenjin Shrine (Kameido Tenjin keidai)**

From the Series *One Hundred Famous Views of Edo*
1856, Utagawa Hiroshige (Japanese)
Color woodblock print, oban
L.:36cm, W.:24.1cm
Collection of the Art Institute of Chicago, Clarence Buckingham Collection (1925.3762)

Here, a display of purple wisteria blossoms falls gracefully along the pole of a trellis. The arched bridge in the distance would have immediately identified the place to an Edo viewer-as indeed it does to any present-day connoisseur of Tokyo's seasonal attractions, for the scene remains little changed. It is, as the title indicates, within the precincts of Kameido Tenjin Shrine, on the eastern fringe of Edo.The dedication is to Tenjin, the deified Sugawara Michizane, patron saint of learning and calligraphy. (AIC)

169 亀户天神社

选自《名所江户百景》
1856，[日本]歌川广重
彩色木版画
纵 36 厘米，横 24.1 厘米
芝加哥艺术博物馆藏

莫奈是日本版画的热情收藏者，尤其喜爱风景版画，并拥有一幅歌川广重的《亀户天神社》。画中的紫藤花沿着花架优雅地垂下，远处的拱桥让人一眼便能认出这是江户时代的著名景致。即便对于熟悉东京季节性景点的现代人来说，这里的景色也几乎未曾改变。正如画作标题所示，这里是位于江户东部边缘的亀户天神社。神社供奉的是被尊为学问与书法之神的菅原道真（Sugawara Michizane），即天神（Tenjin）。（芝加哥艺术博物馆）

170 睡莲池

1900，［法国］克劳德·莫奈
布面，油画
纵 112.4 厘米，横 122.6 厘米
芝加哥艺术博物馆藏

1893 年，莫奈开始将家后的一片沼泽地改造成池塘，并修建日式木桥。他在园中遍植本土与异域植物，紫藤垂坠拱桥，静谧的水面上盛开着睡莲，垂柳和鸢尾掩映池畔，这些景致成为他后半生创作的核心主题。画家在 1899 年至 1900 年间创作了 18 幅类似作品，展现了他对光影、色彩与自然景观的独特理解。（芝加哥艺术博物馆）

170 Water Lily Pond

1900, Claude Monet (French)
Oil on canvas
L.:112.4cm, W.:122.6cm
Collection of the Art Institute of Chicago, Mr. and Mrs. Lewis Larned Coburn Memorial Collection (1933.441)

In 1893, Claude Monet began transforming the marshy ground behind his home into a pond. Over the narrow end of the pond, he built a Japanese-style wooden bridge. The artist added both domestic and imported plantings to his garden, which would be a principal subject of his paintings for the rest of his life. Water Lily Pond was among the 18 similar versions of this motif that he made in 1899-1900; their common theme was the mingling of the lilies with reflections of other vegetation on the water's surface. (AIC)

171 三友轩紫檀透雕松竹梅玻璃花窗

清乾隆（1736—1795）
长 205 厘米，厚 9 厘米，高 140 厘米
故宫博物院藏

岁寒三友，即松、竹、梅，是乾隆帝极爱使用的一组意象。早在乾隆十二年（1747）建福宫花园营建时，乾隆帝就在凝晖堂内专门划分出一方内室，用以保存《十八公图》《君子林图》《梅花合卷》三幅对应三友的画作，并为此室题额“三友轩”。此后，乾隆帝又将长春园含经堂区的一座三间小殿命以“三友轩”之名，并作诗解释：“禁中三友轩，额为藏名迹。园中三友轩，窗外真培植。”乾隆三十五年（1770），在筹划归政后的宫殿时，乾隆帝将长春园三友轩原样搬进了紫禁城宁寿宫中俗称“乾隆花园”的小院里，同时沿用了“三友轩”这个名字。1860 年，长春园罹祸；1923 年，建福宫花园被焚毁。乾隆花园这座三友轩，便成了存世唯一的三友轩。

乾隆花园三友轩三面环绕假山，其中北、南有洞，逶迤通幽；西面为壁，前植青松、古梅、翠竹。为呼应室外景观设计，三友轩西山墙上开洞，洞口嵌此紫檀透雕松竹梅玻璃花窗。坐于室内，观者可以透过花窗深色的梅瓣、松枝、竹叶雕刻，窥见室外三友。对此，嘉庆帝有诗云：“静坐西窗对三友，清贞劲节益心身。”而当观者处于室外，在廊下漫步，或在山间小憩时，则可以反观室内，瞥见烛光剪影、锦幛偎红。这种不同层次、不同时空之间的景观互动，便是“借景”。（刘仁皓）

171 Glass Decorative Window with Red Sandalwood Openwork Carving of Pine, Bamboo and Wintersweet in the Bower of the Three Friends (*Sanyou xuan*)

Qing Dynasty, Qianlong Reign (1736-1795)
L.:205cm,Th.:9cm, H.:140cm
Collection of the Palace Museum

The Bower of the Three Friends was built in the Qianlong Garden. Sitting indoors, the viewer could see the ‘Three Friends’–pine trees, bamboo and wintersweets–on the other side of the wall through the openwork carving of the same plants in the decorative window. (Liu Renhao)

暢音

园林是可以听的。

中国园林中有雨打芭蕉、风过竹林的自然交响，也有亭台轩榭之上，与风声、水声相和的琴瑟之音。人们在园林中创造的音乐与自然的声音相融，与园林的景观交织，仿佛可以将赏音的园中人带入又一重园林时空。

白居易以琴音与明月交心，“竹院新晴夜，松窗未卧时。共琴为老伴，与月有秋期”。弗朗茨·李斯特在埃斯特庄园的喷泉旁，以琴声回应水声，创作了《埃斯特庄园的喷泉》。

明清时期，园林与戏曲结缘。《西厢记》《牡丹亭》中无法实现的梦被寄托在园林中、演绎在园林中；在凡尔赛宫，路易十四举办的盛大游园会上，《伪君子》的演出是振聋发聩的戏剧史强音……今天，越来越多来自遥远时代的音乐与戏剧，以回归园林的方式进入当代生活。所畅之音仍宛转悠扬，弥久恒新。

Gardens and Recreation

Gardens are not merely places of visual experiences, but of sound as well. In Chinese gardens, one can hear the natural symphony of raindrops tapping on banana leaves and the whisper of wind through bamboo. Suddenly the sounds of instruments join in. The music created by people in gardens intertwines with the sounds of nature, transporting listeners to another dimension of time and space.

Poet Bai Juyi talks with the moon with the aid of the *qin* (a traditional Chinese string instrument):

In the bamboo courtyard on a clear night,
By the pine-framed window, I sit awake.
My qin is my faithful companion,
And the moon my autumn date.

Similarly, Franz Liszt, by the fountain at Villa d'Este, composed *Les Jeux d'eau à la Villa d'Este* in response to the murmuring waters.

In the Ming and Qing Dynasties, Chinese gardens became stages for traditional opera. *The Romance of the Western Chamber* and *The Peony Pavilion*–dreams that could not be realized in reality–were performed in gardens. Likewise, at the grand garden festivals hosted by Louis XIV at the Palace of Versailles, performances like *Tartuffe* have echoed through the history of theater. Today, music and drama from distant times return to the garden, reviving their timeless sounds. These melodies continue to resonate, eternal and enchanting.

172 倦勤斋小戏台

现代

综合材质

倦勤斋小戏台，建于清乾隆三十七年（1772），在乾隆花园中轴线最北端的倦勤斋内，是一处别具匠心的室内小戏台。戏台为木结构，四角攒尖亭式，柱、枋、垂脊表面油饰彩绘斑竹纹。主台之前增设一方形小台，是宫廷戏台中仅见的双台形制。戏台三面设有斑竹彩绘夹层篱笆，可以作为演员上下场的通道。戏台对面仙楼上下均设有宝座床，以小楼梯相通，是皇帝观戏的座位。据清宫档案记载，乾隆帝曾命南府太监在此演唱御制腔、太平歌词、岔曲等。主台前的小台则是为表演戏法而设，因距离宝座更近，易于皇帝看得真切。

在设计上，这处室内戏台空间别具一格，着意模糊室内外的界限。戏台本身极力还原园林中的方亭构造及形制细节，四壁更以大幅通景线法画满绘室外园林场景，天顶用透视手法绘制繁茂的藤萝架，虚虚实实，制造出室内外交融的空间幻象。这是乾隆帝本人格外偏爱的室内设计手法，在圆明园和紫禁城中有多处运用。此处通景线法画最初的蓝本在圆明园半亩园室内，是宫廷画师、意大利传教士郎世宁的作品。乾隆帝格外欣赏这一创意，曾命郎世宁在建福宫花园敬胜斋室内再次绘制。而倦勤斋的通景画则是另一位宫廷画师王幼学参考郎世宁原稿，带领学徒在倦勤斋内绘制而成。

展览中用当代设计手法再现了这处小戏台，以对角线分割方式，用白色亚克力材质搭建出半边实体戏台，另半边安装镜面屏，实体戏台映入镜面屏幕中，与屏幕上的戏台空间数字影像合而为一，虚实结合，共同构成完整的戏台空间。观众站在小戏台上观看镜面屏幕时，镜中的影子恰好置身于倦勤斋西四间之内，宛若身临其境，亦真亦幻。乾隆帝本人很喜爱以自身形象入画，这处戏台装置设计将观者影像与倦勤斋室内空间结合，在某种程度上构成了对乾隆帝趣味的当代呼应。（刘梦雨）

172 **Opera Stage in the Studio of Exhaustion from Diligent Service (*Juanqin zhai*)**

Contemporary
Mixed media

Exhaustion from Diligent Service (*Juanqin zhai*), which located in the northernmost part of the Qianlong Garden. The Qianlong Emperor instructed eunuchs from the Southern Office to perform *chaqu* (a type of folk opera) and other small-scale theatrical acts here. The smaller stage in front of the main stage was used for magic performances. The exhibition employs a modern design approach to reconstruct this small Opera Stage, using mirrored screens to recreate the interior imagery of the original studio. The fusion of the virtual and the real offers visitors an immersive experience, making them feel as if they are inside *Juanqin zhai*. (Liu Mengyu)

173 乾隆花园第四进院烫样

现代

综合材质

长 77.8 厘米，宽 96.5 厘米，高 31 厘米

故宫博物院藏

这件烫样再现了乾隆花园的第四进院落。四进院的中心是体量宏伟的符望阁，南面布置假山，向西延伸到玉粹轩；北端是一座九间的倦勤斋，与符望阁相对，构成静谧封闭的院落空间，是整座乾隆花园景观序列高潮之后的落点与结束。

符望阁是一座重檐攒尖顶的高大楼阁，其内部格局极尽繁复，装饰琳琅，有“迷楼”之称，其室内空间是乾隆帝着意经营的。登临二层，可以从四面外廊眺望紫禁城内外景色，视野极佳。乾隆帝每年腊月会在此宴请王公大臣。

院落西北角的竹香馆也颇具匠心，外观仿佛坐落在假山顶端的单层亭榭，实际上是一座二层小楼，一楼隐没在四围假山之中，与曲尺形的院墙与花窗构成精致的对景。竹香馆南侧有爬山廊可通玉粹轩，北侧游廊则暗通倦勤斋，构成一条隐蔽的闭合流线。（刘梦雨）

173 Model of the Fourth Courtyard of the Qianlong Garden

Contemporary

Mixed media

L.:77.8cm, W.:96.5cm, H.:31cm

Collection of the Palace Museum

The main structure of this courtyard is the towering Belvedere of Viewing Achievements (*Fuwang ge*), renowned for its intricate interior layout and lavish decorations. To the south of Belvedere of Viewing Achievements lies an arrangement of artificial rock formations. At the northern end stands the nine-room the Studio of Exhaustion from Diligent Service. Together, these structures form a serene and enclosed courtyard, serving as the concluding focal point of the Qianlong Garden's scenic sequence. (Liu Mengyu)

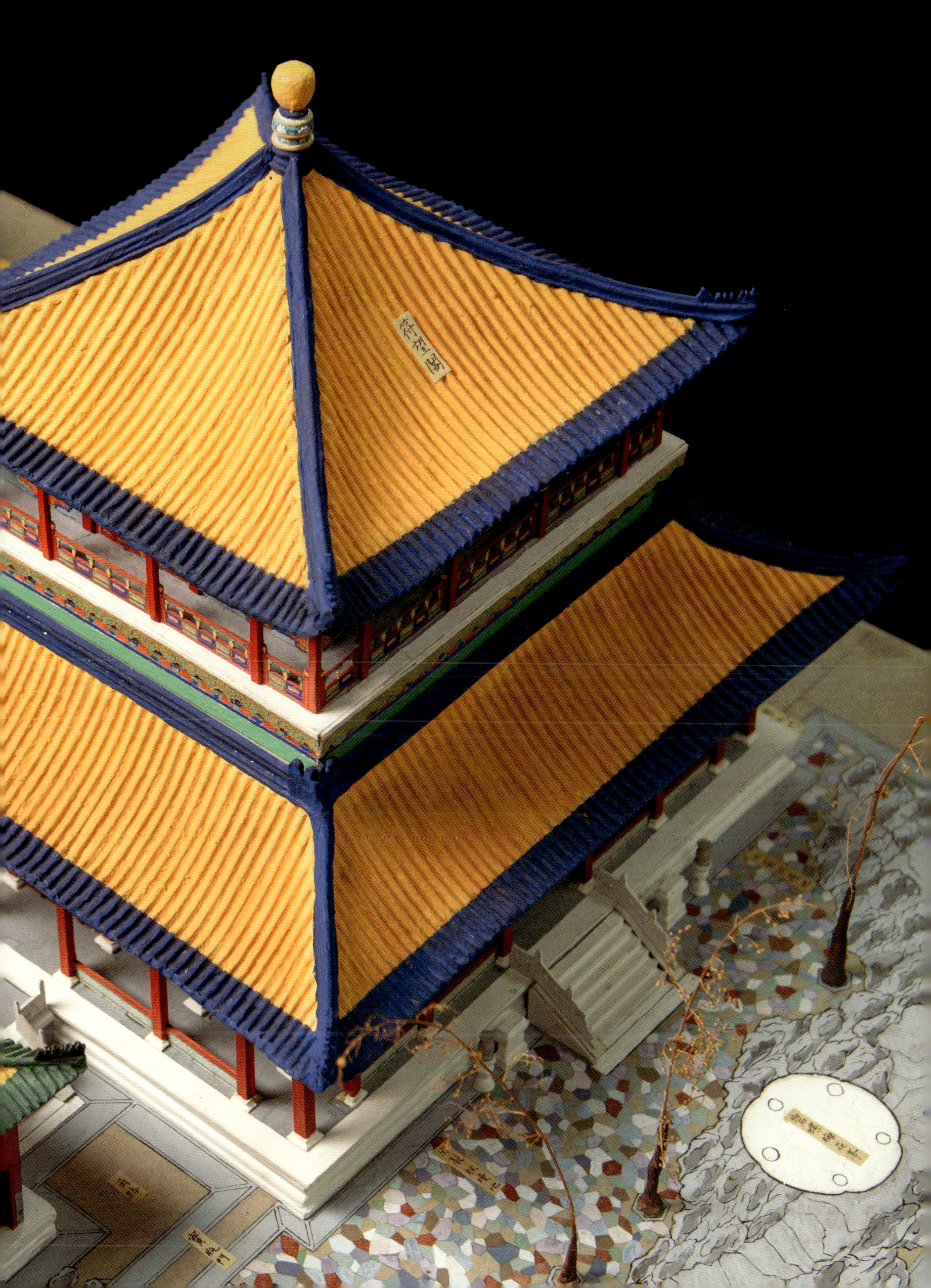
符望閣

174《游园惊梦》(曲谱)

清(1644—1911)
纵 23.5 厘米,横 14.1 厘米
故宫博物院藏

本剧取材于《牡丹亭》,分为“游园”与“惊梦”两出戏。讲述少女杜丽娘带丫鬟春香至花园游赏,见园林美景,春思顿发,百感交集,于园中入睡。梦神指引与书生柳梦梅相见,遂成姻好。戏文中的“则为你如花美眷,似水流年”一句最为经典,流传甚广。(杨国彭)

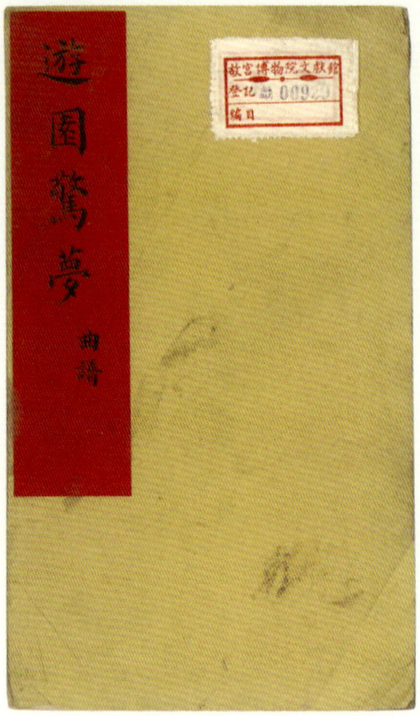

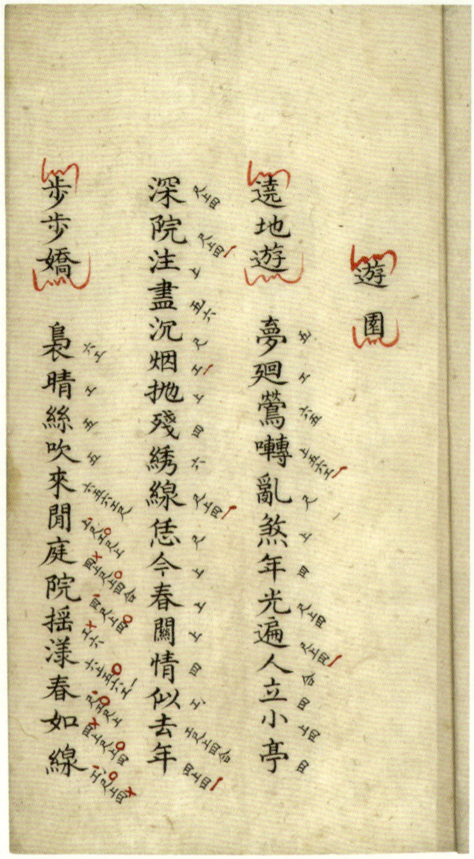

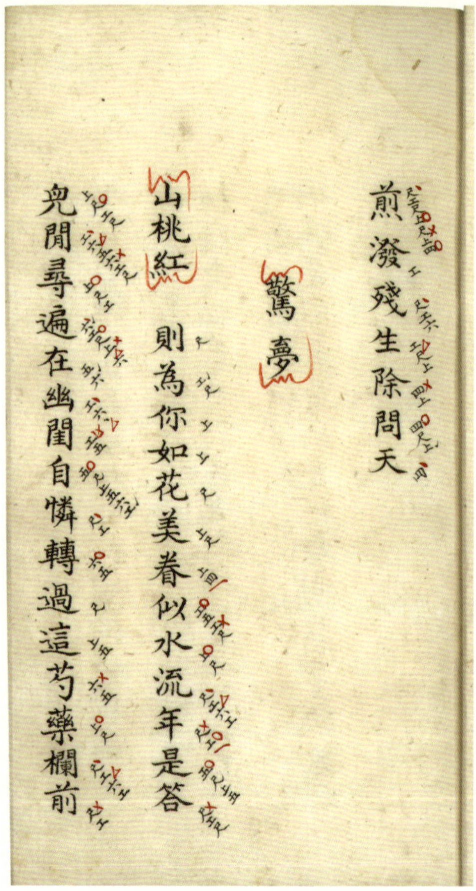

174 *Wandering in the Garden, Dreaming of Love* (Musical Score)

Qing Dynasty (1644-1911)
L.:23.5cm, W.:14.1cm
Collection of the Palace Museum

Wandering in the Garden and Dreaming of Love are two scenes from the opera *The Peony Pavilion*. It tells the story of the young maiden Du Liniang, who strolls through a garden accompanied by her maid Chunxiang. Enchanted by the beauty of the scenery, she is overcome with emotions and drifts into slumber. In her dream, a deity guides her to meet the scholar Liu Mengmei, with whom she falls in love. (Yang Guopeng)

175《万花向荣御苑献瑞》（题纲）

清（1644—1911）
纵 21.1 厘米，横 12.7 厘米
故宫博物院藏

本剧为皇太后万寿、皇后千秋和上元日承应剧目之一。剧情为花神奉玉帝旨，邀约散花玉女及福神来到瑞蓂阶畔、朱草庭前，议定御苑献瑞事宜。花神职司花部，玉女献四时吉庆宝花。众玉女各持玉梅花、山茶花、水仙花、瑞香花、畹兰花、任春花、樱桃花、望春花、春菜花、红杏花、白李花、碧桃花、棣棠花、蔷薇花、海棠花、雪梨花、木兰花、荣桐花、瑞麦花、杨柳花、牡丹花等，歌舞献花。(杨国彭)

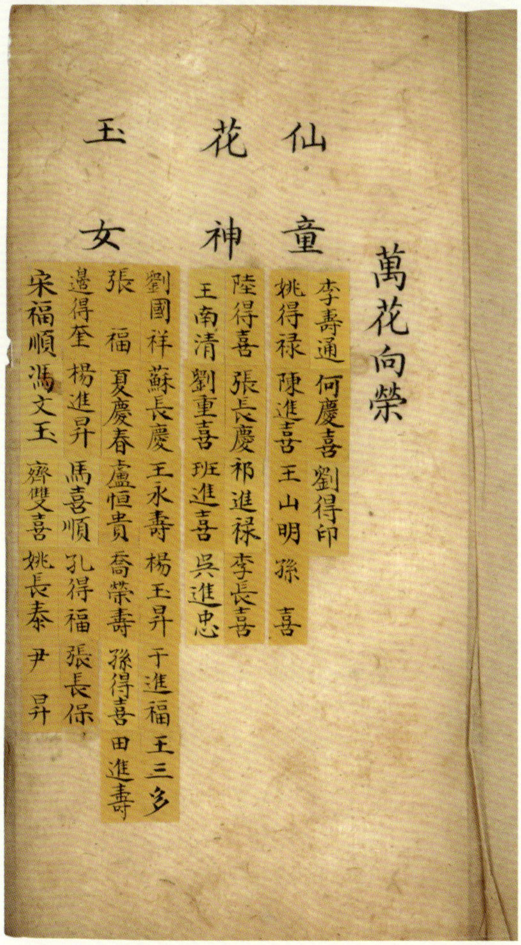

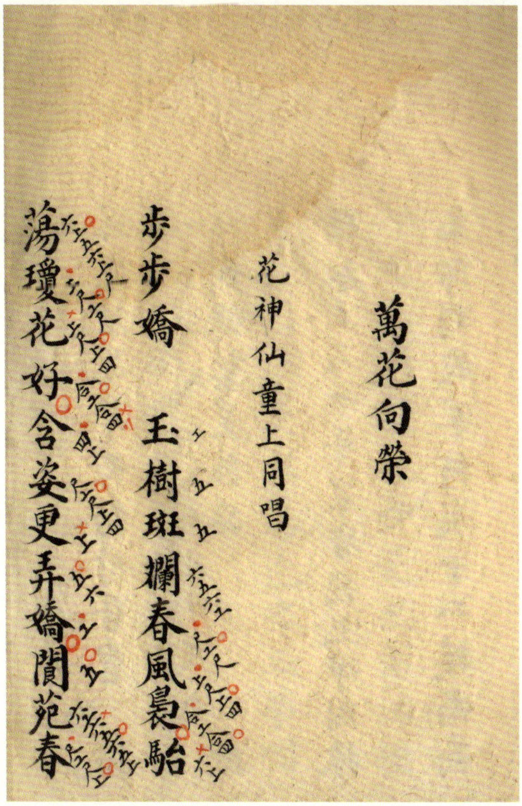

175 ***Presenting Auspicious Blessings at the Imperial Banquet* (Opera Script)**

Qing Dynasty (1644-1911)
L.:21.1cm, W.:12.7cm
Collection of the Palace Museum

This opera was one of the performances staged for the Empress Dowager's longevity celebration, the Empress's birthday, and the Lantern Festival. The story follows the Flower Goddess, who, by the Jade Emperor's decree, gathers the Flower-Spreading Maidens and the God of Fortune at the Auspicious Terrace to discuss presenting auspicious omens in the imperial garden. (Yang Guopeng)

176 木黑漆斗笙

清（1644—1911）
底径 3.5 厘米，通高 50 厘米
故宫博物院藏

笙，又称芦笙，是簧片与竹管耦合共振而发声的吹奏乐器。此笙木制笙斗，上面列植 17 根竹管笙苗，右口稍缺，左右分列，中管最长，两旁渐短，形如凤翼。每根竹管下端嵌紫檀木笙脚，笙脚末端镶接薄铜页簧片。笙苗外束以竹箍，笙嘴镶饰以牛骨。

笙是中国流传已久的乐器，是世界上现存大多数簧片乐器的鼻祖，传说女娲发明了笙。甲骨文中“和”字即是当时小笙的象形，源于笙的发音遵循和声原则，声音饱满，悦耳动听。笙为“八音”中匏之属的乐器，源于笙斗本用匏制，唐代以后多改用木制。笙为昆曲伴奏乐器之一。清李斗《扬州画舫录》载，昆曲清唱“以笙、笛、鼓板、三弦为场面”。（李文洁）

176 Black-lacquered Wooden *Sheng*

Qing Dynasty (1644-1911)
Diam.(at base):3.5cm, H.(in total):50cm
Collection of the Palace Museum

This version of the *sheng*, also known as *lusheng*, features a wooden wind chamber. The *sheng* is a traditional Chinese musical instrument with a long history and is considered the ancestor of most modern reed instruments worldwide. It is one of the accompanying instruments in the *Kunqu* opera. (Li Wenjie)

177 竹镶骨笛

清（1644—1911）
通长 66.3 厘米，底径 2 厘米
故宫博物院藏

笛，吹奏乐器，俗称笛子、横笛。竹制，横吹，上开吹孔一、膜孔一、按音孔六，尾部出音孔二。笛身上用丝线缠绕以加固，两端镶骨。

笛是昆曲伴奏的主要乐器，在昆曲乐队中处于领衔地位。“平均孔”曲笛以雌、雄二笛为一对。清李斗《扬州画舫录》叙昆曲场面云：“笛子之人在下鬼门，例用雌雄二笛。”雄笛管稍粗，音色浑厚，为伴奏“阔口戏”（净、丑等角）所用；雌笛管略细，音色清亮，为伴奏“生旦戏”所用。此笛即为雌、雄二笛中的雌笛。（毛劼）

177 Bamboo Flute Inlaid with Bone

Qing Dynasty (1644-1911)
L.(in total):66.3cm, Diam.(at base):2cm
Collection of the Palace Museum

The flute is a primary accompaniment in *Kunqu* opera, holding a leading position in the orchestra. The 'even-holed' flutes are paired as male and female flutes. The male flute has a slightly thick tube with a rich and deep tone while the female flute, with a thinner tube, produces a bright and clear sound. This particular flute is the female counterpart of the pair. (Mao Jie)

178 紫竹画花卉纹题字箫

清（1644—1911）
通长 73.8 厘米，底径 2.1 厘米
故宫博物院藏

箫，吹奏乐器，也称洞箫，“八音”中竹之属。单管，竖吹，上端利用竹节封口，在封口处开半椭圆形吹孔，管身开六个按音孔（前五后一）。此箫于按音孔与吹孔之间彩绘兰花纹，书“今日花开又一年”。

箫的音色圆润柔和，为昆曲伴奏乐器之一。昆曲清唱又称桌台清唱，被视为文人雅事。明嘉靖年间，魏良辅改革创制昆腔“水磨调”，其所著《曲律》载：“清唱，俗语谓之‘冷板凳’，不比戏场借锣鼓之势。全要闲雅整肃，清俊温润。”这一时期昆曲清唱的伴奏乐器以箫管为主：“箫管以尺工俪词曲，犹琴之勾剔以度诗歌也。”明清之时，苏州人于“吴中第一名胜”虎丘举行曲会，明代万历年间著名文学家袁宏道在《江南子》一诗中记载了苏州虎丘清唱昆曲的盛况，其曲唱伴奏乐器即为“一拍一箫一寸管”：“蜘蛛生来解织罗，吴儿十五能娇歌。旧曲嘹厉商声紧，新腔啴缓务头多。一拍一箫一寸管，虎丘夜夜石苔暖。家家宴喜串歌儿，红女停梭田畯懒。”（毛劼）

178 Purple Bamboo *Xiao* with Floral Pattern and Inscription

Qing Dynasty (1644-1911)
H.(in total):73.8cm, Diam.(at base):2.1cm
Collection of the Palace Museum

The *xiao*, also known as *dongxiao*, is a woodwind instrument characterized by its smooth and mellow sound. It is one of the accompanying instruments for *Kunqu* opera, which was regarded among the literati as a refined pastime. (Mao Jie)

179 匏描金仙人纹三弦

清（1644—1911）
上宽 2.4 厘米，下宽 3 厘米，音箱长 17 厘米，音箱厚 8 厘米，通高 97.8 厘米
故宫博物院藏

三弦柄为紫檀制，上端凿空以绾弦，弦有三根，左二右一，弦轴亦为紫檀制。槽为匏制，上冒虺（音悔）皮，皮面设柱，柄贯槽中。槽背面饰四仙女，手持笙、笛、箫、拍板等乐器。槽下系弦扣，下垂黄绒紃（音寻）。清代宫廷宴飨庆隆舞乐、番部合乐用之。（刘国梁）

179 Gourd Three-Stringed Lute (*Sanxian*) with Gilt Immortal Pattern

Qing Dynasty (1644-1911)
W.(upper part):2.4cm, W.(lower part):3cm,
L.(sound box):17cm, Th.(sound box):8cm, H.(in total):97.8cm
Collection of the Palace Museum

This *sanxian* lute features a red sandalwood neck, the head of which is hollowed for fastening the three strings. The back is decorated with images of four female immortals holding flutes, pipes and clappers, below which the strings are attached. From the end hangs a yellow silk tassel. *Sanxian* lutes were played at court banquets to accompany court dance or as part of ethnic minority ensembles. (Liu Guoliang)

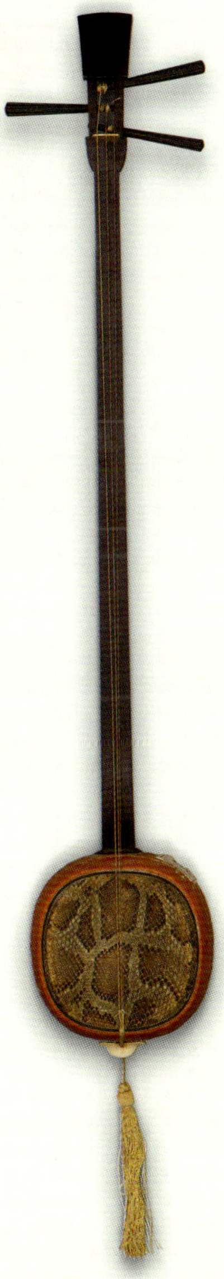

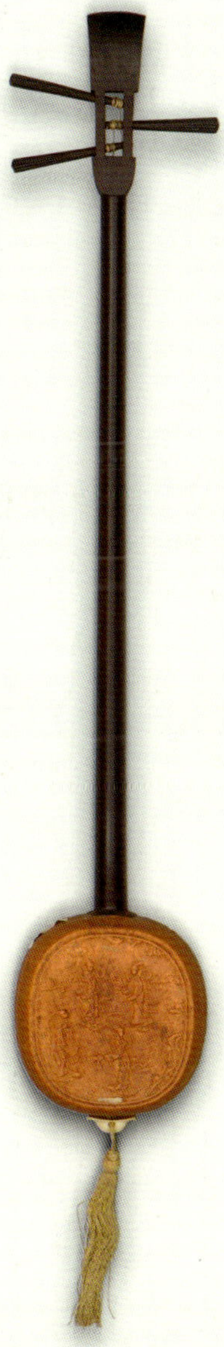

180 “花园中的音乐会”刺绣画

1647，[英国]佚名
帆布，丝线刺绣
纵 34.9 厘米，横 54 厘米
大都会艺术博物馆藏

音乐演奏派对是 17 世纪英国乡绅阶层流行的社交活动。画作描绘了在规则式布局的花园里，女士们手持乐器进行演奏，展现出优雅的社交氛围。背景中，精心布置的花坛、喷泉和廊架等典型元素，生动再现了英国乡间花园的特色景观。（雷浩）

180 Musical Party in a Garden, Embroidered Picture

1647, unkonwn craftsman (British)
Canvas worked with silk thread, tent stitch
L.:34.9cm, W.:54cm
Lent by The Metropolitan Museum of Art, Gift of Irwin Untermyer, 1964 (64.101.1314)

Music performance parties were popular social activities among the English gentry in the 17th century, and gardens became the primary space for performances. The carefully arranged flowerbeds, fountains, and pergolas all vividly recreate a classic English countryside garden. (Lei Hao)

181 天堂园中的舞者献艺

18 世纪晚期，[印度] 佚名，印度北方邦阿瓦德
纸本，不透明水彩和金箔
纵 32 厘米，横 21.9 厘米
芝加哥艺术博物馆藏

画作描绘了一座宫殿和花园，展现了莫卧儿时期园林的典型特征。十字形水渠确立了花园的空间结构，水渠内大大小小的喷泉静静流淌，象征着《古兰经》中提到的天堂花园。画面中心的亭子被精巧设计的烛台环绕，统治者端坐其内，沉浸于轻盈舞者的曼妙舞姿之中。（曹旭卿）

181 Ruler Entertained by Dancers in a Paradise Garden

Late 18th century, unknown painter (India), India Uttar Pradesh, Avadh
Opaque watercolor and gold on paper
L.:32cm, W.:21.9cm
Collection of the Art Institute of Chicago, Lucy Maud Buckingham Collection (1919.953)

This painting depicts a palace and garden, showcasing the typical characteristics of Mughal-era landscaping and symbolizing the gardens of paradise mentioned in the *Quran*. The ruler sits majestically in the pavilion at the center of the cross-shaped water channels, immersed in the music and graceful movements of the dancers. (Cao Xuqing)

182 泥人张宫衣何仙姑像

清（1644—1911）
宽 12 厘米，厚 11.5 厘米，高 40 厘米
故宫博物院藏

本件泥塑外施彩绘，为八仙之一的何仙姑立像。何仙姑鹅蛋脸，弯眉细目，眼角上挑，面带微笑，面容秀丽。头发后梳，绾为发髻，戴流苏状耳饰。头戴金属做成的过翘，装饰红色圆球。身穿宫衣，上穿红色圆领对襟阔袖衣，袖口镶八条彩色花边，左右各有白色团寿纹。下着百褶长裙，前后有平幅裙门，上绘海水江崖和祥云纹，外裹各色缎带组成的凤尾裙。袖口花边和裙身缎带均绘花卉纹。披云肩，立领，绘牡丹纹，周围装饰网子穗。整体衣饰华美，体态婀娜。此像是清宫八仙庆寿剧目形象，具有吉富贵祥、福寿康宁之美好寓意。

何仙姑为八仙中唯一的女仙。有关八仙的戏曲起源于宋元时期，明代吴远泰小说《东游记》始将八仙定为：汉钟离、吕洞宾、何仙姑、曹国舅、蓝采和、韩湘子、张果老、铁拐李，从此八仙结伴同行、扶危济困、造福于民。在清代皇家戏剧演出中，八仙剧目占有很大的比例。主要演出时间为燕九和帝后生日。燕九指正月十九日，据传这天是道教全真派丘处机的生日。届时，京城白云观会举行隆重庆典，清宫内则演出《群仙赴会》《圣母巡行》等剧目。遇帝后生日，演出剧目则更为繁多。自乾隆朝开始，遍及紫禁城及各处皇家园林的大小戏台，每年都会上演祝寿大戏。如遇皇帝、皇太后整寿，还会举行更为盛大的万寿庆典活动。其中八仙庆寿是寿诞日必备节目，所演曲目多为吉庆长寿之剧目，主要有《佛仙会议》《群仙祝寿》《八仙祝寿》等。（赵芸格）

182 Painted Clay Statue of He Xiangu, One of the Eight Immortals

Qing Dynasty (1644-1911)
W.:12cm, Th.:11.5cm, H.:40cm
Collection of the Palace Museum

This painted clay figure depicts a character from the *Eight Immortals Celebrating Longevity* performance in the Qing imperial court. Starting from the Qianlong reign, grand longevity operas were performed annually, with even more elaborate celebrations held for the emperor and empress on milestone birthdays. *The Eight Immortals Celebrating Longevity* was a staple performance on such occasions, with plays often themed around auspiciousness and longevity. (Zhao Yunge)

183 桃红色缎绣团荷鸟纹八仙衣

清（1644—1911）
身长 131.5 厘米，两袖通长 207 厘米
故宫博物院藏

八仙衣大领，斜襟，阔袖，上衣下裳相连，衣长及足，缀水袖。上衣桃红色地，绣六团荷花鹭鸶纹，周围以金线做方格状排列，内填兰花及十字花，领、袖边以玫瑰紫缎缘饰。腰际饰革带，左右各缀侉子，上绣荷花、花篮纹。下裳月白色衬裙，缀双层如意状腰饰及各色飘带，前后正中饰以如意佩，并缀五彩网穗。此衣色彩斑斓，装饰繁复，为八仙中的何仙姑穿用。何仙姑为八仙中唯一的女仙，其衣形制似宫衣，因以荷花为法宝，故何仙姑衣以荷花纹样为主要标志。（张利群）

183 Peach-pink Satin Robe Embroidered with Circular Lotus and Bird Pattern for He Xiangu, One of the Eight Immortals

Qing Dynasty (1644-1911)
Robe l.:131.5cm, Sleeves l.(in total):207cm
Collection of the Palace Museum

This robe is made of peach-pink satin, with gold thread in a grid pattern, collar and sleeve edges trimmed with rose-purple satin, a leather belt around the waist, a blue-white lower skirt, and a central apron-like panel at the front and back. He Xiangu is the only female among the Eight Immortals, and her attire resembles court robes, with a lotus flower as her sacred item. Therefore this garment features lotus patterns as its primary motif. (Zhang Liqun)

184 绿色缎绣桂兔金皮球花纹花神衣

清（1644—1911）
身长 114 厘米，两袖通长 186 厘米
故宫博物院藏

花神衣立领，领两侧外沿饰有弧形镶缘，似翻领。衣对襟，宽身阔袖，袖口缝有水袖，周身饰有镶缘。衣身布满形态各异、憨态可掬的白色绒绣兔纹，金皮球花、桂花等纹样散摆其间，镶缘饰有瓜瓞绵绵、枝桂纹。这件花神衣为八月女花神穿用。花神衣多在清宫节令戏、寿戏等仪典戏中使用，如园林赏荷时节承应戏《御筵献瑞》中就有关于十二月花神的内容。（张锐）

184 Green Satin Flower Goddess Robe Embroidered with Rabbit, Osmanthus, and Golden Ball Flower Pattern

Qing Dynasty (1644-1911)
Robe l. :114cm, Sleeves l.(in total) :186cm
Collection of the Palace Museum

This robe is adorned with endearing, variously posed rabbit motifs embroidered in white velvet. Interspersed among them are golden ball flowers, osmanthus blossoms, and other floral designs. This robe was worn by the August Flower Goddess. Such robes were often used in seasonal and celebratory ceremonial performances and theatrical productions in the Qing court. (Zhang Rui)

185 五彩开光西厢记故事图盖罐

清（1644—1911）
口径 17.6 厘米，足径 17.7 厘米，通高 22 厘米
故宫博物院藏

罐直口，短颈，深腹，圈足。附一平顶圆盖，盖出沿，盖顶置一金属宝珠形钮。罐外壁及盖面以五彩为饰，外壁作四面方形开光，内绘《西厢记》人物故事图四幅，分别为墙角联吟、红娘请宴、莺莺听琴、锦字传情。盖面作四面扇形开光，开光内绘婴戏习武图。颈部及肩部绘龟背锦纹、几何纹等。外底无款识。

元人王实甫作杂剧《西厢记》一经问世，即“天下夺魁”，在社会上迅速风靡。除戏曲表演及文本传播外，以《西厢记》人物故事为主题的图像逐渐成为一类固定图式，这类“西厢图式”在版画及不同材质的工艺品装饰中广泛流行。“西厢图式”在明清时期民间陶瓷装饰上尤为盛行，甚至衍生出专门的“西厢瓷”。（蒋艺）

185 *Wucai* Lidded Jar with *Romance of the Western Chamber* Illustrations

Qing Dynasty (1644-1911)
Diam.(at mouth):17.6cm, Diam.(at foot):17.7cm, H.(in total):22cm
Collection of the Palace Museum

The Romance of the Western Chamber, a Yuan Dynasty play written by Wang Shifu, became an instant sensation upon its release. This lidded jar features a metallic knob shaped like a treasure bead on top of the lid. The outer walls and lid are decorated with four square reserved panels, each depicting a scene from *the Romance of the Western Chamber*. (Jiang Yi)

186 西厢记图册

清康熙（1662—1722），叶逋
绢本，设色
纵 30 厘米，横 30 厘米
故宫博物院藏

在中国古典戏曲、小说中，花园往往是男女主人公邂逅、相会、私订终身的场地。《西厢记》里，崔莺莺和张生的幽会便发生在普救寺后花园中。《西厢记》的故事最早可追溯到唐代李绅所作《莺莺歌》，以及由元稹署名、与之同题材的传奇小说《莺莺传》。其后，金人董解元将之改编为《西厢记诸宫调》。至元人王实甫则又将其改为杂剧，故《西厢记》有董、王之分。图册绘《西厢记》故事，共十二开，对开曲词分别取自《董西厢》和《王西厢》。全剧讲述了才子张君瑞与相国小姐崔莺莺在仕女红娘的帮助下，冲破重重阻挠，终成眷属的故事。

图中女子皆眉眼弯弯，面带笑意，作者用线绵细，布景精细，设色淡雅。本幅无款，仅钤“叶逋”白文方印。对开为康熙五十六年（1717）吴焯题词。叶逋，活动于康熙年间。（邢陆楠）

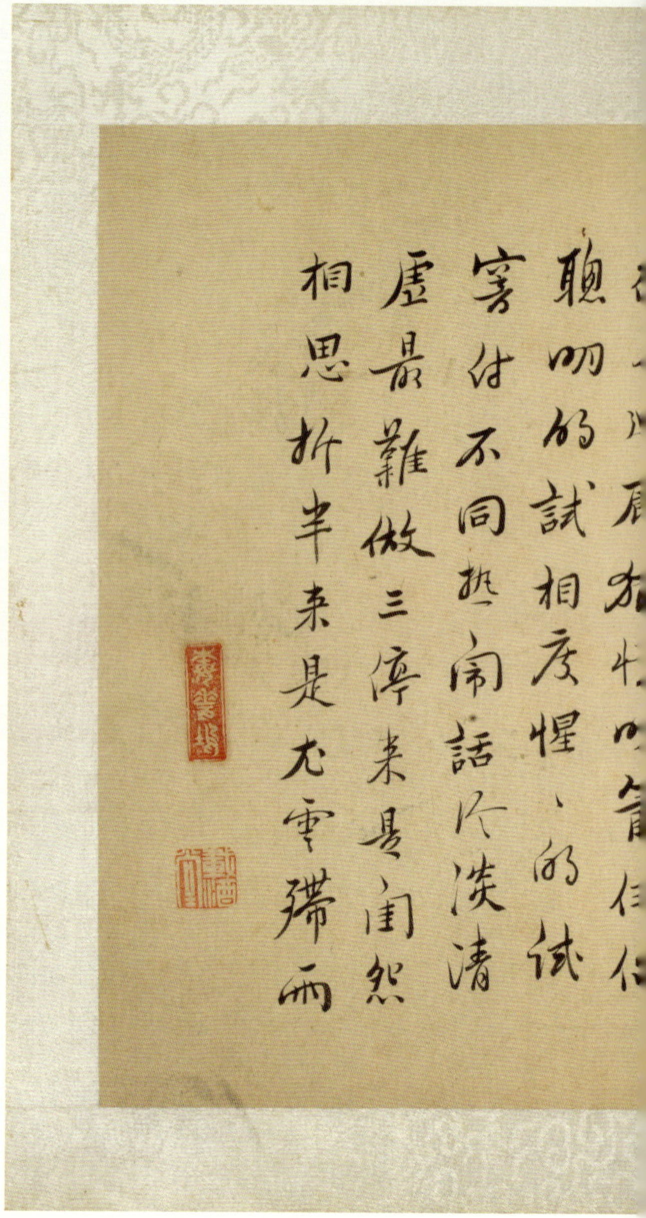

186 *Romance of the Western Chamber*

Qing Dynasty, Kangxi Reign (1662-1722), Ye Bu
Album leaf, color on silk
L.:30cm, W.:30cm
Collection of the Palace Museum

In Chinese classical opera and literature, gardens often serve as the setting for fateful encounters and lifelong commitments between male and female protagonists. In *Romance of the Western Chamber*, the secret meetings between Cui Yingying and Zhang Sheng take place in the rear garden of Pujiu Temple. This album comprises four illustrations depicting key moments from the story. (Xing Lunan)

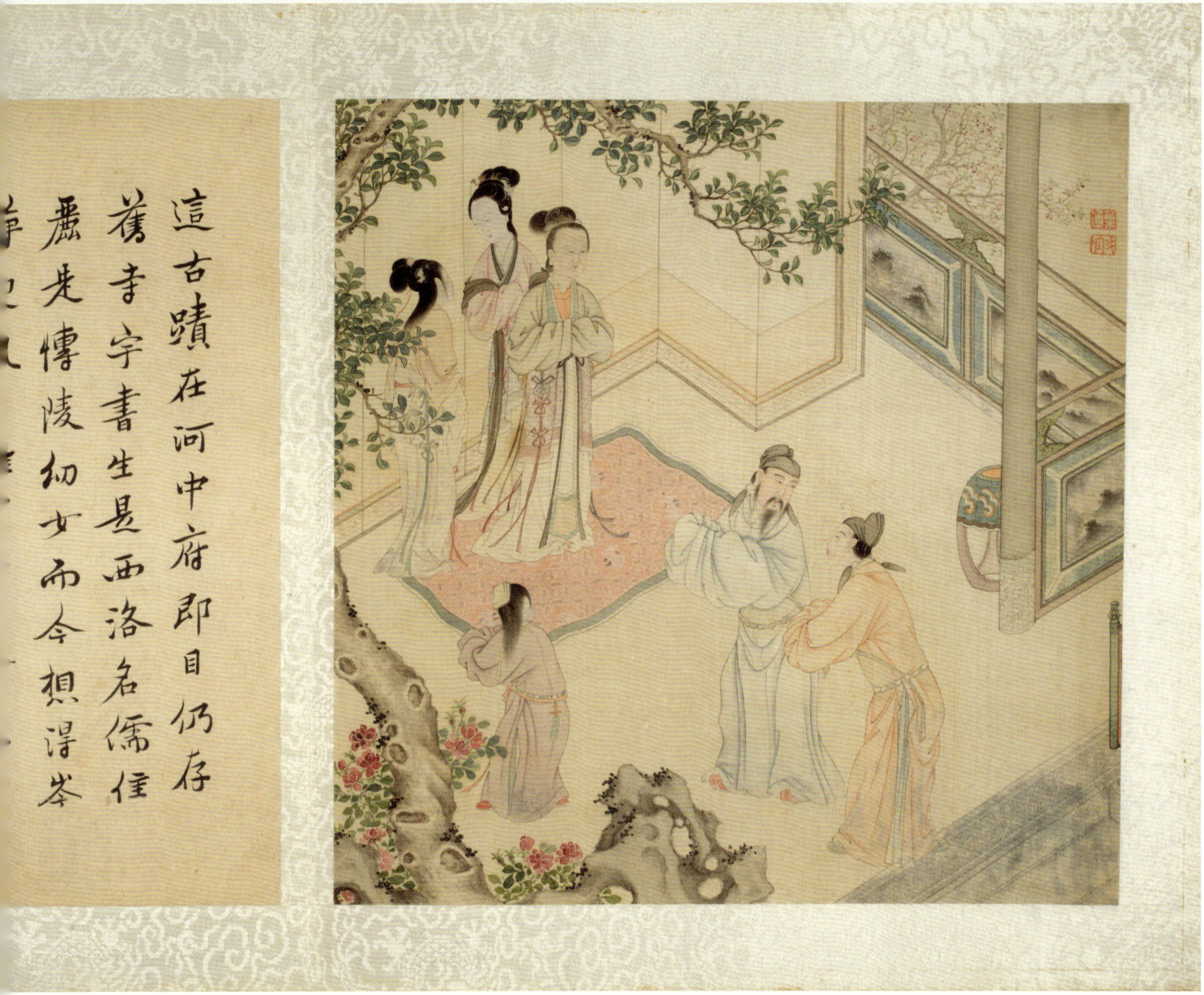

本页绘故事开端，崔夫人携子女欢郎及莺莺，扶亡夫崔相国灵柩返乡途中，在山西河中府普救寺内停歇、借宿。对开题："这古迹在河中府，即目仍存旧寺宇。书生是西洛名儒，佳丽是博陵幼女。而今想得，岑静迎风户，唯有旧题句。空留待月回廊，犹忆吹箫伴侣。聪明的试相度，惺惺的试窨付。不同热闹话，冷淡清虚最难做。三停来是闺怨相思，折半来是尤云殢雨。"曲词出自《董西厢》卷一。(邢陆楠)

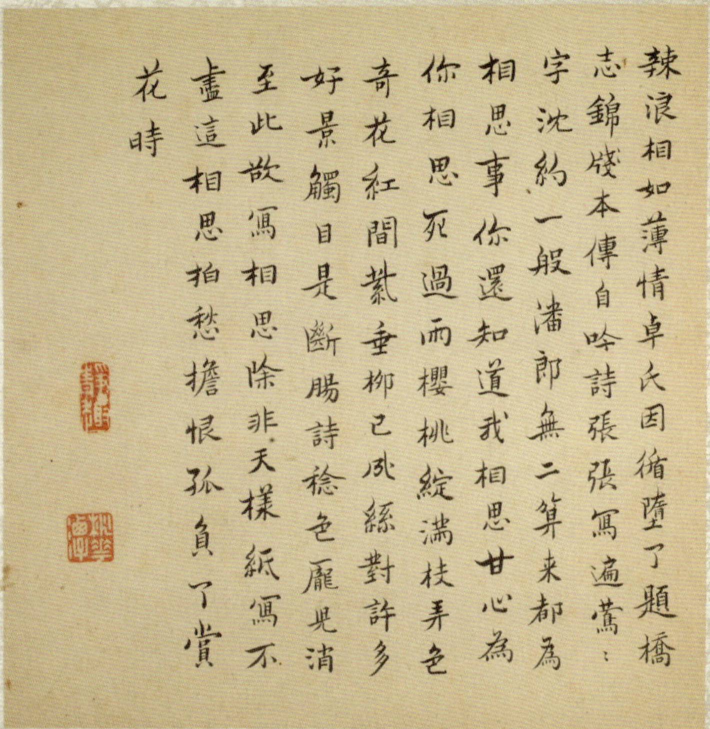

上页绘莺莺读张生来信，红娘偷窥。对开题：“辣浪相如，薄情卓氏，因循堕了题桥志。锦笺本传自吟诗，张张写遍莺莺字。沈约一般，潘郎无二，算来都为相思事。你还知道我相思，甘心为你相思死。过雨樱桃绽满枝，弄色奇花红间紫。垂柳已成丝，对许多好景，触目是断肠诗。稔色庞儿消至此，欲写相思，除非天样纸，写不尽这相思，拍愁担恨孤负了赏花时。”曲词出自《董西厢》卷五。（邢陆楠）

右上页绘张生书童传捷报，莺莺、红娘伏案读信。对开题：“裙染榴花，睡损胭脂皱。纽结丁香，掩过芙蓉扣。线脱珍珠，泪湿香罗袖。杨柳眉颦，人比黄华瘦。”曲词出自《北西厢秘本》（《王西厢》）卷五第一折。图中菖蒲、投壶、青铜等物展现了闺中女子的清玩用品。（邢陆楠）

右下页绘故事结尾，莺莺与张生终成眷属，二人于庭院内对谈。对开题：“从今至古，自是佳人合配才子。一个文章天下无双，一个稔色寰中无二。合欢带连理枝，题彩扇，写新诗。趁了文君深愿酬，了相如素志。将军满满劝金卮，今日极醉休辞欢，撞倒郑恒那村子。”曲词出自《董西厢》卷八。（邢陆楠）

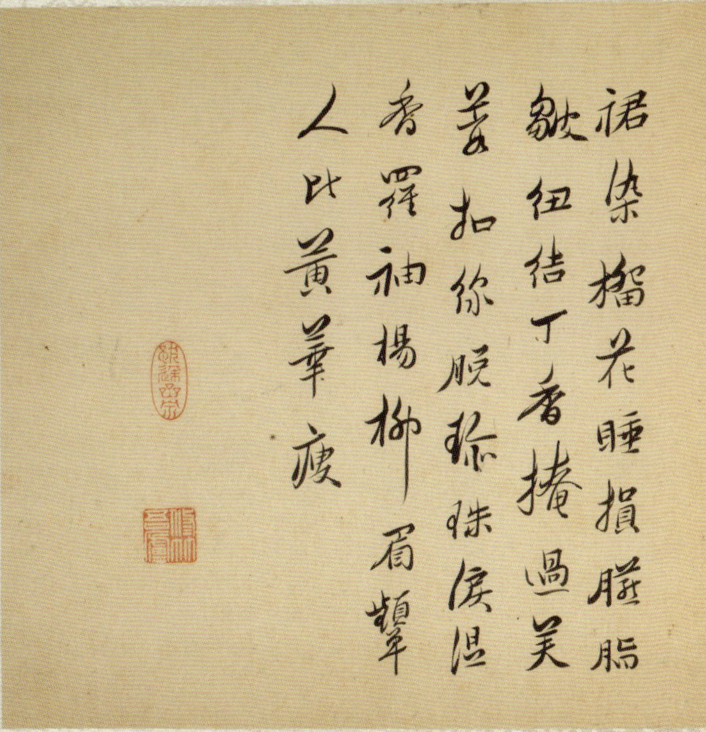
裙染榴花睡損臙脂
皺紐結丁香掩過芙
蓉扣線脫珍珠淚濕
香羅袖楊柳眉顰
人比黃花瘦

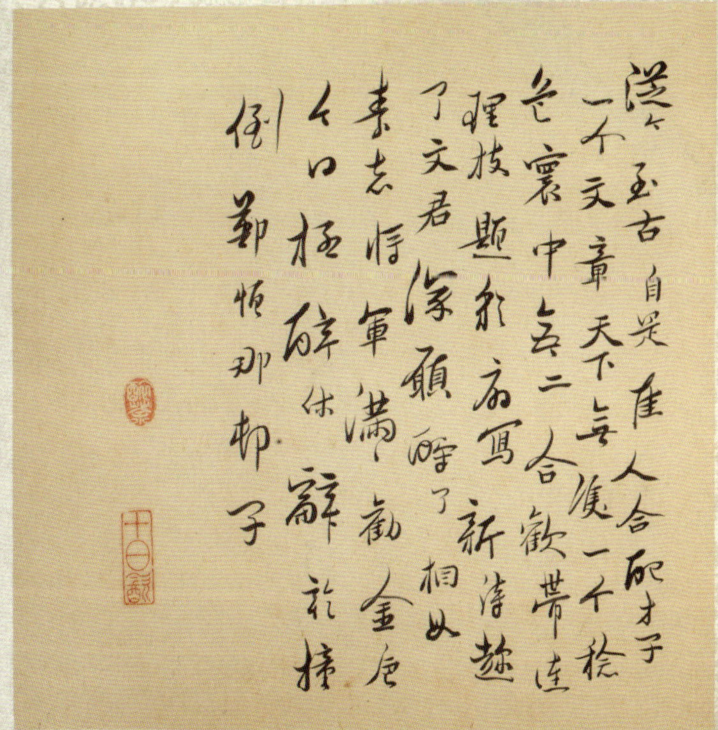
從今至古自是佳人合配才子
一个文章天下無雙一个稔
色寰中無二合歡帶連
理枝題系扇寫新詩燕
了文君深願醉了相如
素志將軍滿勸金卮
口口極辭休辭辭於捧
倒那怕那柳子

佛置身于一场华丽的宫廷演出，为观者带来无尽的惊喜。季节性的喷泉表演，更为其增添了流动的戏剧魅力，仿佛自然与艺术在此共舞。（雷浩）

188 恩克拉多斯丛林园

1700—1725，[法国] 佚名
布面，油画
纵 72 厘米，横 142.6 厘米
凡尔赛宫和特里亚农宫国家博物馆藏

恩克拉多斯，是希腊神话中泰坦神族中的一员。他在与宙斯及奥林匹斯众神的战斗中被打败，被压在岩石之下。根据神话，恩克拉多斯的身体仍在不安分地挣扎。喷泉雕塑以此为主题，池中，恩克拉多斯的雕像半沉，口中喷涌的水柱仿佛他的尖叫一般，象征着挣扎和反抗。在凡尔赛宫的音乐喷泉表演期间，喷泉的水景展示尤为引人注目，水柱高达 23 米，是凡尔赛宫花园中最高的水柱之一。（雷浩）

188 The Enceladus Grove

1700-1725, unknown painter (French)
Oil on canvas
L.:72cm, W.:142.6cm
Collection of the National Museum of the Palace of Versailles and Trianon (MV 7370)

Enceladus was a member of the Titan race in Greek mythology. During the battle against Zeus and the Olympian gods, he was defeated and buried beneath massive rocks. However, his body still struggles restlessly. The fountain sculpture is inspired by this myth. In the pool, Enceladus' statue appears half-submerged, with a jet of water gushing from his mouth, reaching a height of 23 meters, symbolizing his struggle and defiance. (Lei Hao)

189 凡尔赛宫花园中的方尖碑喷泉

1740—1757，据［法国］雅克·里戈作品创作
蚀刻和雕刻版画，水彩及水粉手工着色
纵 31.2 厘米，横 45.9 厘米
耶鲁大学英国艺术中心藏

画中描绘了凡尔赛宫花园的方尖碑喷泉，以其精妙绝伦的水景设计著称。画面中心，巨大的水柱喷涌而上，形成象征力量与永恒的方尖碑造型。四周是台阶式水池，水流逐级而下，营造出动静相宜的层次感，增强了视觉与听觉上的沉浸体验。画面中的贵族们身着华丽服饰，漫步于喷泉四周，展现出凡尔赛宫廷生活的奢华与优雅。（雷浩）

189 The Obelisk in the Garden at Versailles

1740-1757, after Jacques Rigaud (French)
Etching and engraving hand-color in watercolor and gouache
L.:31.2cm, W.:45.9cm
Collection of the Yale Center for British Art (B1995.13.72)

This painting depicts the Obelisk Fountain in the Garden at Versailles. The massive water column shoots upwards, forming the shape of an obelisk, symbolizing power and eternity. Surrounding it are tiered water basins, with water flowing down in stages, creating a harmonious balance of movement and stillness, enhancing both visual and auditory immersion. (Lei Hao)

190 帝王岛丛林园

1700—1725，［法国］佚名
布面，油画
纵 71 厘米，横 129 厘米
凡尔赛宫和特里亚农宫国家博物馆藏

帝王岛丛林园由皇家池（Bassin de l'Île royale）和半圆形的“水镜池”（Miroir d'Eau）组成。皇家池于 1671 年至 1674 年间开凿，广阔的水面可容纳小船，供宾客泛舟游览；水镜池位于其东侧，平静如镜，倒映园林景观。园内布满大小喷泉，水流声交织成自然的交响乐，为宾客带来视觉与听觉的双重享受。（雷浩）

190 The Royal Isle Grove

1700-1725, unknown painter (French)
Oil on canvas
L.:71cm, W.:129cm
Collection of the National Museum of the Palace of Versailles and Trianon (MV 7363)

The Royal Isle Grove consists of the Royal Basin (*Bassin de l'Île royale*) and the semicircular Mirror Pool (*Miroir d'Eau*). The Royal Basin was excavated between 1671 and 1674, and its vast surface could accommodate small leisure boats for visitors. Located to the east, the Mirror Pool is calm and reflective, mirroring the garden landscape. (Lei Hao)

191 沼泽丛林园

1700—1725，［法国］佚名
布面，油画
纵 65.5 厘米，横 136.6 厘米
凡尔赛宫和特里亚农宫国家博物馆藏

沼泽丛林园是凡尔赛宫早期园林的重要组成部分，其主景是一个模仿自然池塘的水景，中央矗立着一株由青铜与锡打造的大树雕塑，树枝末端喷涌出无数水柱，宛如繁茂的枝叶瀑布洒落池中。两侧的芦苇丛中也有水柱喷出，与自然树篱和大树雕塑相互辉映，水声潺潺，营造出宛如舞台般的梦幻景观效果。（雷浩）

191 The Marsh Grove

1700-1725, unknown painter (French)
Oil on canvas
L.:65.5cm, W.:136.6cm
Collection of the National Museum of the Palace of Versailles and Trianon (MV 7378)

The Marsh Grove was a key part of the early landscaping design of the Palace of Versailles, with its main feature being a waterscape designed to mimic a natural pond. At the center stands a large tree sculpture made of bronze and tin, with countless streams of water gushing out from the tips of its branches. On both sides, water columns also rise from the reed beds, harmonizing with the sculpture and the natural hedges. (Lei Hao)

192 水剧场

1700—1750，[法国] 佚名
布面，油画
纵 65.3 厘米，横 136 厘米
凡尔赛宫和特里亚农宫国家博物馆藏

水剧场丛林园内，广场由台阶分隔为两个半圆形区域，分别是供观众就座的圆形剧场和供演员表演的舞台，巧妙地将自然景观与舞台艺术相结合。广场后方的壁龛之间，每条林荫道中央都设有装饰贝壳和喷泉的运河，水花沿着台阶流淌，形成动态的舞台背景，使整个园区充满韵律感与戏剧性。（雷浩）

192 The Water Theater

1700-1750, unknown painter (French)
Oil on canvas
L.:65.3cm, W.:136cm
Collection of the National Museum of the Palace of Versailles and Trianon (MV 7162)

The Water Theater's Plaza is divided into two semicircular areas by steps: a circular theater for the audience and a stage for the performers, cleverly blending natural landscapes with theatrical art. Between the niches at the back of the plaza, each tree-lined avenue features canals adorned with shells and fountains at the center, with water flowing down the steps. (Lei Hao)

193　凡尔赛宫花园中的凯旋门

1768—1779，[英国] 雷米・帕尔据 [法国] 雅克・里戈作品创作

雕刻版画，手工着色

纵 31.8 厘米，横 45.7 厘米

193　The Triumphal Arch in the Garden of Versailles

1768-1779, Remi Parr (British) , after Jacques Rigaud (French)

Hand-colored engraving

L.:31.8cm, W.:45.7cm

Collection of the Yale Center for British Art (B1995.13.73)

194 凯旋门

1700—1750，［法国］佚名
布面，油画
纵 65.3 厘米，横 136.3 厘米
凡尔赛宫和特里亚农宫国家博物馆藏

194 The Triumphal Arch

1700-1750, unknown painter (French)
Oil on canvas
L.:65.3cm, W.:136.3cm
Collection of the National Museum of the Palace of Versailles and Trianon (MV7159)

凯旋门丛林园内的主景，是一座巨大的镀金凯旋门，象征着国王的军事胜利与法兰西的辉煌荣耀。华丽的喷泉与瀑布从凯旋门上倾泻而下，形成宏伟壮观的水景。园中布局精巧，分层设有金色的金字塔和喷涌的水柱喷泉，宛如一场宏大的视觉交响乐。（雷浩）

The main feature of this grove in the Garden of Versailles is its massive gilt metal triumphal arch. Magnificent fountains and waterfalls cascade down from the arch, creating a spectacular water display. The garden is meticulously designed, with layered golden pyramids and gushing water fountains forming a visual symphony. (Lei Hao)

195 穹顶丛林园

1700—1725，[法国]佚名据[法国]让·科泰尔作品创作
布面，油画
纵142厘米，横120.5厘米
凡尔赛宫和特里亚农宫国家博物馆藏

穹顶丛林园的布局仿若圆形剧场，中央设有一个六角形水池。园中雕塑以神话故事为蓝本，生动再现了仙女沐浴及她们用花环装饰阿波罗战车的场景。此外，黄道十二宫的标志被巧妙地嵌入上方的云层之中，增添了神秘色彩。路易十四自诩“太阳王”，凡尔赛宫花园中布满了太阳神阿波罗的形象，寓意光辉与权威，映照着君主的荣耀与宏愿。（许愿）

195 **View of the Grove of the Domes**

1700-1725, unknown painter (French), after Jean Cotelle (French)
Oil on canvas
L.:142cm, W.:120.5cm
Collection of the National Museum of the Palace of Versailles and Trianon (MV 7055)

The layout of the Grove of the Domes resembles an amphitheater with a hexagonal pool at its center. The sculptures in the garden, based on stories from mythology, vividly depict scenes such as nymphs bathing and decorating Apollo's chariot with flower garlands. (Xu Yuan)

196 阿尔多布兰迪尼庄园的风之厅

1660—1678, [意大利] 乔瓦尼 · 巴蒂斯塔 · 法尔达
蚀刻版画
纵 39.2 厘米，横 27.9 厘米
大都会艺术博物馆藏

意大利的弗拉斯卡蒂小镇（Frascati）坐落着多座文艺复兴时期的经典庄园，其中阿尔多布兰迪尼庄园（Villa Aldobrandini）因其巧妙的水景设计尤为引人注目。图中风之厅（Stanza dei Venti）又名阿波罗厅（Sala d' Apollo），厅内设有一座象征帕纳索斯山（Mount Parnassus）的微缩山体，山上环绕着阿波罗、手持乐器的缪斯女神以及天马的神话形象（见后对页照片）。水力驱动系统可以模拟出雷声、风声和雨声等音效，将神话与自然融为一体，营造出充满奇幻色彩的空间氛围。（郭欣悦）

196 **The Room of Winds at Villa Aldobrandini**

1660-1678, Giovanni Battista Falda (Italian)
Etching on paper
L.:39.2cm, W.:27.9cm
Lent by The Metropolitan Museum of Art, Bequest of Phyllis Massar, 2011 (2012.136.112)

The Italian town of Frascati is home to several Renaissance gardens, with the Villa Aldobrandini being especially eye-catching for its ingenious hydraulic engineering. The 'Room of Winds' depicted in this work features a miniature representation of Mount Parnassus, adorned with figures of Apollo, the Muses holding musical instruments, and the winged Pegasus. The hydraulic system simulates sounds of thunder, wind, and rain, blending mythology with nature to fill the space with mythical charm. (Guo Xinyue)

199 玉壶冰琴

南宋（1127—1279）
通长 119.1 厘米，隐间 110.3 厘米，额宽 17.8 厘米，肩宽 19.3 厘米，尾宽 13.3 厘米，厚 4.8 厘米
故宫博物院藏

琴仲尼式，桐木斫，鹿角灰胎，黑漆，蛇腹间流水断。长方形龙池、凤沼，槽腹基本挖成方形，纳音微隆起。金徽，白玉雁足，紫檀木琴轸，紫檀木岳山、焦尾。

琴背铭刻，龙池上方刻小篆琴名“玉壶冰”，池内左右朱漆径寸楷书“宋绍兴二年，公路金道制”。绍兴二年即公元 1132 年，时为宋高宗在位的第六年。金道，字公路，据元人周密《云烟过眼录》载：“金公路，所谓金道者，绍兴初人，琴薄而轻。”玉壶冰琴体现了南宋琴扁平而轻的典型特征。

琴名“玉壶冰”，出自唐代诗人王昌龄“洛阳亲友如相问，一片冰心在玉壶”的诗句，反映了琴之主人不与世俗权贵同流合污，归隐山水、退居园林，以琴书自娱养志的生活境界。（孙召华）

199 Ice in a Jade Vase *Qin*

Southern Song Dynasty (1127-1279)
H.(in total):119.1cm, L.(effective string):110.3cm, W.(upper):17.8cm, W.(middle):19.3cm, W.(lower):13.3cm, Th.4.8cm
Collection of the Palace Museum

This *qin* zither was crafted from paulownia wood. The back bears inscriptions, including its name, *Yu Hu Bing* (Ice in a Jade Vase). This name is derived from a line in Tang Dynasty poet Wang Zhihuan's poem: ‘If my friends in Luoyang ask about me, tell them my heart is as pure as ice in a jade vase’. It reflects the *qin* owner's disdain for worldly power and wealth, seeking a reclusive life in harmony with nature; playing the *qin* and reading books were means of self-cultivation and retreat from society. (Sun Zhaohua)

200 大忽雷

唐（618—907）
音箱宽 20.8 厘米，通高 90.8 厘米
故宫博物院藏

琴龙首，细长直柄，二轴二弦，梨形音箱，上蒙蟒皮。龙首作突目张口状，雕刻精美，表面罩金漆。弦轴雕作六棱形，嵌有牙头，插于龙首两侧。龙颔下刻篆书“大忽雷”三字，字下琴颈表面镶有一段象牙片，柄与音箱相接处嵌有碧玉一段。背面髹朱漆地彩绘描金勾莲纹，极尽富丽，富有大唐开放雄阔的时代气息。

此大忽雷与另一件传世的小忽雷，均为唐代著名画家、晋国公韩滉创制。小忽雷龙首背面刻有楷书“献，臣滉手制恭，建中辛酉春”，建中辛酉，即唐德宗建中二年（781），知其为韩滉于此时进献于宫廷。大、小忽雷的问世颇有传奇色彩，据晚唐段安节《乐府杂录》和北宋钱易《南部新书》记载，建中元年（780）韩滉奉命出使四川，途径骆谷（今陕西周至县西南）偶然发现一坚实有金石之韵的奇树，如获至宝，遂募请名匠制成大、小忽雷。

“忽雷”因其发音忽忽若雷而得名，实即龙首琵琶或二弦琵琶，是我国古代少数民族弹拨弦鸣乐器。元人马端临《文献通考》将其列入琵琶类，称之为“忽雷琵琶”。忽雷在唐代曾相当流行，后世渐趋少见，鲜有记载。（孙召华）

200 Large *Hulei*

Tang Dynasty (618-907)
W.(sound box):20.8cm, H.(in total):90.8cm
Collection of the Palace Museum

This large *hulei* (a lute-type instrument), along with another surviving piece known as the small *hulei*, was created and presented to the imperial court by Han Huang, the renowned Tang Dynasty painter. The three characters 'da hu lei' (large *hulei*) are inscribed in seal script. The term hulei includes the Chinese character for thunder, due to its deep, thunder-like sound. It was popular during the Tang Dynasty but gradually became rare in later periods, with few historical records remaining. (Sun Zhaohua)

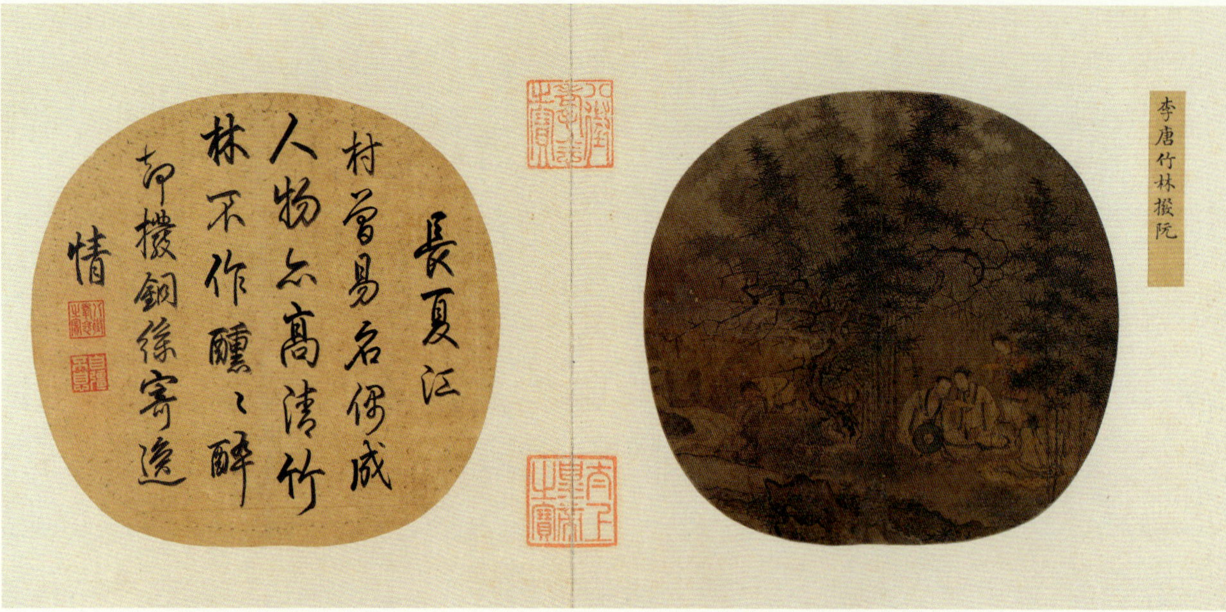

201 竹林拨阮图页

宋（960—1279），佚名
绢本，设色
纵 22.7 厘米，横 24.5 厘米
故宫博物院藏

图绘竹篁苑囿中，三位高士拨阮清娱，文宴雅集之景。其一抱阮拨弦，其一捧瓶斟浆，其一执卷清吟。三人坐于兽皮之上，神情恬淡适然，高标清韵，似有出尘之貌，尽显风雅。

高士所抱乐器为阮，相传创制于汉武帝时期，唐前称“琵琶”。西晋学者傅玄《琵琶赋》对其音色记载甚详：“弱腕忽以竞骋兮，象惊电之绝光。飞纤指以促柱兮，创发越以哀伤。”阮于古代素有“仁智之器”“乐之雅者”之谓，范晔、阮咸等皆为知名的弹阮名士。旁有奚童侍奉左右，一奉盘、一舀水。溪畔者持匜，当作“奉匜沃盥”之用。图中所绘与辽宋白釉瓷匜形制近似，同类器型也见有银制者。

兹图旧题“李唐竹林拨阮”，对开有乾隆帝题诗：“长夏江村曾易名，偶成人物亦高清。竹林不作醺醺醉，却拨铜丝寄逸情。”唐代诗人杜甫有《江村》诗：“清江一曲抱村流，长夏江村事事幽。”“曾易名”指乾隆内府曾先后收藏李唐山水长卷《长夏江寺图》及《长夏江村图》，乾隆帝认为后者图中所绘“塔势涌出，是寺非村”，遂“正其讹误”，因此“长夏江村曾易名”并非指本幅《竹林拨阮》。（喻理）

201 *Playing the Ruan in a Bamboo Grove*

Song Dynasty (960-1279), unknown painter
Album leaf, color on silk
L.:22.7cm, W.:24.5cm
Collection of the Palace Museum

This painting depicts a serene scene in a bamboo grove, where three refined gentlemen play the *ruan* for their enjoyment. The *ruan*, a plucked string instrument, was regarded as the 'instrument of benevolence and wisdom' and the 'elegant instrument'. This painting's depiction of music played amidst mountains and bamboo mirrors the elegant gatherings of literati in later Chinese garden culture. (Yu Li)

202 仕女图轴

明（1368—1644），陈洪绶
绫本，设色
纵 119 厘米，横 45.3 厘米
故宫博物院藏

此图绘二仕女坐于芭蕉叶上，正欲抚琴吹箫，神情肃穆，凝神静思。其一人微侧身，横琴于腿上，琴尚包裹于锦布中，身后石几上，瓶插菊花一枝；另一人侧颜，已作吹箫状，身后放置有蜷曲老干。二人衣带缓垂于地，衣袖、下摆褶皱层层，有长裾曳地的美感。仕女发式似为牡丹头发髻，这种发髻流行于明末清初，是一种高耸的发髻。清初尤侗诗称“六宫争学牡丹头”即指此。

据陈洪绶自题，此图画于西湖定香桥。定香桥位于今西湖“花港观鱼”附近，原为拱桥，今已改为平桥。宋朝时桥旁有定香庵，桥因此而得名。此地又为南宋内侍卢允升宅邸所在。卢氏在此开凿池塘，栽花养鱼。清康熙时在定香庵故址上筑建园亭，并有御题“花港观鱼”石刻。西湖自古为风流蕴藉之地，明末文人多在此泛湖聚饮、吟诗作画。张岱《陶庵梦忆》曾记载过陈洪绶夜游西湖，与女子载舟饮酒的韵事。（施含牧）

202 *Court Ladies*

Ming Dynasty (1368-1644), Chen Hongshou
Hanging scroll, color on silk
L.:119cm, W.:45.3cm
Collection of the Palace Museum

This painting depicts two women seated on plantain leaves, poised to play the *qin* and blow the *xiao*. Their robes are elegantly draped, with long trailing skirts that sweep the ground, creating a graceful and flowing effect. According to the artist's own inscription, this work was created at the Dingxiang Bridge near West Lake. During the late Ming Dynasty, many literati frequented the area to enjoy the lake, drink, compose poetry, and create paintings. (Shi Hanmu)

203 秋鸿图谱册

明（1368—1644），佚名
谱：纸本，墨笔
图：绢本，淡设色
纵 24.3 厘米，横 62 厘米
故宫博物院藏

明宫旧藏浙派徐门所传古琴曲《秋鸿》图谱册，先图后谱，为绘图与曲谱的合装本。图谱共四册，册面签题“琴谱”二字，并分别标注“平”“沙”“落”“雁”四字于其下。每册均钤“乾隆御览之宝”印。曲谱为纸本，每段一开配有标题；图为绢本，亦每段一开按标题内容绘制，无作者款印。

本册为《平册》。第一开是“清商调”和“夹钟清商意”两首练习曲；第二开为序曲“飞冥吟”，其后进入“秋鸿”主题。依次是“渡江”“宾秋”“依渚”“呼群”“呼芦”“悲秋”“聚沙”“宿芦”段落。在主题“秋鸿”二字下，有小字注明：“妙品夹钟清商曲，世谓清商楚望谱。瓢翁、晓山翁累删。”“瓢翁”为南宋浙派古琴名家徐天民别号，“晓山”乃其孙徐梦吉之号，梦吉子徐和仲名列洪武元年（1368）“文华堂”名琴家，并曾觐见过燕王朱棣。故此《秋鸿》图谱册可能为徐和仲所进献，成书年代早于朱权《神奇秘谱》，为存世最早的《秋鸿》版本之一。

图谱所配画作完全依照琴曲意境绘制，表现了秋日群雁在江畔、沙渚、芦苇间的活动。鸿雁或呼朋引伴，或展翅翱翔，动静间与琴音相合。作者采用横向构图展示水边沙岸、坡坂汀渚等近景，画幅虽小却引人入胜，既表现了曲谱所传达的山林意象，又反映出作曲、听曲人对于天籁之音的追求。（邢陆楠）

203 Album of *Autumn Wild Geese* with Illustrations and Scores

Ming Dynasty (1368-1644), unknown painter
Score: ink on paper
Illustrations: ink on silk
L.:24.3cm, W.:62cm
Collection of the Palace Museum

This album, originally part of the Ming court collection, features *Autumn Wild Geese*, an ancient piece of music for the *qin*. Illustrations precede each musical score, with the paintings created to faithfully reflect the imagery and mood conveyed by the composition. The illustrations depict autumn geese engaging in activities along the riverbanks, sandy shores, and reed marshes. Despite the small scale of the paintings, they are captivating, evoking the natural imagery conveyed by the music and reflecting both composer and listener's pursuit of the pure and harmonious sounds of nature. (Xing Lunan)

喜在嘉生兆有丰

研究篇

Research Chapter

重视 Garden　重视园林

朱育帆　清华大学建筑学院

园林（Garden）作为人类与自然互动的核心载体，在历史演进中经历了从中心到边缘的转变。19 世纪以来，现代主义与城市化浪潮使 Landscape Architecture 逐渐取代 Garden design 的话语主体性，传统贵族园林的宏大叙事让位于实用主义与公共空间需求。然而，园林的本质植根于人类最早的园艺（Gardening）行为——新石器时代女性通过围合空间培育植物，既满足生存需求，又寄托审美理想。这一行为催生了园林作为"理想世界"的空间实验，融合自然秩序与文化表达，成为人类文明的重要基因。

中国园林的独特性源于其"山水精神"，魏晋时期的文人雅集将自然山水升华为哲学与艺术的实践，形成"景面文心"的特质。明清计成、张南垣等人以山水画论为蓝本，创造"虽由人作，宛自天开"的叠山艺术，奠定中国园林的巅峰范式。全球视野下，欧洲的伊甸园理想、波斯伊斯兰的天堂园与中国山水园共同构成古典园林三大体系，反映不同文明对"乐园"的多元诠释。

"园事视角"为园林研究提供了动态的文化切片。故宫"乐林泉"展览以雅集、鉴藏、游山等园事活动为线索，串联起中外园林的对话。园事不仅是空间使用的载体，更是文明互鉴的媒介，揭示不同文化如何通过园林活动诠释人与自然的关系，凸显园林作为"异托邦"的避世属性与自由的精神追求。

在当代，园林（Garden）既是回溯 LA（风景园林学）本源的钥匙，也是人与自然"谈判"的场所。从社区花园到文化遗产，它承载着跨越时空的情感链接与文化基因。重视园林（Garden），需在传统智慧与现代需求间寻求平衡，以开放姿态探索人与自然的共赢之道。

一　被边缘化的 Garden

在 1937 年巴黎召开的第一届国际园林建筑师协会（AlAJM）年会上，时任大会主席阿奇尔·杜谢内（Achille Duchêne）先生语出惊人，发出了“园林艺术已死”的哀叹。这位时年 71 岁、20 世纪前 30 年法国最具代表性的园林设计师，一生坐拥 6000 余座园林设计作品，经历了法国美好时代（Belle Époque）的辉煌，与其父[1]一起复兴了勒诺特法国古典园林的艺术传统。而在一战后社会环境的巨大变迁面前，他们感受到了现代主义时代降临的阵阵压迫。阿奇尔在他自己所主持的法国行业调查报告面前不得不作此推断，因为“它”（Art of the Garden）已经从公共话语中消失了，花园（Garden）转入地下，成为个人生活环境的非官方领地。[2]

阿奇尔说的是园林艺术已死，因为他知道园林还会活下去，只是根据他的判断，世界大战和大萧条之后以往靠为贵族设计大型园林的职业是没有前途的。相反，花园必须更小、更容易维护和更实用。他非常担心未来园林建设所关心的重点已不再是他所最为珍视的那种艺术呈现。在他看来，这样的园林也就会失魂落魄，而始作俑者便是业态公共属性的主流化——那种昂贵的大型贵族园林从资本到趣味上都无法适应的属性。尽管他立刻做出了回应，他在 1935 年出版了一本名为《未来花园》[3]的书，书中试图用设计展示说明传统巴洛克风格的贵族品位略加调整是可以适应新局面的，这样古典审美便获得时空延伸的可能。但是他没有想到的是，更大的取代浪潮还在后面——现代主义艺术的降临，使他书中的那些“未来花园”没成为那种“未来”。

回首 70 年前，雄心勃勃的奥姆斯特德发起了对 Garden 的“夺权”行动，尽管找不出最满意的词汇，但他能肯定的是 Garden 一词已不能满足他所处时代城市化新兴产业拓展的方向，他需要的是一个更有力度的新词汇以便在上位博弈中争取更大的专业权益空间。Garden 被认为具有私人的和封闭的固有属性，而 Landscape 则可以成为公共的和开放的化身。Landscape Architecture 虽然不完美，但至少要比 Landscape Gardening 合适。今天的 Landscape Architecture，从专业概念角度来看，实质上是 Architecture 对 Landscape 的修辞；而 19 世纪之前的专业代名词 Landscape Gardening 中的 Landscape 则是对 Gardening 的修辞；今天 Landscape design 也取代了 Garden design 话语上的主体性。一句话概括，从专业名称可以看出，Garden 被 landscape 的主流话语所取代。另外还有一个比较有趣的现象，是 19 世纪后半叶专业主体人——Landscaper 的运用涵盖了传统的 Gardener 的专业范畴。

1　亨利·杜谢内（Henry Duchêne）（1841—1902），阿奇尔·杜谢内之父，法国著名古典园林修复专家。

2　With the ascendency of modernism, the French landscape architect Achille Duchêne (1866–1947) wrote in 1937 that the 'art of the garden was dead because it had disappeared from public discourse.The garden went underground into the unofficial territory of the personal living environment.' Laurie Cluitmans (E), *On the Necessity of Gardening: An ABC of Art, Botany and Cultivation* [M]. Amsterdam: Valiz, 2021:9.

3　《未来花园》在 1935 年出版。书中强调未来花园的社交属性，但并没有提出新时代的设计语言，并没有产生影响。

Garden（园林）作为 LA 专业和学科中最为原发性的核心，为何在进入 19 世纪之后的历史演进中屡被边缘化？是否能够通过重新审视获得新的意义？

二　Gardening 与 Garden

Garden 在中文中主要可以翻译为花园或园林，显然两种翻译对应着不尽相同的语义和所指，花园指代更加具体，园林的含义则更加广泛。而 Gardening 可以翻译为造园，这样的语言导向是我们所熟知的规划设计的语境，但在英语中 Gardening 则更接近于 Horticulture，对应的中文是园艺。

英国学者海姆斯[4]将 Gardening（园艺）和 Garden（园林）联系起来，为园林（Garden）之源的假说提供了建设性的观点，他认为人类 Gardening 的行为可以追溯至新石器时代。史前人从野外挖掘植物或采集种子，将它们种植在自己的种植园（Plantation）里，照顾它们，并在此过程中不知不觉地确保了固定和改进他们选择这些特殊植物的品质。海姆斯研究的启示让我们回到人类 Gardening 行为的核心是对于植物的种植和维护。作为栽培植物的方式之一，园艺充满了人对植物的干预动机，从收集、筛选到修正，符合了人的观赏意志。他提出园艺行为（Gardening）的首创者几乎可以肯定是妇女[5]，围篱的设置是为了保护种子的安全。围合及其内部土地经营的本初就是为这些植物的存在提供环境条件，而这个条件本身也是人工营造的。这种空间的围合正是 Garden 的空间原型，就像雷普顿定义的 Garden，围合（Enclosure）、内部培育（cultivated）和娱乐。[6]

就结果而言，园林（Garden）营造的目标在于对自身观念中的理想世界的现实再现，也是人类践行理想世界（乌托邦或异托邦）诸多方式中最具可行性的实验途径，因此它本身必然是文化性的。园林（Garden）这种人类打造的理想世界以三维空间的形式，整体上呈现出自然的样貌，布局上表现出可以判别的形式秩序，并在成熟期以时代风格的方式体现出来。与其他诸多理想世界的表达形式不同的是，植被是最具标识性的特征，植被的角色耦合了形式秩序的空间需求，也使得空间具备了显著的时间属性，而这个特点源于人类 Gardening 行为的本质。

通过 Gardening 的劳作，人类建立起与植物的特殊情感，从而获得其他行为中难以获得的愉悦和幸福。这是一种近乎本能式的园艺欢愉，但并不一定要依赖成熟的园林（Garden）环境。从 Gardening 的行为发展到专门营建用于娱乐的园林（Garden）[7]，

4 Edward Hyams（1910—1975），英国园丁、园艺家、历史学家、作家和无政府主义者。他大胆提出果树种植的历史可能要早于谷物农业的种植，种植园（Plantation）是最早的园林（Garden）类型，史前新石器时代的园丁（Gardener）是人类文明的奠基人。

5 Edward Hyams, *A History of Gardens and Gardening* [M], Westport: Praeger, 1971:9.

6 ［英］Tom Turner 著，林箐等译：《世界园林史》，中国林业出版社，2011 年。

7 笔者非常认同佩内洛普·霍布豪斯（Penelope Hobhouse）女士在《造园的故事》一书中对于园林范畴的界定。参见［英］佩内洛普·霍布豪斯著，童明译：《造园的故事》，清华大学出版社，2013 年。

还存在一个漫长的过程。与园艺植物栽培由实用（食用—药用）转向观赏相通的是，园林营建的目的由实用（果蔬等获取）转向游憩娱乐[8]，并使得园林（Garden）演化出独立完整的观念、理论和实践技术体系。纵观今天如此多元复合的LA学科内部，Garden体系仍是最具专业自主性的核心。从当代LA学科教育的发展看，主要有两条脉络，一条是从农学的园艺演化出来，总体强调植物的功效；另一条由建筑学庭园分支演化出来，总体强调空间意义的规划设计。两者在19世纪后半叶的欧美近代教育中率先完成融合。

三　重视园林

回到中国的文化语境，放之世界任何一个园林体系进行参照，中国古典园林都可谓独树一帜。尽管大众层面对于中国园林的认知仍旧大多是“小桥流水”或是关于“花花草草”，但中国造园所关注的认知核心并非仅仅是Gardening所关注的植物。直至今日，如何传承与发扬传统文化精髓，始终是中国当代风景园林学科与专业发展的元命题。

简而言之，中国园林之所以与众不同，因为其背后所笼罩的巨大的身影：山水精神。

中国山水思想发肇于先秦，定形于魏晋，成熟于唐宋。中国园林则源起于先秦之囿台，其自然山水园风格形成于魏晋南北朝时期。那时“竹林七贤”的时代刚刚谢幕，西晋政权遭“永嘉之乱”，匈奴袭都，皇族遭诛殆尽。五年后，西晋为后赵所灭，士族纷纷南渡，江南会稽成为最重要的移民中心。习惯于平原风景之大开大合的北方人，一时竟不禁为禹禅之所、鉴湖烟雨等江南山水的俊秀温润所震撼，终日慰藉，不能自拔。于是便铸就了中国文化“寄情山水”的特质，这也是中国园林文化的特质。

353年，王右军领衔41名高士的一次集会发生在会稽山阴的一条自然溪流旁，在所谓“崇山峻岭，茂林修竹，清流激湍，映带左右”的环境中，行“修禊事”。自此，兰亭雅集名垂千古，并留下了天下第一行书《兰亭序》，骈散结合，尽显魏晋风度，标榜后世。其中“流觞曲水”作为雅集的代名词流传深远，并通过造园的方式赓续不绝。

“流觞曲水”源于先秦的“祓禊”，是先民在天然水际涤身的一种祈福行为，后演化为节律性的社交聚会，并出现了“禊觞”的宴饮形式。兰亭雅集最大的不同在于并非单纯的酒肉推杯，而是在山水间以和诗的方式进行即时即兴的文学创作[9]，催生出

8　另外一点相通的是，从初始阶段的人和植物通过土地的高度合一，发展到园林中欣赏（能主之人）与劳作（能者）可能的分离，即标志着那种原发性的关系：人与植物通过身体接触的劳作方式不再是唯一的和必需的。

9　创作包括诗文、书画和乐曲等。

可以流传千古的文学作品，此所谓集之“雅”，赋予了山水以人的格调精神。750 年后，北宋刊行的崇宁本《营造法式》中出现了“流杯渠图”，可见“曲水流觞”以官方认定的方式由一处自然风景转换为既定的造园形式和内容，而我们这里看到的是一种典型的中国文化中自然的人化与人化的自然的过程，其结果是中国园林“景面文心”特质的体现。

5 世纪上半叶，一批代表性山水文学作品开始涌现，如陶潜创作《归园田居》（约 405）、《桃花源记》（约 421）；谢灵运开创山水诗派，于元嘉初年（426—428）隐居会稽创作《山居赋》；宗炳晚年隐居江陵，潜心佛道，游历山水并从事绘画创作，在《画山水序》（520—550）中提出“澄怀味象”“山水以形媚道”，首次将山水提升至“体道”的境界，标志着山水艺术理论的诞生。这些文学经典以文本的方式塑形出中国独到的山水理想，成为后世自然山水园创作的“圣经”，而所有的山水实践都最终使得“山水”成为植根于中国人心中的一种精神图腾。

如果说园林是一种观念自然的呈现，那么自然山水园就是呈现山水精神最为理想的空间载体。以自然山水造园，首要考虑、也是最根本的，是关于山水的营造，如何去经营山水？郭熙说：“山得水而活，水得山而媚。”从工程角度来看，园林中挖湖与堆山通常是个孪生行为，而中国造园更加追求山水之间的生动关系。因此，必须从构园底层确立山水环境的逻辑关系。山水营造即地貌营造，鉴于水形的容器性特征，如何造山（堆山 / 叠山 / 掇山）成为自然山水园最受关注的内容。

史学家曹汛先生将中国叠山艺术分为三个阶段[10]：自然主义，浪漫主义和写实主义。从早期的规模真山，到中期小中见大的缩摹写意，再到局部的写实创作，认为写实主义阶段标志着中国叠山艺术的最后成熟，也标志着中国造园艺术的最后成熟。

在园林中试图建造一个比拟自然界真山尺度的人工山体，无论从经济还是技术上看，都是不可行的，需要经过设计上的微缩和抽象。因此换个角度看，中国园林叠山具有两种基本的设计意识和创作思路，第一种是形态上[11]摹写真山，包括源于对自然山水的感悟而进行的写形、写神或写意的自由创作；第二种是依据间接理论进行的自由创作。其实两者并不矛盾，可能是叠加的。比如艮岳整体肖形余杭凤凰山，但并没有妨碍它依据画论展开的复合山水园林的在地创作。

第一种以真山真水为蓝本的模山范水，自古有之，东西也相通，是世界范围造园的朴素共性。例如据公元前 7 世纪末西拿基立棱柱的记述，这位亚述王在都城尼尼微无双宫旁建设了一座模仿阿马努斯山（Amanus）的高峻花园（kirimāhu）[12]；公元前

10 曹汛：《中国造园艺术》，北京出版社，2019 年。

11 平面或空间动势。

12 I made those pavilions look just right. I raised the height of the surroundings of the palace, to be a Wonder for All Peoples.I gave it the name 'Incomparable Palace'. A high garden (*kirimāhu*) imitating the Amanus mountains I laid out next to it, with all kinds of aromatic plants, orchard fruit trees, trees that enrich not only mountain country but also Chaldaea (*Babylonia*), as well as trees that bear wool, planted within it. Stephanie Dalley.*The Mystery of the Hanging Garden of Babylon*: An Elusive World Wonder Traced [M], Oxford: Oxford University Press, 2013 : 63.

2 世纪汉武帝开昆明池，筑土山以象东海蓬莱仙境的做法开启了中国园林中经典的“一池三山”模式。

第二种通过间接媒介和理论指导创作，这点相当具有中国特色。与中国自然山水园最为相关的媒介是山水文学（包括山水画论）和山水画。艺术史与园林史学界基本形成了先有山水文学，再有山水画，再有山水园的递进关系共识，反映出不同艺术形式之间存在互动关系。山水文学提供山水园创作的观念框架和叙事文本，山水画论提供了创作所需的更加具体的布局关系和建构原则，而山水画则提供的是造园叠山（尤其是石山）所需的最为直接的视觉气质和形式语言。

17 世纪的中国晚明出现两位造园巨匠：计成和张南垣，他们共同扭转了中国造园的整体风格和面貌。那个理学时代，儒道释已深度融合，中国山水画艺术也已过宋元之巅。董其昌以禅喻画提出左右中国画坛三百年的南北宗论，推崇南宗摩诘董巨、龙眠元章、大痴云林、黄鹤梅花等文人画之天真自然，后世山水创作以宗之效仿为耀。山水画风格类型化趋势至《芥子园画传》的刊行而达到高峰，画家获得了跳过真实山水的感悟转而通过间接媒介直接进行山水画创作的途径。尽管清代山水画也是异彩纷呈，但总体而言，经典艺术语言的类型化并盲目推崇导致陷入艺术上僵化的桎梏，这不免让我们想起 16 世纪欧洲文艺复兴时期的手法主义（Mannerism）艺术的历史定位。而又何其相似的是，欧洲文艺复兴时期最经典的代表作品却都是 16 世纪的手法主义园林（Mannerism Garden）。

对应中国明末艺术品位上的变化，尤其是类型化的结果为自然山水园的创作提供了更加直接而且有效的蓝本。计成自述自幼“最喜关仝、荆浩笔意，每宗之”，而曹元甫感慨计成的园林叠山“以为荆关之绘也，何能成于笔底”。这说明曹元甫通过他的假山作品可以直接判断出其背后的创作依据和风格蓝本。这也是中国山水画与中国叠山之间非常奇妙的通感对应关系，平面的绘法——山石皴法确实可以转换为真实叠山的基本语言和空间语法。通过间接媒介而不是直接仿效真实山水，恰恰使得园林叠山获得了中国特质和身份。当然，这也是最令人费解的地方，从而影响了传播和普及。因为要想真正理解并读懂中国山水园，必须理解和读懂中国山水画。

计成《园冶》所述“虽由人作，宛自天开”是对中国古典园林成就的极为精辟的概括，将独步千古。而张南垣以“平冈小坂”为代表的截山法，有似园林版的“马一角”[13]，其叠山实践重新定义了江南园林的视觉基因，也重新定义了中国自然山水园的艺术高度。发展到 19 世纪初，中国传统造园叠山以戈裕良颇具王蒙笔意的苏州环

13 这里指南宋马远取山水一角的创作手法。

秀山庄假山收官封神。

如果我们再跨越到18世纪的英国，那时一场轰轰烈烈的如绘（Picturesque）运动正在欧洲兴起一场具有颠覆性的美学革命：巨大地理尺度的自然风景园建设逐渐在欧洲风起云涌，这场变革正是源于对普桑和洛兰等人的风景绘画的刻意模仿。

四　引论

从开篇对Garden的重视，到前述对中国园林的重视，可见研究包含了对于两个基本点的观察，即园林产生的源点特征和对于风格（特质—文化基因）形成尤其是转折点的专注。从实物园林看，世界古典园林总体可以划分为三大类型：欧洲园林、波斯伊斯兰园林、东方园林（中日园林）。从园林形成源头看，尽管出现时间上略有差别，但园林（Garden）无论是在人与植物关系的本源属性上，还是在早期园林的内容特征上都展现出高度的同频和类同，而各大园林风格特质点的形成时间和动因则各不相同。

有趣的是，与中国园林的魏晋变革几乎同时，古罗马皇帝狄奥多西一世在380年颁布了《萨洛尼卡敕令》，规定基督教为罗马帝国唯一合法宗教。392年，禁止一切异教祭祀，关闭异教神庙，基督教一跃成为罗马帝国的国教。在此后基督教统治的世界里，《圣经》成为影响人类历史和文明发展最为重要的书籍之一，伊甸园（Eden）的理想也通过《圣经》确立了其在欧美园林的核心地位，成为西方园林的代名词。随后中世纪的封闭花园（Hortus Conclusus）成为最能代表伊甸园思想的园林类型。在那里，信徒们通过园艺劳作获得救赎，通达彼岸。直到18世纪自然风景园的出现。

而波斯伊斯兰园林则较为复杂。伊朗帕萨尔加德发掘出土了迄今为止最早的规则式园林，时间在公元前6世纪。居鲁士王宫波斯花园，一座反映琐罗亚斯德教《阿维斯塔》教义的四分园（Chabagh），也被公认为迄今为止考古学意义上西方园林规则式园林的源头。7世纪始，伊斯兰教势力影响到从北非到中东再到印度的广大地域，作为伊斯兰教最高经典的《古兰经》是穆斯林生活和信仰的指南，书中关于来世乐园（Jannah）的记述，既是信仰的终极目标，也是道德实践（造园）的驱动力。《古兰经》将天堂之水描绘为水、乳、蜜、酒的交融，完美承接了波斯花园四分园（Chabagh）的传统，形成“伊斯兰天堂花园”范式，普及到更加广阔的地域中。

从重视Garden，到重视园林，以全球视野拓宽比较和交流的维度，可以为我们提供更加深刻地理解园林发展本质的多元和互补性视角，这也是开展世界园林文化交流的意义。

园林营建主要提供的是容纳园林活动的空间容器，为的是生活其中人的福祉。园林活动（简称园事）是园林空间与人关系建构的纽带，从这个意义上讲，园事自身便带有探究造园中何时（when）、何地（where）、何人（who）、何为（how）、为何（why）与为何（what）的多重属性，而从可比性的园事来并行比较多元的世界园林，将为重视世界园林文化带来新鲜和有趣的视角。

故宫博物院“乐林泉：中外园林文化展”就是这样一种展览。

“乐林泉”，取自故宫博物院藏清代乾隆帝为李世倬《皋涂精舍图》诗塘的题写。林泉，象征着园林的精神内核，可谓园林的一种雅称，脱胎于农业文明中的中国园林精神。其背后的“身影”是山水，身披的“袈裟”是田园，而内心流露的却是林泉。此所谓山水天地间，田园稻粱外，林泉笔墨中。“乐林泉”准确表达出中国园林的道法自然、物我交融、景面文心和天人合一，在隐逸世界中追求精神自由的特质。

“乐林泉”即园事之乐。中国园林的园事最具代表性的是文人雅集，包括了流觞曲水、赋诗挥毫、鉴古清玩、品茗焚香、抚琴拨阮、辩经参禅、坐隐设谜等多元活动。本次展览以体现紫禁城园林[14]园事特征的雅集[15]、鉴藏、游山、静修、观花和畅音六大主题，串联起整条中外园林文化的展示动线，并以与紫禁城花园相关的展品为主线，分板块、有针对性地精选可以匹配主题的世界园林内容。如与雅集相应的波斯诗人萨迪和哈菲兹诗集及波斯细密画所体现的天堂图景；与鉴藏相应的梵蒂冈望景楼古罗马雕塑和美第奇家族园林的名人雕塑收藏；与游山相应的埃斯特台地庄园与美第奇园林洞窟场景；与静修相应的中世纪修道院园林与圣哲罗姆的野境修行；与观花相应的莫奈睡莲池与葛饰北斋的浮世绘花鸟画；与畅音相应的印度天堂园中的舞者献艺和凡尔赛宫丛林园的恩克拉多斯泉池。这些为观者带来了一次跨越东西、贯通古今的通景对话之旅。

在北京大力推动践行花园城市建设的当下，作为基础理论研究，对于 Garden（园林 / 花园）进行重新审视也具有特殊的意义。沿着海姆斯的观点即将 Gardening 行为视作构建人类文明的基础，直至发展出代表人类文明瑰宝的古典园林，再到 21 世纪的新型社区自治花园，Garden（园林 / 花园）始终是传达人与自然最为真切的情感世界的窗口和载体。正如它天然的避世属性，是置入人世间的异托邦（Heterotopia），无

14 主要包括宁寿宫花园（乾隆花园）、建福宫花园，慈宁宫花园和御花园（后苑）。

15 狭义雅集，指曲水宴饮和诗。

论是个人的、人人的还是众人的。

迪奥戈（Maria Paula Diogo）[16] 女士有个颇为深邃的概括："把园林（Garden）比喻成人和自然谈判的场所，正如所有的谈判告诉我们的一样，它们的结果不一定是双赢。"[17] 重视园林（Garden），我们既不能忘却学科和专业自身的本源，是通过什么让我们建立起与自然的情感连接，又要反复审视、理解专业固有的文化基因，在开放的和不断自我驱动更新的语境中走向人与自然的共赢之道。

16 1957 年生，葡萄牙里斯本新大学（NOVA）科技学院的技术史教授，《人类世的园林与行动力》主编。

17 Maria Paula Diogo, Ana Simões, Ana Duarte Rodrigues, Davide Scarso, *Gardens and Human Agency in the Anthropocene* [M], London: Routledge, 2020:14.

Re-examining 'Garden' and 'Landscape Architecture'

Zhu Yufan School of Architecture, Tsinghua University

Abstract: In 2023, the People's Government of Beijing Municipality issued the 'Beijing Garden City Plan (2023-2035)', and the semantics of 'Garden City' as opposed to Chengdu's 'Park City' are provoking in-depth professional thinking. Different languages often cause information attenuation and deviation in cross-linguistic translation and conceptual conversion, as seen in the Chinese translation of 'Garden' as either '花园' or '园林', and 'Gardening' as '造园' (garden-making) or '园艺' (horticulture). Beginning with Achille Duchêne's declaration on 'Death of Garden Art', this paper re-examines the historical trajectory of Landscape Architecture (LA) by tracing the evolving professional identity of the Garden by establishing the intrinsic connection between 'gardening' and 'garden'. It discusses the value source of the direct emotional relationship between people, plants and land that the Garden should have and why garden thinking forms a system of recursive inertia, and proposes that professional development should embrace an open, self-renewing paradigm to achieve a harmonious coexistence between humanity and nature.

Keywords: Landscape architecture, Garden, Gardener, Gardening, Vegetation, System Gradation, Garden City, The Spirit of Chinese Landscape, Shanshui Construct, Literati-scape in Classical Gardens, Rejoicing in Woods and Springs

从仙山到人境
——中国古代绘画中的园林意象与表达

邢陆楠　故宫博物院书画部

在中国传统绘画的分科中，以园林、庭院或花园为表现的绘画并没有形成独立的画科门类。东晋顾恺之在《论画》中，曾提到“台榭”一词，指代描绘园林风光和建筑形象的画作；宋人郭若虚《图画见闻志》则用“界画”囊括以精细手法展现亭台楼阁的作品。这些词汇概念与园林画有所重合，但无法覆盖或阐释这一主题的多样性与复杂性。借由此类作品所绘对象的主次，它们或作为人物画、行乐图的背景出现，或被并入山水、屋宇、楼阁门下，呈示出难以界定的特征。

颇具意味的是，此类题材的演变与中国古代实体园林的发展走势相契合。园林绘画在唐代发展，彼时士大夫酷爱山水，他们“移天缩地”，对自然之景加以提炼并纳入自家宅院，更亲自参与设计规划，王维的辋川别墅、白居易的白莲池，皆为后世造园典范。其中，传为王维所创《辋川图》中有文杏馆、辛夷坞等 20 景，景物“无不可游、可玩、可忘世”[1]。白居易提出“聚拳石为山，环斗水为池”的理念也成为明清时期以小见大、塑造写意小园的基本纲领。[2]受风气影响，唐代以降，描绘园居场景的画作日趋丰富，它们以点景的形式穿插错落于人物或山水画中，代表作如卫贤《高士图》、郭熙《早春图》、钱选《幽居图》等。明代中后期，文人从山野隐居、中隐[3]向市隐生活转变，江南地区尤其是吴门（今苏州）造园风气之盛，使园林由点景变为了画面的主要表现对象，相关作品层出不穷。

除此之外，中国园林和园画在艺术处理上也有着相似性。园林通过借景、障景、框景等手法，创造出层次感；绘画则通过远近、虚实、疏密等构图技巧，营造画面的深度和空间感。“虽由人作，宛自天开”[4]，园林和绘画都属于人造的自然景观，二者既相互依赖，又相互影响，许多山水画家同时也是优秀的造园师，如倪瓒之于狮子林、

1　见《石渠宝笈续编》第二十一册，元代柯九思题《宋人临辋川图》。

2　曹汛：《略论我国古典园林诗情画意的发生发展》，《中国造园艺术》，北京出版社，2019 年，第 19 页。

3　“中隐”一词出自唐代白居易《中隐》诗：“大隐住朝市，小隐入丘樊。丘樊太冷落，朝市太嚣喧。不如作中隐，隐在留司官。”

4　（明）计成著，李世葵、刘金鹏编著：《园冶》，中华书局，2018 年，第 27 页。

文徵明之于拙政园，园林设计师以画入园，画家又以园入画，园画相合，形成了独特的艺术形式。

可以说，当园林作为一个构图元素出现在画面中时，其作用在于映射人们的情感、思想和心绪，它不仅是实体园林的再现，也是人们内心对于隐逸、天人合一等境界追求的载体，更是园主和画家借以展开的艺术表达。诚然，园林画并未形成独立画科，但其意象、表达形式却广泛存在。本文将选取并关注中国古代绘画中具有园林意象的作品，以仙山洞府类绘画为引，希冀通过分析、列举不同审美形态下的画作，以及它们与园林的关联，对此题材进行阐释与梳理。

〔图1〕
东王公在昆仑山上，沂南汉墓墓门东立柱画像拓片

一　仙山与洞府

园林的营造乃至山水画的创作，其根源可追溯至秦汉时期人们对于仙境的朴素追求。彼时由于方士、儒生群体的活跃和对长生不死理念的宣扬，神仙观念、仙山信仰风行。[5] 以天界仙山（昆仑体系）和海岛仙山（东海体系：方壶、瀛洲、蓬莱）为载体的早期仙境逐步成形。然而，西极昆仑和海外仙岛毕竟遥不可及，于是“洞天福地”类的陆地仙境应运而生。这一体系融合了“洞穴崇拜”、道教“地仙”及东晋陶渊明后尤为兴盛的“隐逸”思想，构建了一个可为凡人所达，又极具神秘感的人间仙境。[6]

自此，仙山、洞穴的形象便在历代图像系统中不断出现，成为了一种特有的视觉元素和符号。

（一）东海三山与造园

先秦文献《列子》描述了五座仙山，岱舆、员峤、方壶、瀛洲与蓬莱，“其山高下周旋三万里，其顶平处九千里……其上台观皆金玉，其上禽兽皆纯缟”。[7]《十洲记》则记载了昆仑形貌：“方广万里，形似偃盆，下狭上广，故曰昆仑。”[8] 这些描述文字，顶平、上广下狭、禽兽纯缟、台观宫阙等，为仙山的主要特征，上广下狭者尚有佛教须弥山，五峰顶平、其间生长白色兽类者则有五台山，它们在图像表达方面具有一定的相似度，成为世人营造圣山、辨识仙境的关键因素。

在汉画像石刻中，频繁出现的昆仑山即呈上广下狭的柱状形象〔图1、图2〕。唐代杜甫曾作《戏题王宰画山水图歌》赞美一幅《昆仑方壶图》：“壮哉昆仑方壶图，挂

5　顾颉刚：《秦汉的方士与儒生》，上海世纪出版集团，2005年。

6　吴会、金荷仙：《“洞天福地”题材绘画中的仙境景观流变驱动力与呈现要素》，《中国园林》2022年第1期。

7　杨伯峻：《列子集释》，中华书局，1979年，第159—160页。

8　（北魏）郦道元著，陈桥驿校证：《水经注校证》卷一，中华书局，2007年，第12页。

君高堂之素壁。”[9]至明代吴门画坛仍流行着仙山图像的绘制，学者林圣智因此提出“吴派仙境”的概念，文伯仁《方壶图》是这一脉络的代表〔图3〕[10]，可见此题材的流传。

在文伯仁之后，画家吴彬也绘制过方壶山。有别于文氏笔下的三角状尖山群，吴氏方壶上部重峦耸峙、峰高而尖，下部隐于海浪、底窄而狭。这一造型看似奇诡，实有本所依。继《列子》后，晋人王嘉《拾遗记》指出：“三壶则海中三山也。一曰方壶，则方丈也；二曰蓬壶，则蓬莱也；三曰瀛壶，则瀛洲也。形如壶器。此三山上广、中狭、下方，皆如工制，犹华山之似削成。”[11]吴彬的图像综合了《列子》与《拾遗记》，山如壶器、上广下狭、群峰如刀削而成，山间云气缭绕、楼阁仙人分布其间，正似汉代昆仑图与博山炉的组合。与文伯仁等吴门画家倾向于将苏州本地的人文地理融入仙山主题不同[12]，吴彬的表现手法是复古的：其仙山图式可上溯至汉代，居中而立的方壶则继承了北宋“大山堂堂”的构图。

而上广下狭的图式又和造园时的叠石理念颇相合，与吴彬生活时代相近的计成，在其《园冶》中提到过一种叠石方式：“理宜上大下小，立之可观。或峰石两块三块拼缀，亦宜上大下小，似有飞舞势。”[13]这种手法显然源自人们对仙山的想象。参考吴彬曾以米万钟所藏怪石为本绘制长卷[14]，此《方壶图》或也与某家园中、案头奇石造型有关。

除叠石手法外，东海三山还为园中的水

〔图2〕
西王母在昆仑山上，沂南汉墓墓门西立柱画像拓片

〔图3〕
明文伯仁《方壶图》，台北故宫博物院藏

9 仇兆鳌：《杜诗详注》，中华书局，1979年，第754页。

10 林圣智：《游历仙岛——文伯仁〈方壶图〉的图像源流》，《故宫文物月刊》2018年第8期。

11 （晋）王嘉：《拾遗记》卷一，明万历二十年新安程氏刻汉魏丛书本。

12 前揭《游历仙岛——文伯仁〈方壶图〉的图像源流》。

13 前揭《园冶》，第166页。

14 （清）于敏中修，（清）窦光鼐纂：《钦定日下旧闻考》卷四十四，“太仆（米万钟）常以所蓄石，令闽人吴文仲绘为一卷”，清乾隆五十三年武英殿刻本。

〔图4〕
清袁江《蓬莱仙岛图》，故宫博物院藏

系景观提供了意象参照。最典型者便是“一池三山”模式，一池指太液，三山即方壶、蓬莱、瀛洲。据目前可考的资料来看，秦始皇首次运用人工建筑象征大海及蓬莱，汉武帝也曾于东莱郡黄县遥望蓬莱山，并筑三山于太液池中。[15] 这种布局形式为历代皇家园林继承、发展，影响深远。鉴于统治阶层对蓬莱的格外向往与追求，这一缥缈海岛跻身为东海三山中最流行的题目，至明清时期尤为富商巨贾群体所爱。

从宋人《蓬瀛仙馆图》到清代袁江、袁耀笔下的多幅大型立轴和屏风〔图4〕，各个时期的画家总倾向于在蓬莱仙山中营建楼阁。《史记·封禅书》曾专文记述方士公孙卿“仙人好楼居”的说法，将楼阁与仙人联系起来，又或与“一池三山”的造景形式相关，总之仙山楼阁逐渐成为表现蓬莱的经典图式，且这类画作中的楼阁建筑异常华美宏大，从总体布局到形体、细节均不合法度，往往并非真实，乃画家杂采历代画作中的建筑形象组合而成。[16]

（二）“洞天福地”与静修

作为仙境宇宙观的重要组成部分，“洞天福地”实际包括十大洞天、三十六小洞天和七十二福地，它们大多分布在中国境内的大小名山之间，并与现实世界有着各种各样的联系（如世人有时误入洞天，洞天居民时有造访人间）。正是这种相对隔绝、隐蔽的神秘感和在现实中可通达的真实属性，“洞穴”和“福地”成为人与仙之间沟通的桥梁，也成为修行者梦寐以求的修炼场所。

15 陈刚：《唐前蓬莱神话流变考》，华中师范大学博士论文，2011年，第70、89页。秦始皇于兰池筑蓬莱一说大多见于后期文献，虽有疑问，但此事屡见于古籍引述，被广泛接受，此处暂引为参考。

16 傅熹年：《中国古代的建筑画》，《文物》1998年第3期。

〔图 5〕
明仇英《玉洞仙源图》，故宫博物院藏

〔图 6〕
明仇英《桃源仙境图》，天津博物馆藏

《太平经》有载："洞者，其道德善恶，洞洽天地阴阳，表里六方，莫不响应也。皆为慎善，凡物莫不各得其所者。"表明洞是宇宙的象征、通往仙境的通道，能让凡人进入有别于俗世的天地，具有沟通天地、连接地脉的作用，因而被当作修仙得道的场所，信道徒能在里面谈道、修身和炼丹。[17] 明清两代，"洞天"被纳入园林造景中，尤其是标榜隐居理想的文人士大夫园林，山洞景观俨然成为他们静修的特殊空间。《园冶》中"掇山"的要领就有"岩、峦、洞、穴之莫穷，涧、壑、坡、矶之俨是"，书里更论及园中"理洞法"、洞上应堆土植树，或作台、置亭屋等景观搭建办法。[18]

仇英《玉洞仙源图》〔图 5〕《桃源仙境图》〔图 6〕便兼具园林、洞天、静修多重意涵。图中隐士们或在溶洞前停琴静坐，或在山洞中听琴静思。荀子《乐论》指出："君子以钟鼓道志，以琴瑟乐心。"抚琴听曲在士人生活中既是精神享受，又能升华心灵境界，作者以之点明修心，此为其一。其二，两图涉及了广泛流行的艺术题材《桃花源记》，文中捕鱼人通过山洞进入世外桃源，其描述正如洞天、福地之所在，后世画家绘制时，常用山洞表示两个时空的连接口。其三，两图远景绘有仙山楼阁，近景处侍童们忙着煮茶、端盘、陈设古玩，画家看似描画仙境，实则将文人雅士在庭院中品古、弹琴、雅集的活动直接挪到了"洞天福地"布景中，本质上表现的还是现实中人与园的互动。

明代"吴派仙境"多有"洞天"主题创作，代表作还有陆治《仙山玉洞图》、文嘉《二洞纪游图》等；清代宫廷绘画如《胤禛行乐图》〔图 7〕，则更为直接地表现了人与洞的关系，画中，喇嘛装的胤禛正手持念珠，在洞里打坐修行。

作为洞穴通达的目的地，福地具有度世太平、使人不死之功能。葛洪《抱朴子》有言："又按《仙经》可以精思合作仙药者，有华山、泰山……地肺山……此皆是正神在其山中，其中或有地仙之人。上皆生芝草，可以避大兵、大难，不但于中以合药也，

17 前揭《"洞天福地"题材绘画中的仙境景观流变驱动力与呈现要素》。

18 前揭《园冶》，第 154、169 页。

若有道者登之则此山神必助之为福药。”[19]后期道经如《南岳九真人传》则探讨了福地与修行的关联，“修道者若非其地，如植五谷于砂石之间，则不能成矣。虽有升飞之骨，当得福地灵墟，尔后可以变化”。文中将修道择地与农夫播植相提并论，可见修行之人对于福地的重视。[20]

〔图7〕

清人《胤禛行乐图·喇嘛图页》，故宫博物院藏

明末画家崔子忠的《藏云图》〔图8〕即与福地有关，图绘李白赴地肺山以瓶瓿贮藏浓云事。作者自题：“史称李青莲安车入地肺，负瓶瓿而贮浓云，归来散之卧内，日饮清泉卧白云，即此事也。”此地肺山正是道教“洞天福地”体系中“七十二福地”第一地，乃适合修炼的太平场域和中级仙人地仙的居住地，具有浓厚的修道、隐逸色彩。画家绘此图时正值明朝天启年间，阉党横行、政治黑暗，福地“过不逢兵戈之乱，不为豪强之侵，不近往来之冲”的属性正符合世人所追求的美好愿景，且崔子忠很有可能是道教净明道的修行与追随者。[21]由此可见，福地不仅是修行者的理想居所，更是文人雅士心中避世隐逸的象征。

〔图8〕

明崔子忠《藏云图》，故宫博物院藏

仙山和“洞天福地”对于造园理念、手法的影响深刻且绵长，同时它们与绘画的联系也分外密切。人们对山的崇拜与喜爱，体现了朴素的自然观和祈求长生的美好愿景，而可通达的福地又将山与人活动的空间连接在了一起。所谓园林就是在人居住的地方拥有一片山林意象，那么大到山川河流，小到一花一石，皆可承载其意，赏石、观花莫不如此。

按不同的审美意趣和所有者身份而言，园林可分为皇室、私人与寺庙三种，当中又以皇家园林和文人私园成就最为突出，其遗存及相关画作数量颇为可观。

二　地上的天宫

由于皇宫被视为天上宫殿在人间的映射，故有“地上天宫”之谓，本身就具备神圣空间属性，皇家园林、宫中御园更是以布局精巧闻名，往往带有模拟天宫、仙境的做法。前文所提“一池三山”即典型的皇家造园模式。帝王在这处不对外界开

19 （晋）葛洪：《抱朴子内外篇》内篇卷四，四部丛刊景明本。

20 张广保：《唐以前道教洞天福地思想研究——从生态学视角》，参见《道教教义与现代社会国际学术研讨会论文集》，上海古籍出版社，2003年。

21 邢陆楠：《〈藏云图〉研究——崔子忠的道教情思与对李白形象的再造》，《湖北美术学院学报》2020年第2期。

放的空间内宴乐、赏景、品古、静修，其造园艺术既体现了皇权至高无上、富贵享乐，又寄托着他们对长生不老的永恒向往。

其中，奇石、古玩和花木是皇家园林中的三大观赏对象，对它们的赏玩行为进而影响到了宫廷花园的营建和图绘。

（一）赏石

文震亨《长物志》提到“石令人古，水令人远”[22]，水和石是园林构建最基本的元素，造园必然离不开石；李渔认为通过垒石“能变城市为山林……自是神仙妙术”[23]，可见石对于园的重要性。赏石之乐自宋延续至清，其间王公贵族、文人雅士都对奇石特别钟爱，他们不仅在造园时置石叠石，还为小型奇石搭配精巧底座，置于厅堂、案几以作清供之用。石是他们“卧游”、隐居时的主要欣赏对象，由此生发的咏石、画石行为比比皆是。历史上耳熟能详的奇石有米芾“苍雪堂砚山”、苏轼“雪浪石”“仇池石”、宋徽宗“祥龙石”、乾隆帝“青芝岫”等。

北宋皇家御苑中最为关键的审美来源便是石。宋徽宗在位期间，曾于汴京（今开封）宫城东北隅建造艮岳，又在大内后苑营造瑞应之状的微缩景观，两处造景的基础都是石。艮岳之石大多采自洞庭西山的太湖石，共计65块，均经赵佶品题并绘入《宣和石谱》。北宋灭亡后，金人将艮岳中上好的太湖石运至中都（今北京），点缀于琼华岛上，元、明相沿不改，今北海公园琼岛上部分湖石即当年艮岳旧物。[24]

表现大内后苑奇石景观的遗存则有《祥龙石图》。此作被认为出自徽宗手，图上题识：“祥龙石者，立于环碧池之南，芳洲桥之西，相对则胜瀛也。其势腾涌，若虬龙出，为瑞应之状，奇容巧态，莫能具绝妙而言之也。”又据《大宋宣和遗事》中蔡京所记宣和殿“后有沼曰环碧”可知，画中所绘湖石并非艮岳之物，而是置于大内西北角后苑附近的一块造景石。[25]宣和殿乃徽宗起居之所，亦是其法帖绘画、图书金石收藏之要地。《宋史·地理志》云：“宣和殿，在睿思殿后，绍圣二年四月殿成，其东侧别有小殿曰凝芳，其西曰琼芳，前曰重熙，后曰环碧。”[26]环碧池与宣和殿相近，祥龙石静立于二者之间，为构成花园景致的要点。

宋代皇宫的赏石之乐、掇山之法乃至奇石本身，被明清宫廷保留、承继。北京紫禁城中的御花园、宁寿宫花园等，均巧妙采用掇山之法，以石作景增加了园林观赏的高度与长度，并为花园营造出山水意境。清宫旧藏一幅钱维城《御花园古柏图》〔图9〕，作者以写实手法表现了花园东北隅的堆秀山之景。堆秀山始建于明，垒石为山，

22 （明）文震亨撰，胡天寿译注：《长物志》卷三，重庆出版社，2017年，第65页。

23 （清）李渔：《闲情偶寄》，知识出版社，2015年，第217页。

24 单国强：《赏石与中国绘画》，《故宫博物院院刊》1999年第2期。

25 司徒元杰：《一片石亦有深处——从宋徽宗〈祥龙石图〉看中国宋代文人的赏石文化》，参见陈燮君主编：《上海文博论丛》第34辑，上海辞书出版社，2010年，第29页。

26 （元）脱脱等撰：《宋史》卷八五至卷九七，中华书局，1977年，第2099页。

〔图 9〕
清钱维城《御花园古柏图》，故宫博物院藏

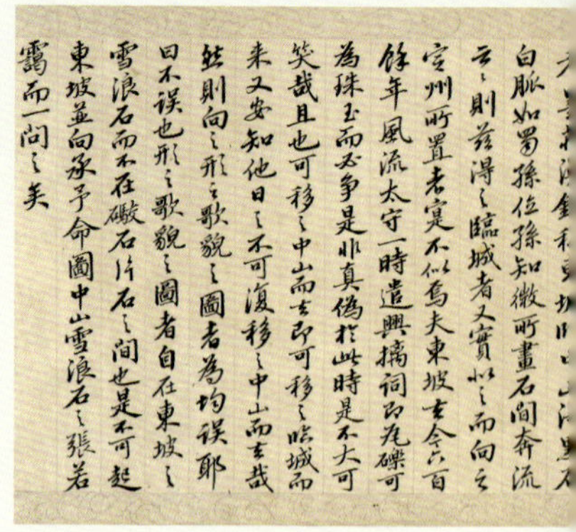

〔图 10〕
清钱维城《画雪浪石并书乾隆帝雪浪石记》（局部），故宫博物院藏

石峰造型玲珑、形貌苍古，山南营造洞穴景观，洞门上刻“堆秀”二字；图中另绘出藏书之所摛藻堂，以及凝香亭和浮碧亭，这些建筑皆现存完好。

除园林造景外，在赏玩特定奇石方面，乾隆朝宫廷也继承了宋人遗风。最具标志性的案例当为乾隆帝对雪浪石的寻找与图绘。宋哲宗元祐八年（1093），苏轼在定州（今保定）衙署后花园觅得一石，黑质白脉、中含水纹，仿若山水之中隐现浪花奔流之态，故命名“雪浪”。他对此石喜爱有加，先琢汉白玉为盆，后建“雪浪斋”亭安放，又为之赋诗作记，一时引起众多文人响应唱和。乾隆帝深爱此石，曾先后六次过定州，观赏题咏，另命张若霭绘制《雪浪石图》。乾隆三十一年（1766），直隶总督方观承又进献了一块刻有“雪浪”二字的奇石，乾隆帝看后认定是好事者所为，将其运至定州与东坡雪浪石为伴，作《雪浪石记》并让张若霭之弟张若澄绘《后雪浪石图》。[27]

现存故宫博物院的钱维城《画雪浪石并书乾隆帝雪浪石记》〔图 10〕全面记录下这段历时多年、围绕前后二雪浪石展开的君臣风雅互动。画家先仿二张笔意绘两石，后仿苏轼书风录乾隆帝《雪浪石记》，最后作跋赞颂乾隆帝考证之精。可视为清代皇家赏石之乐的最佳例证。

27 随着岁月流逝，雪浪石与盆均湮没不见，直到明万历年间重现；清康熙年间被定州知州韩逢庥移入众春园内。张若霭图现藏河北承德避暑山庄博物馆，张若澄图现存台北故宫博物院。

（二）品古、观花

品古，即鉴赏、品评古玩书画，这一雅趣与赏石、观花往往相辅相成。在精心

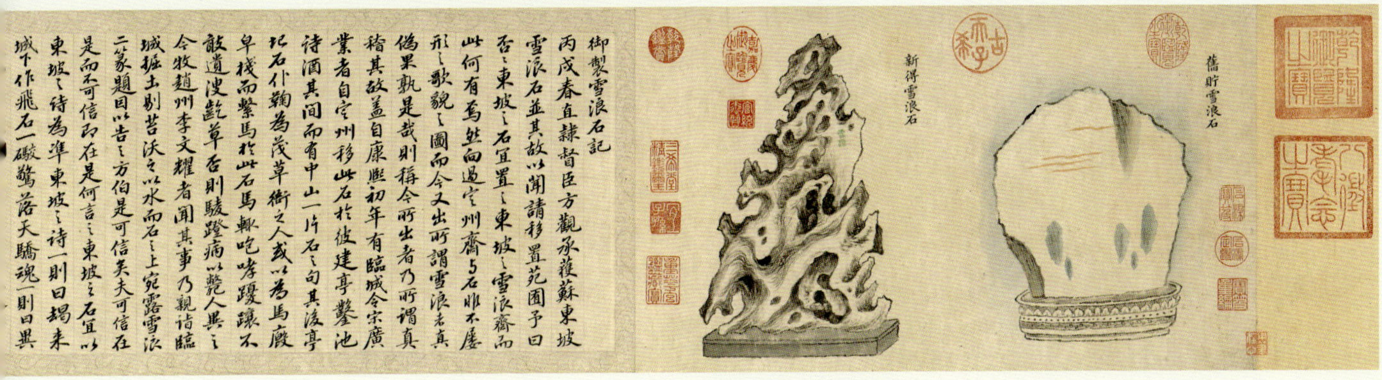

打造的景观旁，大多为宫中贮藏法书字画的重要场所，如祥龙石与宣和殿、堆秀山与摛藻堂、叠翠峰与三友轩，可见皇家造园，观赏与功能缺一不可。从皇帝贵族、文人墨客到普通民众，当他们置身庭院中时，最直观的景色就是眼前佳趣，移步异景，“一花一石，无不妙有位置”[28]。画家们则通过细微观察、师法造化，描画各类树植，从微观视角展现花园、庭院的盎然生机。

乾隆十二年（1747），乾隆帝敕命作櫝匣，将内府所藏《曹知白十八公图》《元人君子林图》及《宋元人梅花合卷》一并收贮于建福宫花园之“三友轩”中。在为三友轩赋诗时，乾隆帝写道：“乔松恒落落，新笋已亭亭。虽是梅子候，无妨通体馨。允宜诗著句，况有画传形（轩中弆曹知白十八公图、元人君子林图、宋元梅花合卷，因以命名，而窗外亦植松竹梅三种）。即景思三益，堪为砥锡型。”[29]可知，为贴合所藏书画“岁寒三友”的主题，三友轩室外遍植松竹梅三种植物。作为建福宫花园中的一处景观和赏景之处，其建筑本身、内里所藏古玩与轩外的花木、假山，共同构成了帝王观赏的对象。

清代宫廷多有表现帝王后妃在园中品古、赏花的图绘，乾隆帝常作为主角出现。《弘历观荷抚琴图》〔图11〕便描绘了乾隆帝身着汉装，在一临水亭中抚琴观花，水中荷花亭亭玉立，远景处的高山流水喻示着画中人高逸超脱的心境。另在《薰风琴韵图》〔图12〕中，乾隆帝身处庭院一角，端坐榻上，膝上横琴轻抚，周遭花团锦簇、绿柳成荫、荷花满池，侍从手捧书籍古玩，像主显然正享受着来自多层面的赏玩之乐。值得注意的是，画中屏风之上是乾隆帝御笔松梅双株，为画中之画。类似的作品还有表现他在无背景空间中赏玩各类古物的《是一是二图》屏〔图13〕，作为画中画的屏风上同样是一幅御笔墨梅。

这些作品的创作与帝王的收藏、品古活动密不可分（《薰风琴韵图》《是一是二图》分别仿自宋人《琴书乐志图》和宋人《人物图册》）。基于多方位的观赏来源，乾隆帝在表现他本人品古、赏花的画作中，加入了亲笔所绘花木。据学者考证，乾隆帝

28（明）祁彪佳：《越中园亭记》，参见陈从周、蒋启霆选编，赵厚均校订注释：《园综》下，同济大学出版社，2011年，第100页。

29（清）弘历：《御制诗三集》卷七十四，《清高宗（乾隆）御制诗文全集》第5册，中国人民大学出版社，2013年，第449页。

〔图 11〕
清人《弘历观荷抚琴图》，故宫博物院藏

〔图 12〕
清人《薰风琴韵图》，故宫博物院藏

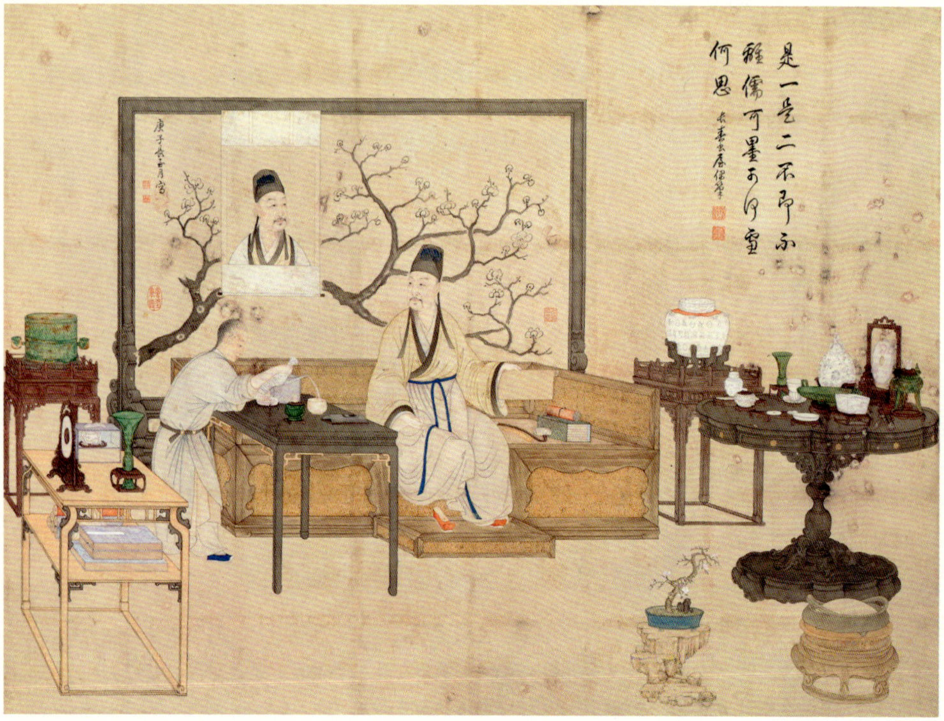

〔图 13〕
清人《是一是二图》，故宫博物院藏

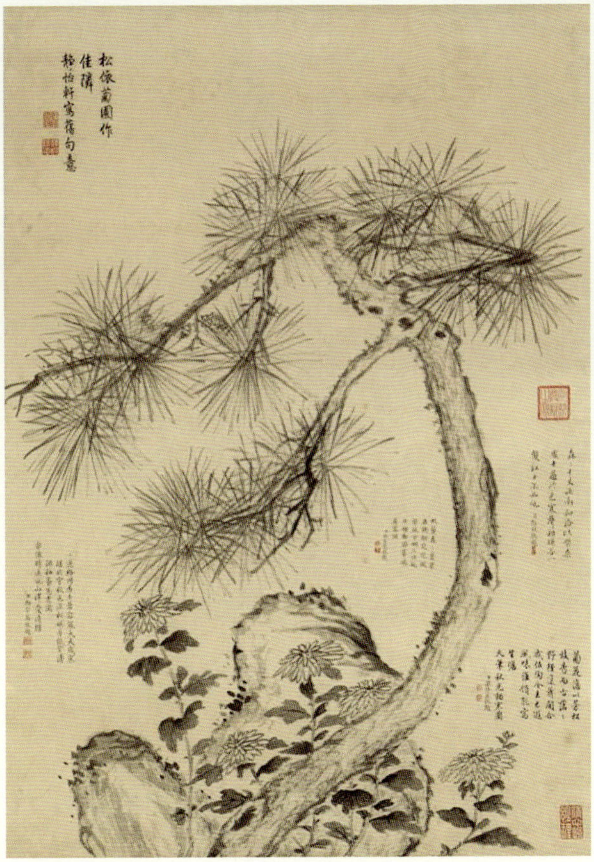

〔图 14〕
清弘历《松依菊圃图》，故宫博物院藏

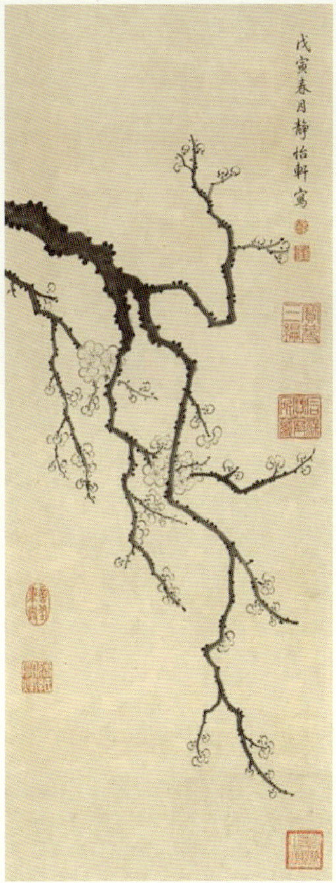

〔图 15〕
清弘历《梅花图》，故宫博物院藏

在收藏“三友”后，自己也开始大量涉猎松竹梅题材作品，《石渠宝笈》初编到三编的多幅著录就是文献佐证。[30] 故宫博物院所藏弘历《松依菊圃图》〔图 14〕、弘历《梅花图》〔图 15〕、弘历《墨梅图》〔图 16〕等作品则为实证，且这几件作品都绘制于建福宫花园的静怡轩和三友轩之中，当是乾隆帝在赏玩“三友”之余，对照轩前所植松竹梅临写而成，可视为窥探花园景色的一扇窗口。

按石之奇状、花草树植之容色本属天然造化，经人们赏鉴、掇放、栽种后，便从自然之物衍化为可游、可赏的美学主体了。而画面中有限的奇石、花木所承载的山林意象，正是作画者与观画者对于游山、游园趣味的无限想象与追求。帝王赏玩古物之所需有山林之景，他们观看的对象也是丰富立体的，这种设置与文人在园林中雅集、品古的传统一脉相承。

三　城市之山林

城市山林，是指文人雅士在城市中隐居的住所。作为拙政园的参与设计者，文徵明形容此园“虽在城市，而有山林深寂之趣”，是于极致喧嚣中寻求寂静的表征。

30 段勇：《乾隆“四美”与“三友”》，紫禁城出版社，2008 年，第 130 页。

〔图 16〕
清弘历《墨梅图》(局部),故宫博物院藏

明代以苏州为中心的江南地区,私家园林鼎盛,同一时期,吴门画派涌现出众多园林题材作品。这两者的契合,与城市经济的繁荣有关,但究其根本,在于园林与园画都是隐居文人(或有隐居山林之心的为官文人)卧游的媒介,体现了他们身在城市、心在山林的意趣。[31]

作为修身养性的空间,园林同时也是展开交游的场所,通过召集志同道合者进行以诗画创作或古玩赏鉴为主题的聚会,文人士大夫群体相互交流、议论学问、结社成团,园景与活动相互合融,更能体现园主品位。

(一)雅集

雅集是古代文人钟爱的活动,从西晋金谷园、东晋兰亭,到唐代香山、北宋西园及元末玉山,这些聚会被反复记诵,并制成各类艺术品流传于世。最著名者兰亭,其聚会方式与园林造景亦有关联。

永和九年(353)三月初三,东晋书法家王羲之、谢安等人在兰亭饮酒赋诗、结成诗集,由王羲之作序。当日正值上巳,按习俗人们要到水边举行仪式以求消灾除病。《后汉书·礼仪志》载:"是月上巳,官民皆洁于东流水上,曰洗濯祓除,去宿垢疢,为大洁。"[32] 故此兰亭雅集采用了"曲水流觞"的方式,众人列坐环曲水边,将斟满酒的觞置于水上任其漂流,酒杯停则饮酒赋诗。

这种宴饮方式在魏晋时期颇为流行,魏明帝就曾建立过一座流杯亭,以石渠引水流觞。明清宫廷亦有延续,明代宫后苑(今紫禁城御花园)乐志斋前的曲流馆、雍正帝在圆明园所建流杯亭[33],以及乾隆朝宁寿宫花园禊赏亭,都是园林景观的重要组成部分。有鉴于此,历代有关兰亭曲水流觞的图绘也十分丰富。《绘事备考》记南宋赵士遵曾绘《曲水流觞图》二,另有元赵原初《兰亭觞咏图》、明益王刻《兰亭流觞图》等。现存文献虽未见有李公麟绘《兰亭雅集图》的记载,但赵原初和益王刻

31 卢辅圣主编:《中国山水画通史》上,上海书画出版社,2014 年,第 469—471 页。

32 (晋)司马彪撰,(梁)刘昭注补:《后汉书志》志第一至第一八,中华书局,1973 年,第 3110 页。

33 此亭后被乾隆帝改为重檐八角攒尖亭,并在亭中立石柱八根,摹刻唐虞世南摹、褚遂良摹、冯承素摹《兰亭序》,柳公权书《兰亭诗》,董其昌临柳公权《兰亭诗》,戏鸿堂刻柳公权《兰亭诗》,于敏中补戏鸿堂刻柳公权《兰亭诗》所缺部分,以及乾隆帝御临董其昌仿柳公权《兰亭诗》。传拓后装裱成册,即《清拓兰亭八柱帖册》。

〔图 17〕
明尤求《修禊图》(局部)，故宫博物院藏

本均言仿自李公麟，且二者构图、画法接近，当依同一底本制成。[34] 明代流觞图盛行，故宫博物院就保存有文徵明《兰亭修禊图》、黄宸《曲水流觞图》等多本，这些作品在传为李公麟本的基础上，呈现了彼时文人在园中雅集的形态，尤求《修禊图》卷〔图 17〕是其中代表。

与李公麟本在卷首表现王羲之水亭观鹅不同，尤求本呈现了四位围案而坐的文士，他们或展卷，或倚栏，或交谈，案上置青铜觚；随后进入众人曲水流觞段落，文士们凝思觅句，身侧放着卷轴和书册；结尾处，两人从桥上走过，身后二童子带着成捆的卷轴、书册。这些新加入的元素如觚、卷轴、书册等，正是当时文人品古所观赏的对象。与前文所提《玉洞仙源图》的处理手法相似，画家直接将园林雅集的现场移到了兰亭主题中。

明代中期以后，许多致仕还乡的官员和隐逸文人热衷于营造私人书斋、草堂或园林，并在其内雅集。如董其昌所言“骨董非草草可玩也，先治幽轩邃室”[35]，品古当先治园，园林、雅集、品古彼此相连。而展现古今文人在园中品古图像的流行，正是画家对于现实生活的一种反映与记录。以仇英《竹院品古图》〔图 18〕和尤求《品古图》〔图 19〕为例，两图均描绘了庭院内、屏风前，头戴东坡帽的高士倚案而坐、展阅字画。在高士周围，文士们或一同赏玩字画，或赏爵、炉等物；梳双髻的童子或持扇侍立，或捧古物前来；院内几案上另摆放有各式瓷器、茶盏、铜鼎；园中湖石嶙峋，翠竹掩映。二者的构图模式和元素非常接近，充分体现了人、物、园之间的关联与互动。

与兰亭、品古同具吸引力的雅集题材，尚有表现北宋苏轼、米芾、黄庭坚等人在驸马都尉王诜西园集会之事，吴派画家如仇英、尤求等人都曾参与绘制。现存故宫博物院的晚明陈洪绶未竟之作《西园雅集图》，由清代画家华喦续补完成，它既是跨越时代、珠联璧合的佳作，也见证着西园雅集题目的传播与延续。

34 王中旭：《兰亭题材图像样式的传承与演变——以故宫藏兰亭题材绘画为中心》，参见故宫博物院编：《兰亭图典》，故宫出版社，2012 年，第 305 页。

35 （明）董其昌：《骨董十三说》，参见严文儒、尹军主编：《董其昌全集》第 3 册，上海书画出版社，2013 年，第 195 页。

〔图 18〕
明仇英《竹院品古图》，故宫博物院藏

〔图 19〕
明尤求《品古图》，故宫博物院藏

（二）丘园养素

宋人郭熙在《林泉高致》中曾提及："丘园养素，所常处也。"所谓养素，即指颐养生命之天然本性。这一观念多少受到老庄思想的影响，如《老子》便有"见素抱朴，少私寡欲"之句。不过，在郭熙的语境中，"养素"应不只包含道家思想，也有儒家涵养心性之意。从这个层面而言，园林不只为园居者提供了栖居之所、享乐之地，更为他们营造了一个意义完满的精神世界。

这些园子的旧貌究竟如何，现今唯有仰赖留存下来的、具有实景意义的园画窥视一二。明代苏州画家钱榖所绘《求志园图》便为我们保存了一幅已然消失的庭园影像。求志园乃明代文人张凤翼的居所，园主与吴派书画家交游密切，除请钱榖作画外，还分别请文徵明和王穉登作引首题。按卷后王世贞《求志园记》所述，画家以怡旷轩、风木堂、尚友斋为中心，其后绘出采芳径、文鱼馆、香雪廊，重点园景皆按记可索，当参考实景绘成。图中采用了略高于平视的角度描绘庭院景致：堂屋庭中以竹篱为架，藤花攀缘其上，形成天然隔墙；后院则是一片开阔的人造湖，红栏环绕，花树葳蕤，水禽嬉戏。

与《求志园图》同类型的作品还有沈周《东庄图》、文徵明《东园图》《拙政园图》等，它们大多采用俯瞰或平视的角度展示园内景观，庭院大多处于画面的中景或近景位置。除此之外，还有表现处在更为广阔背景中的庭院形貌，如《洪崖山房图》就绘出了一座处于山水间的庭院远景。图中山峦起伏，江面开阔，院落位于画面左侧一角。此别墅为胡俨怀着归隐之情在家乡江西所筑，并请画家陈宗渊绘出。图制成后，深得胡俨珍爱，并多次题写，以抒发其身居庙堂、心怀归思的情感。

在表现园景的作品中，别号图是十分特殊的类型。从作品诗文题记或相关文献资料看，凡斋室之名，往往也是斋室主人的别号。如唐寅《毅庵图》的毅庵之名，即来自朱秉忠名号；文徵明《真赏斋图》是为其好友华夏所作，华夏以“真赏”命名其斋。在这些作品中，画家的用心之处不在于表现斋室建筑和园林景致，而是根据斋室、别号的寓意，撷取可以突出主人喜好的物象（如室内陈设、古玩用具、花木等），同时着力渲染室外环境，最终呈现出具备园林意象且能够体现园主品格的庭院小景画。[36] 杜琼《友松图》是标准的别号图，画家以魏本成的别号为题作画，描绘了园中文士书写对谈、品评画卷的场景，屋外遒松三株盘旋而上，正契合“友松”之名。

不论实景画，抑或是别号图，园画所承载的园林意象，正是官员、文人所追求的隐逸理想，通过赏画、游园，他们可以暂时从政治生涯、社会价值的得失中解脱出来，而在日常的审美体验中重拾本真。总而言之，在那个时代，园林和园画不仅承担着居住和欣赏功能，更是充满了文人意趣的精神寄托。

四　余论

明代江南造园之热，不仅豪门贵族垒石筑洞、栽花植木，连小民之家也会用盆景装饰居所，以求园林之态。时人有记：“至今吴中富豪竞以湖石筑峙，奇峰阴洞，至诸贵占据名岛以凿，凿而嵌空妙绝，珍花异木，错映阑圃。虽闾阎下户，亦饰小小盆岛为玩。”[37] 这股风气与构筑模式，对清代皇家园林的修建产生很大影响，而这点又与乾隆帝的喜好、倡导有关。对于文人间的雅集、品古与诗文书画创作，乾隆帝十分向往，园林成为他施展这一乐趣的理想场合。他以皇帝的身份亲自主导，从园的整体构想、文化定位，再到景的设计，乃至陈设其中画的绘制，全方面渗透，处处都是他美学品位和文化功底的体现。

苏州的狮子林和千尺雪是他格外爱重的两处景观。狮子林始建于元至正二年

36　孔晨：《聊以画图写清居——谈以园林、庭园为题材的吴门绘画》，《故宫博物院院刊》1990年第3期。

37　（明）黄省曾：《吴风录》，明隆庆至万历刻百陵学山本。

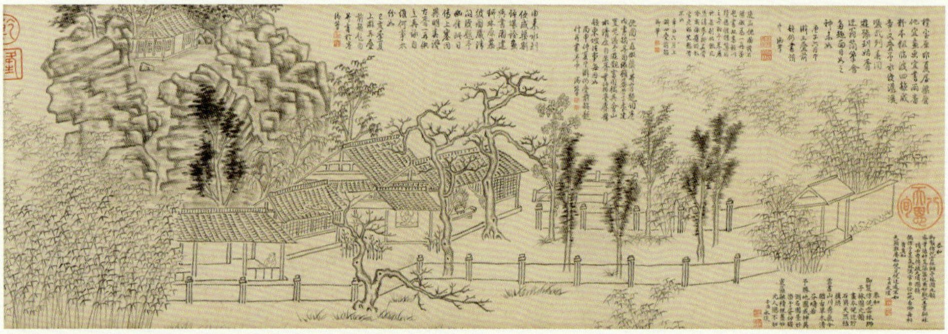

〔图 20〕
清弘历《再仿倪瓒狮子林图》(局部)，故宫博物院藏

〔图 21〕
清弘历《盘山千尺雪图》(局部)，故宫博物院藏

(1342)，乃天如禅师门人为延请其来苏州讲道而修，因园中奇石状如狮子，且天如禅师得法于天目山狮子岩，故命名“狮子林”。当乾隆帝得到一件倪瓒款的《狮子林图》后，不仅对其重新装裱、反复题写、多次临摹〔图 20〕，甚至据画寻景。在找到狮子林后，又于圆明园、避暑山庄等处仿建。千尺雪则是苏州名士赵宧光在寒山别业营造的景观，乾隆帝曾多次造访，后在北京西苑、热河避暑山庄、蓟州静寄山庄三地仿造，更亲绘《盘山千尺雪图》〔图 21〕，并命张宗苍画苏州寒山、董邦达绘西苑、钱维城绘热河的千尺雪景观，各为四图，每处合贮。

这两个经典案例充分阐释了园林与园画的关联。可以说，园林主题绘画是画家借园林的表面形象和要素进行表达的艺术形式，它既反映真实又超越真实；这种绘画形式再次反哺到园林的设计和意境借鉴上，二者互为因果。

从仙山到人境，中国古代绘画中具有园林意象的作品范围非常广泛，其中所蕴含的是人类各种美好的情感表达：理想中的仙山洞府是度世太平、长生不死愿景的载体；形状奇巧的太湖石代表着祥瑞、吉兆与山林之趣；对古董字画、珍花异木的占有、品鉴体现了皇家权威和享乐；真实或虚构的园景则是文人雅士们追求隐逸、释放天性的渠道，它们共同构建了一个立体、丰富的文化景象，并通过造园者和画家之笔呈现于世，为我们留下了绚丽多彩的园景与画作。

From Celestial Mountains to Human Realms:

The Garden Imagery and Expression in Ancient Chinese Painting

Xing Lunan　Painting and Calligraphy Department, The Palace Museum

Abstract: This paper explores the garden imagery and its expression in ancient Chinese painting, tracing its evolution from the fantastical Celestial Mountains to the human realm. By analyzing the aesthetic forms and symbolic meanings embedded in these paintings, this study examines how gardens, as both physical spaces and metaphors, reflect the cultural values, philosophical ideals, and aspirations of ancient Chinese society. Through a detailed investigation into representative works, this paper aims to provide a comprehensive understanding of the garden imagery in ancient Chinese painting and its significance in the broader context of Chinese art and culture.

Keywords: Garden Painting, Celestial Mountains and Caves, Terrestrial Palaces, Urban Mountains and Forests, Stone and Flower Appreciation, Antiquarian Gatherings

乾隆花园衍祺门，陈敬哲摄

耄期致勤倦　颐养谢尘喧

——乾隆花园里的归去来兮

李越　故宫博物院古建部

这一年是乾隆三十五年（1770），六十岁的天子走过了人生甲子。当六旬庆典的余音渐散，那位曾在登基之初许下“不敢上同皇祖纪元六十一载之数”誓言的帝王，已着手将诺言化为现实。在紫禁城内廷东路的静谧处，一场持续十载的营造即将展开——这里将诞生中国历史上最独特的太上皇宫：宁寿宫。

历代太上皇宫的营建，多是继任者草草为之的权宜之作。从唐高祖的大安宫到宋高宗的德寿宫，形制局促，意趣匮乏。乾隆帝却以“十全老人”的自信，亲手擘画了一座空前绝后的退隐之城。自乾隆三十五年始，至四十五年(1780)告竣的宁寿宫区，不仅是紫禁城建筑群中璀璨的明珠，更成为帝王权力美学的巅峰之作。今天，当晨光照向九龙影壁，这座融合前朝后寝、礼乐修身的复合空间，在宫城东隅焕发出超越时空的辉光。

一　超越前代的太上皇宫

宁寿宫，原本建于康熙二十八年（1689），是康熙皇帝奉养孝惠章皇后的寝宫。也许为了彰显“禅让”者的崇高品德，并冀其福寿绵长，乾隆帝在沿用其名的同时，以“移天缩地”的魄力重塑格局。

前朝区以皇极殿为中心，沿中轴线复刻乾清宫规制，以宁寿宫、皇极殿构建礼仪轴线，萨满祭祀的袅袅青烟与朝贺仪典的钟磬之声在此交融。后寝区纵向分为三段，中路养性殿、乐寿堂层层递进，将帝王起居升华为“颐养天和”的生命哲学，既延续了养心殿的帝王威仪，又融入了长春园里淳化轩的退隐雅趣，而景祺阁室内戏台的弦歌更添了几许生趣；东路建筑群自畅音阁至佛日楼，将观剧之乐、礼佛之诚熔铸一炉；而西翼狭长而幽致的花园，则以文人意趣编织出“身居庙堂，神游江湖”的精神秘境。〔图 1〕

这种“一宫四境”的空间结构，恰如乾隆帝晚年心境的物质投射——既执着于权力的仪式化存续，又渴慕文人化的精神超越。当暮年的帝王漫步于宁寿宫长廊，前朝的金戈铁马、后寝的笔墨春秋、东路的梵音钟鼓、西路的林泉幽趣，皆化作一曲复调的生命乐章，在紫禁城的天空下久久回荡。

二　凝固的山水诗：乾隆花园的空间叙事

在宁寿宫的整个营建过程中，乾隆帝可谓倾注了无尽的心血。从图纸到烫样，从备料到施工，每一个环节无不亲自操办——太上皇的宫殿必须奢华、气派，必须有严谨的格局，华丽的装饰，处处彰显皇家的威仪与至尊的权力。那么，花园呢？花园的设定或许不必受此束缚，或许它更为灵动与自由，它可以是生机盎然的，也可以是幽静深邃的。在这里，主人可以将他无尽的灵感与期望寄托在这片自由流淌的空间里。尽管它坐落在深宫之内，却如同主人心灵的世外桃源，充满了主人的爱好、经历与理想。这里既是他冥想静思的净土，也是他思绪飞扬的江湖。正如《园冶》所言：“世之兴造，专主鸠匠，独不闻三分匠、七分主人之谚乎？非主人也，能主之人也。”乾隆花园，对乾隆帝而言，不仅仅是一个游憩的后花园，它承载着多重功能，成为了一个多用途的复合体：倦勤斋用于观戏，符望阁用于宴集，养和精舍供奉佛事，玉粹轩供读书，遂初堂供起居……它几乎成为了皇帝园居生活不可或缺的重要组成部分。

这座花园南北长达 160 米，东西宽度却不足 40 米，研究者通常将其划分为从南至北的四进院落〔图 2、图 3〕。然而，事实上，花园与宁寿宫后中路并无明显的隔断，近十处出入

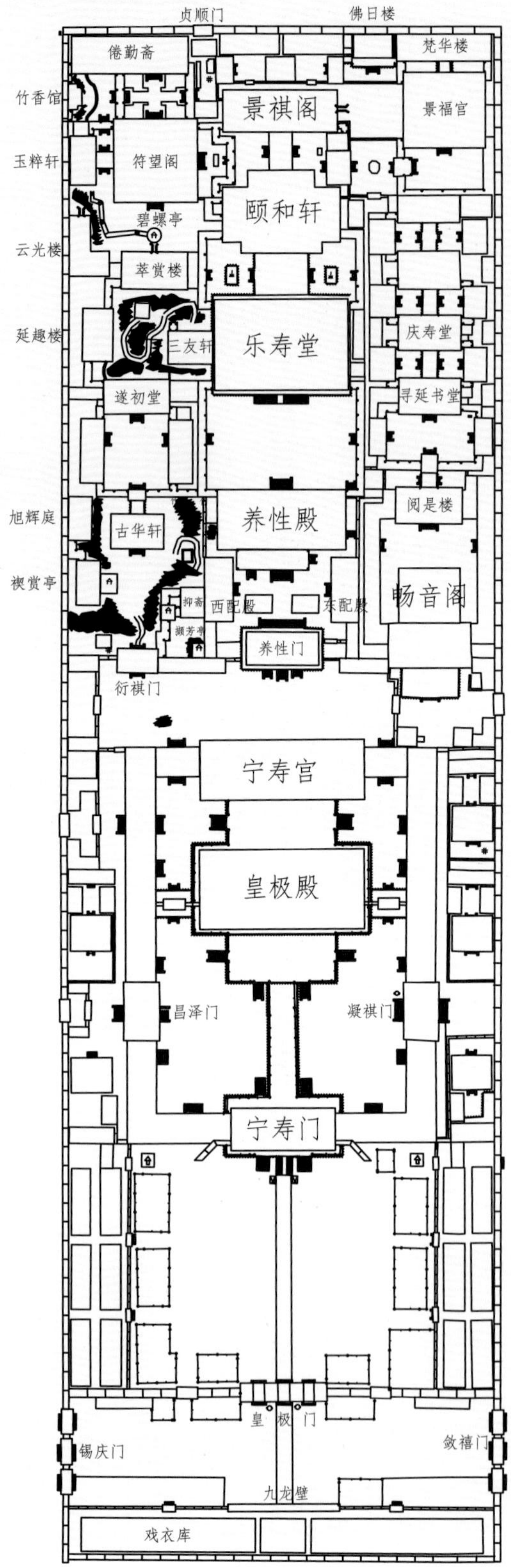

〔图 1〕
宁寿宫区总平面图

口通向中路，这些横向的通道打破了南北长轴线的单调与拘谨。这些相连的游廊与交错的门口，将中路与西路紧密相连，构建起寝宫与花园的互动与连接。

由衍祺门入内，迎面太湖石堆砌的曲径将视线引向古华轩〔图 4〕。这座敞轩因西侧古楸而得名，其建筑形制呈现出独特的“内外双相”美学：外部红柱黄瓦的卷棚歇山顶尽显皇家气象，内部黑漆描金落地明罩与木本色井字雕花天顶透出文人雅趣〔图 5〕。轩内环顾，每槽隔扇皆框景成画，形成移步换景的视觉韵律。

古华轩前西侧禊赏亭三面出抱厦〔图 6〕，东抱厦内流杯渠暗藏水法玄机，人工汲水经暗渠注入，再现《兰亭序》曲水流觞的魏晋风雅〔图 7〕。与之呼应，东侧叠山上的承露台占据制高点，形成“左山右水”的空间对仗。轩西堆石成山，旭辉庭凌山而建，其面东取势的布局暗合晨光时序，将自然天象转化为建筑语言〔图 8〕。

衍祺门右转自成独立庭院，此处通过养性殿西配房的佛堂暗门与抑斋小院形成空间渗透。抑斋书房内楠木框“玻璃眼”的改造堪称点睛之笔——乾隆帝命工匠拆去传统窗棂，代以尚属稀有的平板玻璃，既增强采光又拓展观景视野〔图 9〕。组成小院的建筑还有矩亭和蜿蜒的游廊，东南角的撷芳亭与抑斋书房相得益彰，成为绝佳的对景〔图 10〕。

从康熙八年（1669）《皇城宫殿衙署图》〔图 11〕到乾隆十五年（1750）《京城全图》〔图 12〕，可见乾隆花园格局的演变轨迹。原仁寿宫西花园的对称式布局在乾隆帝改造中升华为非对称的灵动空间：流杯亭由四角攒尖改为重檐抱厦，药栏花墙化作曲廊假山，既保留古楸树等历史遗存，又通过旭辉庭、斜廊等新建元素重构了空间节奏。

进入古华轩后，便是乾隆花园的第二进院落——一座典型的四合院。院中散置湖石，正殿名为“遂初堂”，寓意乾隆帝在位六十年后“归政得遂初愿”〔图 13〕。四合院似以民居形制承载帝王心迹：歇山顶的五开间正殿、抬高的五色石台明等细节悄然打破民间尺度，形成“皇家四合院”的特殊语汇；而东厢房“惬志舒怀”的匾额，则昭示着太上皇对归隐生活的期许。建筑群通过隐秘的侧门与宁寿宫中路相连：经东厢房屏门可直抵乐寿堂西廊，这种空间勾连既维持了帝王居所的功能性，

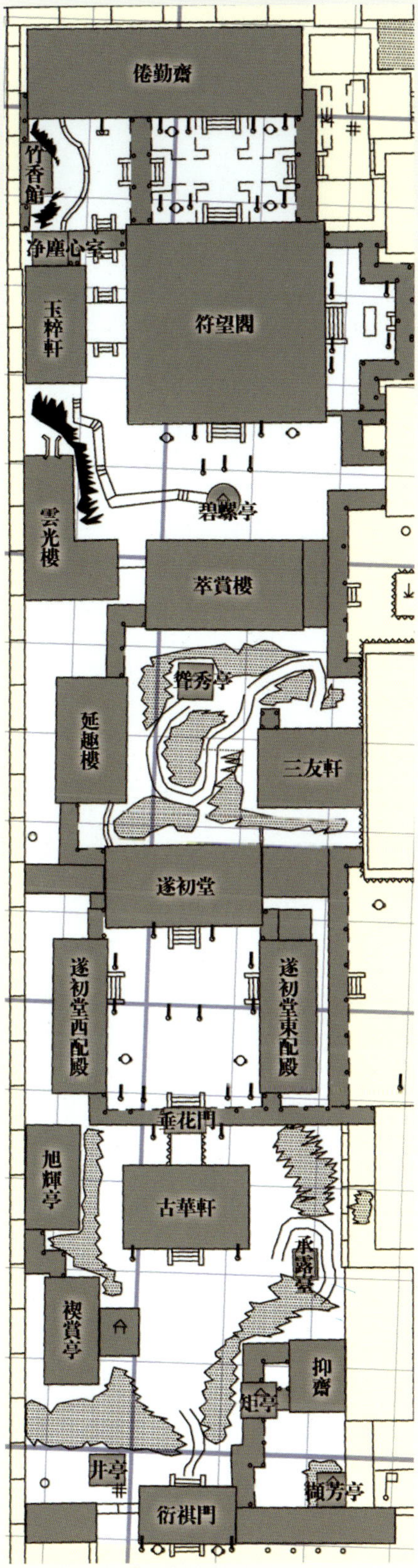

〔图 2〕
乾隆花园总平面图

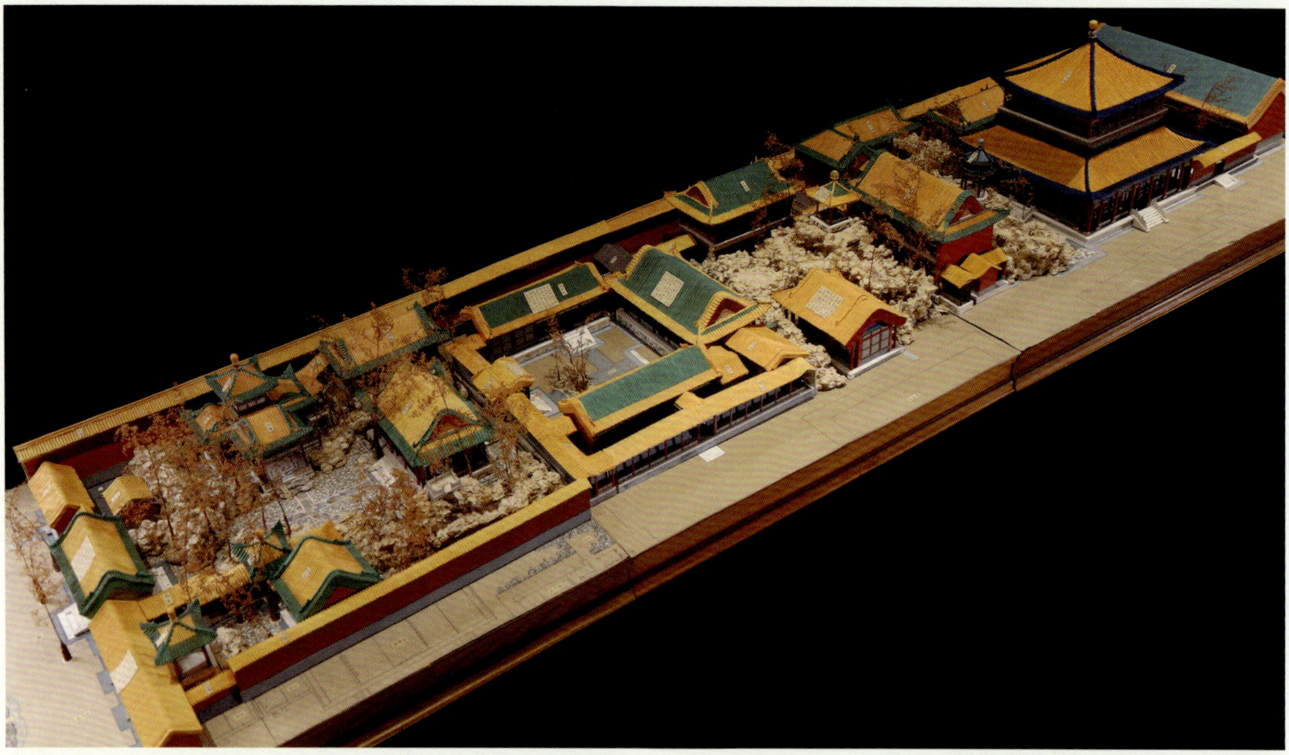

〔图 3〕

乾隆花园烫样全景，任玉丹摄

〔图 4〕

古华轩与古楸树，李越摄

〔图 5〕

古华轩内外装修，李越摄

图 3

图 4 | 图 5

图 6	图 7
图 8	图 9

〔图 6〕
禊赏亭，李越摄

〔图 7〕
禊赏亭内流杯渠，李越摄

〔图 8〕
旭辉庭，李越摄

〔图 9〕
抑斋院落，李越摄

〔图 10〕
撷芳亭，李越摄

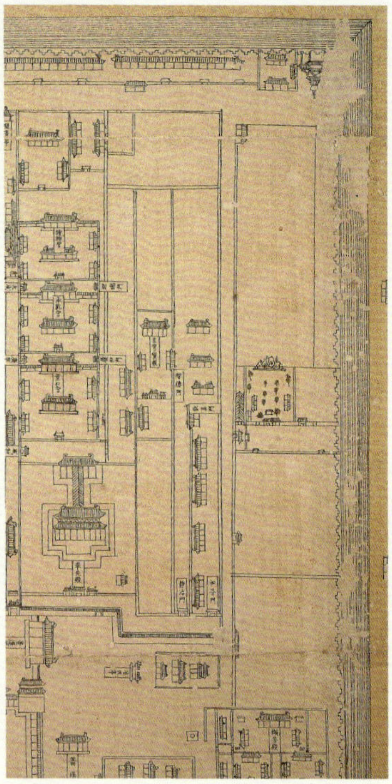

〔图 11〕
康熙八年《皇城宫殿衙署图》所示宁寿宫区风貌

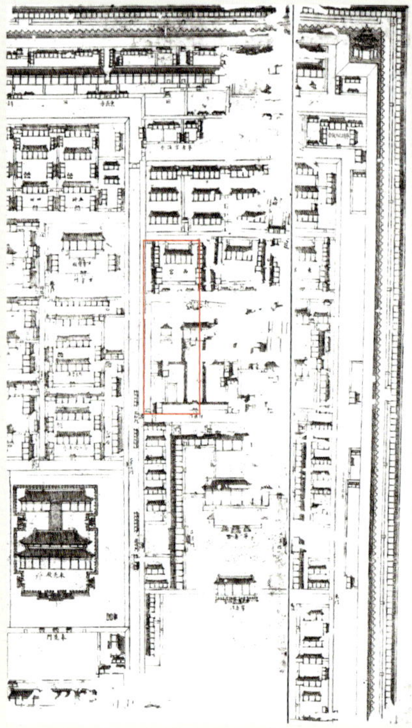

〔图 12〕
乾隆十五年《京城全图》所示宁寿宫区风貌

又在空间错位中实现了礼制空间向园林空间的过渡。可惜遂初堂各座建筑室内历经清末和近代的改造，只留下清晚期工艺略显粗糙的木装修，以及零星的早期通景画残片[1]，往昔的浮华已消失在朝代的更迭中。

遂初堂北的第三进院落堪称立体山水长卷。满庭叠山，耸秀亭〔图 14〕踞七米高山之巅，其下洞壑纵横，延趣楼〔图 15〕、萃赏楼〔图 16〕如峭壁悬阁般贴山而筑。东南隅三友轩却独辟蹊径〔图 17〕：紫檀透雕的松竹梅虬曲而洒脱〔图 18〕，将院外真景与室内雕饰融为一体，内外景致“互文”的艺术效果恰似《园冶》“纳千顷之汪洋，收四时之烂漫”的造境理想。尤为精妙的是，建筑与山体的光影游戏——萃赏楼下层终日荫翳，上层豁然开朗；三友轩却四面沐光，这种明暗辩证法在方寸间拓展了空间感知。

第四进院落呈现明显的建福宫花园克隆特征：符望阁仿延春阁，倦勤斋仿敬胜斋，玉粹轩仿凝晖堂，竹香馆仿碧琳馆，云光楼仿玉壶冰……对照着乾隆十三年（1748）宫廷画家丁观鹏的《太簇始和图》和故宫藏样式房图样〔图 19〕，可以更加形象地了解当年复制的情况。

作为空间序列的高潮，符望阁统摄全局〔图 20〕。楼阁底层四面各面阔五间，带周围廊。阁内“中心发散式”布局颇具深意：四方宝座构成的建筑核心，向四个方向发散出结构完全一致的“凸”字形空间〔图 21〕，仿佛将四座一字型殿宇背对背组织

1 “乾隆三十九年二月二十三日接得郎中德魁等押帖一件，内开本月十三日，太监胡世杰传旨宁寿宫遂初堂东配殿五间内，着艾启蒙照玉玲珑馆林光澹碧殿内西洋景改正线法，着王幼学等画，钦此。”中国第一历史档案馆藏《内务府活计档》，胶片 132。

〔图 13〕
遂初堂，陈敬哲摄

〔图 14〕
耸秀亭，陈敬哲摄

〔图 15〕
延趣楼，陈敬哲摄

〔图 16〕
萃赏楼，李越摄

〔图 17〕
三友轩，李越摄

〔图 18〕
三友轩紫檀透雕松竹梅窗，李越摄

图 13		图 14
图 15	图 16	图 17
图 18		

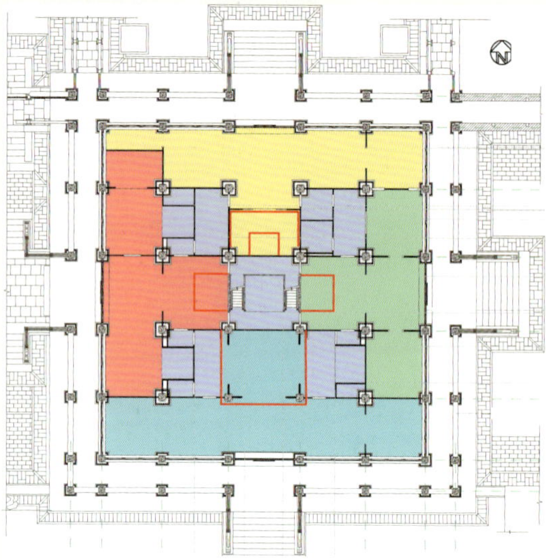

图 19

图 20 | 图 21

在一起，以各种不同工艺的隔罩巧妙地分隔空间，真真假假变幻其中，身临其境难辨东西。底层的幽暗迂回仿佛置身“迷宫”，又与顶层开敞的大空间形成强烈反差。

阁北侧倦勤斋暗藏玄机：穿过真假难辨的镜门，鸣响着岔曲余音的小戏院竟容纳着三层视觉系统——天顶藤萝画、墙面通景画与方亭戏台共同构建沉浸式剧场〔图 22〕。这种空间嵌套手法，恰如乾隆帝晚年心境的隐喻：在权力核心营造出世幻境。

阁南峰峦上，建有五瓣梅造型的碧螺亭。该亭突破了传统亭式，通过柱础、栏板、天花、彩绘等的全要素梅花意象，完成了从建筑形式到装饰母题的完整叙事〔图 23〕。

碧螺亭西翼的云光楼以曲尺形平面布局破解方正格局，下层养和精舍承载日常起居与礼佛功能，上层通过 16 尊黑漆地嵌白玉罗汉围屏构筑圣殿般的空间场域。这种垂直向度的功能划分，将世俗生活与宗教崇拜熔铸于同一建筑内，又通过独立的室外楼梯相分隔，形成独特的空间氛围〔图 24〕。

玉粹轩明间后檐整墙通景画突破了二维局限，以全景式宫廷生活画卷实现空间延展〔图 25〕。南次间山墙方窗匠心独运，将乾隆帝《得闲室》诗意具象化为“窗含湖石”的立体画框，营造出虚实相生的空间意向。这里恰如诗中描绘的场景：“真树

〔图 19〕
建福宫立样图

〔图 20〕
符望阁，李越摄

〔图 21〕
符望阁一层平面图

〔图 26〕
竹香馆，陈敬哲摄

制高点，各个单体建筑错落有致地分布，营造出独特的空间序列。在这片错动的轴线上，乾隆花园如同一部交响乐，从序曲到高潮，随着高低、疏密的变化演奏出一曲壮丽的乐章。

三　叠山理水，尽在方寸

在皇宫大内营建花园面临诸多制约，其中最致命的便是缺乏自然景观——无山可依，无景可借，甚至连水源都难以引入。

首先是叠山。掇山置石是我国独有的造园手法，始于秦汉的“一池三山”仙人思想。古代园艺家常根据山石的形态、纹理、色泽、质地等特点，塑造出峰峦、壑洞、涧流等假山形式。通过对真山真水的抽象提炼，便可在庭园内一览山河之景，足不出户，便可游历万里。

乾隆帝对此有极高的造诣。他曾自诩道：“山水之乐，不能忘于怀。”在他漫长的帝王生涯中，曾六度巡幸江南，命宫廷画师随行。所到之处，皆以江南私家园林为题，园中的叠山理水技法深深影响了他日后营建的皇家园林。

然而，要造出真假难辨的假山，首先要解决一个问题——石材储备。

在今天的北海公园琼华岛，依然能看到一些形态各异的奇石，分布在白塔山下、永安寺后殿前等处。明宣宗《御制广寒殿记》曾记载：“万岁山在宫城西北隅，皆奇石积叠以成……朕侍皇祖太宗文皇帝万几之暇，燕游于此。天颜悦怿，顾兹山而谕朕曰：'此宋之艮岳也。'"[2] 艮岳是宋代皇家园林，其设计者为徽宗皇帝。为了理想中的园林，徽宗命人搜集江南奇石，这些石头在北宋灭亡后分散各地，部分被运至金中都。其叠山技艺也随之传入北京，最终在北海假山中得以传承。

琼华岛的艮岳石经历了两次大规模的运输。第一次发生在康熙二十年（1681），为修葺瀛台；第二次是在乾隆年间，目的为宁寿宫花园的建设。乾隆帝早在十六年（1751）时便在此处立碑留题，诗文中提到"承光殿之北，孤屿瞰临北海，相传为辽之琼华岛。上多奇石，宋艮岳之遗，金人辇至于此。"他自此亲自挑选琼华岛之石，运至宁寿宫花园，为园内假山增添了艮岳园林的风貌。

〔图 27〕
遂初堂前南北太湖石的合奏，陈敬哲摄

不过，虽然运来数百块湖石，仍然无法满足花园所需的石材数量。如今园中的大量假山石大多来自京西房山的北太湖石，这些石头与来自苏州的南太湖石截然不同。北太湖石呈橙黄色，纹理清晰，适宜构建古朴雄奇的山峦。乾隆帝《文园四咏》曾道："南方石玲珑，北方石雄壮；玲珑类巧士，雄壮似强将。"而南太湖石则玲珑剔透，呈青灰色，质地坚润，常带有独特的孔洞。其中的上等石材，往往嵌空穿眼，婉转险怪，玲珑剔透，具备"透、漏、瘦、皱"之特征，但数量稀少，故显得十分珍贵。因此在乾隆花园内，各景区的单块峰石和孤石，多为琼华岛 377 块南太湖石中品相较好的上等湖石，虽鲜有江南名园奇石那样玲珑剔透、百窍通达的天然妙品，但园中置石立峰的体态风韵和气势，也堪称南太湖石中的佳品〔图 27〕。

乾隆帝对于园林中的山石使用，展现了南北风格的完美融合。园中南北太湖石互为辉映，彼此交织，相得益彰，正如《园冶》所述："取巧不但玲珑，只宜单点；求坚还从古拙，堪用层叠。"每一块石头的选材和布置，都在展示乾隆帝对山水的深刻理解与品味。

穿过花园正门，迎面便是一座玲珑的山屏。山前一条曲折的小径隐于两座山间，引导人走向未知的景致。园中景物被巧妙地隔而不断、露而不显，体现了"曲径通

2 （明）朱瞻基：《御制广寒殿记》，参见（清）于敏中等编纂：《日下旧闻考》卷三十六，北京古籍出版社，2001 年，第 567 页。

〔图 28〕
衍祺门内假山，李越摄

幽”之趣〔图 28〕。前行的过程中，景色不断变换，让人犹如进入一处深邃幽静的秘境。景越藏越深，境越掩越大。而在蜿蜒前行之中，轴线已稍稍偏转，而人在其中却浑然不觉。

走出山屏峡谷，豁然开朗，院落变得开阔，仿佛进入另一个世界。乾隆帝将江南园林中“欲扬先抑”的手法运用得淋漓尽致，层层递进的景致让人眼前一亮。正如李渔所言：“景贵乎深，不曲不深。”园中的每一个转折、每一处景致，都带有深厚的江南山林气息。

古华轩是园中最具代表性的建筑之一，其四面围绕着连绵的假山和建筑，形成了“山包院”的格局。院中假山自北向南，层层叠叠，与亭台楼阁相呼应，气势雄浑〔图 29〕。与假山紧密相连的建筑，形制多样，使空间富于变化，充满了诗意氛围。

院西由假山与禊赏亭、旭辉庭围合。假山自北向南，发起于东北隅，辗转向南堆叠成崖，恰有旭辉庭耸立崖上。随后山势极速转缓，古木林立，山石以散点的形式，

〔图 29〕
从东侧假山望古华轩，李越摄

〔图 30〕
衍祺门内通往储水房的假山，李越摄

续接禊赏亭前的湖石踏跺，经亭檐下几点山石转至禊赏亭南，此时叠山再起，婉转向东，逶迤至衍祺门内。

禊赏亭南假山，巧妙地遮住储水房，而由衍祺门通往储水房的路径，是一条极狭的山谷，仅可容一人侧身通行，仰而观之，似“一线天”〔图 30〕。此并非园中主要山道，然而中国叠山艺术中那种宽狭、主次、虚实的对比和变化，在此处体现得恰到好处。

假山细节也值得细品。如西路假山虽有禊赏亭这一大体量建筑阻隔，却巧妙地运用散点聚合的置石手法，使山势接续，泯然无迹。这里的虚实、断连、结散，颇有章法，整体山势显现出自然发展的脉络和态势。同时，建筑与山体遮挡了西侧 8 米高的宫墙，营造出活泼的园林空间，是对《园冶·掇山》“有假为真，做假成真”的绝佳诠释，从“掇山莫知山假”达到“虽由人作，宛自天开”的效果〔图 31〕。

由古华轩东望，简约空灵的落地明罩勾勒出错落嶙峋的山石组图〔图 32〕。这一带假山起脚于院北五色石墙下，由西向东折为山坳，结合养性殿“香雪”窗景，形成幽闭的竹石小景。“香雪”室内，洁白如雪的宣石，层层叠叠“镶嵌”在满墙的松竹梅贴落上。当年，乾隆帝盘腿坐在“香雪”西窗前，在读书的间歇，难免不会窥视花园的景色！

出幽谷，山势陡增如云起，循石径可直达承露台〔图 33〕。承露台四面栏杆，台

〔图 31〕
堆石成山，陈敬哲摄

〔图 32〕
透过古华轩望东侧假山，李越摄

〔图 33〕
沿山石蹬道至承露台，陈敬哲摄

面与宫墙等高，极好地打破和遮挡了宫墙呆板的线条。承露台以南，山势一折而为幽谷，再渐而成余脉，在院落东南隐入抑斋小院〔图 34〕，形成旷奥交替、大小对比极为强烈的多空间组合。大有十步千寻、百里一瞬之感。而在抑斋小院中，乾隆帝更是运用了极具诗意的石山布局。撷芳亭立于假山之上，亭与山石的融合，仿佛云朵般缥缈。这种横向堆砌的方式被称为“云头皴”，极具江南园林的诗意感〔图 35〕。乾隆帝曾在此留诗：“移石动云根，叠石看云起。”充分体现了他的园林理念。

除了充溢于山石间的诗意，乾隆帝也极懂节奏和留白。游毕第一进院落，由古华轩北进入垂花门，即为花园第二进院落。比起极尽变化的第一进，这里的景物简淡幽静得多，只用极少的湖石加以点缀，似为整个花园的低潮，而这正是园林空间设计序列所遵循的启、承、转、合手法。

随着游园的深入，景致愈发变化多端。第三进院落的假山气势磅礴，山势起伏，石洞幽深，宛如进入深山大壑。假山几乎占据了整个院落，成为园林中最具挑战性的一部分。山势高低起伏、悬崖峭壁、峰峦叠嶂，创造出丰富的空间感和层次感，给人一种深邃广阔的视觉冲击。

〔图 34〕
穿过假山至抑斋，陈敬哲摄

假山几乎占据了院落的全部，山之西北隅为全局高潮——石壁森然耸峙，陡直峭立，直逼院中延趣、萃赏二楼。山崖与楼檐距离仅一米余，又植柏木参天，于逼仄中营造巍峨之势〔图 36〕。山势虽不甚高峻险奇，但占有突出的位置，并在顶峰上建耸秀亭，以亭而增其势〔图 37〕。站在耸秀亭凭高俯视，如临深渊，颇为险峻。如果置身岩下幽谷，仰望抬头，山石层层挑出作悬崖状，似倾似倒，极具视觉冲击力〔图 38〕。正如李渔所说："使尘客仰视，不能穷其颠末，斯有万丈之势。"以近求高，虽只数米，却见万丈。

仰观俯视，大异其趣。若是行走山中，更觉其乐无穷。这里的山石结体，丰富繁缛，空间多变，悬崖峭壁、峰峦叠嶂、洞曲玲珑，山涧峡谷更是沟壑贯通，设计者在极有限的空间里，将虚实、高下、旷奥、明暗之对比发挥到极致。又在上台下洞间〔图 39〕，构建出多路径的"立体交通"——既有崎岖蜿蜒的蹬道汀步，又有开敞疏朗的平台石座；既有险峻的飞梁跨步〔图 40〕，又有小巧的拱桥连壁〔图 41〕。正如《园冶·掇山》中所描述："蹊径盘且长，峰峦秀而古，多方胜景，咫尺山林。"幽谷、深涧、悬崖、峭壁、涵洞、汀步……叠山造势手法之丰富，空间转合变化之巧妙，给原本拥塞的

图 35	图 36	图 37	图 38
图 39		图 40	图 41

〔图 35〕
云头皴，陈敬哲摄

〔图 36〕
萃赏楼与山石的关系，陈敬哲摄

〔图 37〕
于山下仰望耸秀亭，李越摄

〔图 38〕
山石层层挑出，李越摄

〔图 39〕
涵洞，陈敬哲摄

〔图 40〕
萃赏楼后飞桥，李越摄

〔图 41〕
跨步飞梁，陈敬哲摄

空间平添一种琳琅满目、应接不暇的神秘气质。

至第四进院落，假山和景色愈加雄伟壮观，山石的堆叠技艺达到了顶峰，真正展现了“叠山理水”的精髓。碧螺亭位居假山中心，为了形成变化和对比，这里假山东西狭长弯似碧螺，山上石径绵延，贯穿东西，山石之中的洞穴也随之贯通。碧螺亭的添建比整体工程晚了两年，大抵是工程进展过半，因乾隆帝某一次“步辇看工亦趁凉”[3]的造访，才有了这座梅花小亭的创意[4]。而这座小亭之于这座山，真是神来之笔，甚至不能想象，没了碧螺亭的假山，该是多么的寂寥空落。

最令人拍案处，当属第三进院落创造的“旱园水意”。7 米高的假山峡谷中，峭壁以“悬崖式”叠石法凌空挑出，形成深涧幽潭的虚空想象。这种“以实写虚”的造境智慧，暗合倪瓒《狮子林图》中“空亭翼然，群峰俱响”的留白之道。当游人在耸秀亭俯瞰“万丈深渊”，或在碧螺亭穿越“螺壳”状贯通的洞穴，早已忘却足下并无半滴活水。正如计成所言“理涧壑无水，似有深意”，乾隆帝硬是在琉璃瓦的海洋里，用叠石之法劈出了精神的山水。

这座耗费十载心血的花园，终究成了帝王最精妙的“纸上江山”。当八旬老翁漫步萃赏楼，望着楹联“四围应接真无暇，一晌登临属有余”时会心一笑，或许正是领悟了造园如治国的终极奥秘——在重重限制中创造自由，于方寸之处见天地无穷。那些凝固在太湖石纹路里的江南烟雨、燕山风骨，何尝不是另一个维度的《兰亭集序》呢？

四　南北装饰美学的对话

乾隆时期的皇家建筑工程，凝聚了全国顶尖的工艺智慧，尤以乾隆花园为集大成之作。这座皇家园林不仅以丰富的园林建筑和叠山理水的园林技艺著称，更以其室内装饰中蕴藏的江南工艺密码，展现了 18 世纪宫廷与地方的技术互动图景。

在符望阁的百宝镶嵌间，在倦勤斋的竹丝嵌玉中，江南工匠以令人惊叹的技艺解构、重构了传统宫廷装饰语汇。通过两淮盐政李质颖的奏折可知，乾隆三十八年（1773）六月至次年四月间，以扬州为核心的江南工匠团队完成了符望阁、萃赏楼等五处建筑的内檐装修工程。[5]这种以月为计量单位的高效营造，折射出江南地区成熟的产业协作体系——从紫檀、乌木等珍稀材料的统筹调度，到漆器、竹雕、双面绣等十余种工艺的协同作业，展现出工业化前夜手工艺生产的巅峰形态〔图 42〕。

3　转引自章乃炜《清宫述闻》，紫禁城出版社，2009 年，第 843 页。

4　“乾隆三十七年十一月初六日，福隆安等奏修建宁寿宫后路殿座工程估需工料钱粮数目事……东路……再查遵奉谕旨，中路游廊内镶嵌摹刻石，西路符望阁前山上补建亭座、改堆山石……”“乾隆四十年五月二十四日，福隆安等奏修建宁寿宫续添工程估需银两数目事……古华轩、碧螺梅花亭内里添安楠柏木天花……”中国第一历史档案馆藏《内务府奏销档》。

5　乾隆三十八年十月六日《李质颖奏请陛见并交卸盐政印务事》：“李质颖恭请陛见奏：奴才李质颖谨奏，为仰恳圣恩事。窃奴才前蒙皇上豢养望成擢用盐政。前岁在京仰遇，格外荣恩，垂慈教育，至厚至周，刻骨铭心，难酬万一。自叩辞回任今已二年，感激日深，孺恋倍切。伏查六七等月接奉内务府大臣寄信，奉旨交办景福宫、符望阁、萃赏楼、延趣楼、倦勤斋等五处装修。奴才已将镶嵌式样雕镂花纹，悉筹酌分别预备集料，加工选定，晓事商人，遵照发来尺寸详慎监造。今已办有六七成，约计明岁三四月可以告竣。至监务一切应办事件，奴才俱已办理全完。现在正值闲暇之时，仰恳天恩俯准，奴才于十月二十八日将盐政印信照例暂交运使廷抡护理。奴才即束装起身，趋赴阙廷，瞻仰天颜，跪聆圣训。奴才犬马恋主之忱，得以少伸，感沐高厚恩遇于无次矣。为此恭折奏恳伏乞皇上恩允谨奏。乾隆三十八年十月十八日，奉朱批不必来，钦此。”中国第一历史档案馆藏《乾隆朝汉文录副奏折》。

〔图42〕
乾隆花园令人惊叹的各种装饰手法，李越、陈敬哲、吴西羽摄

与其他宫廷建筑的内饰不同，乾隆花园的内檐装修少用金饰，多用古木、绿玉、珐琅、铜、螺钿、雕漆、琉璃、宝石等富有地方特色的装饰材料。仅在个别罩隔的局部，才点缀有铜镏金纹饰。如此多元的材料运用、华丽的装饰表现，是工匠高超技艺的见证，体现了设计者独特的艺术品位。匠师们运用平嵌、凹嵌、浮雕、透雕等技法，根据材料的自然纹理，恰如其分地发挥其质感，将不同的材质合理地结合在一起，巧妙地呈现在空间之中。薄如蝉翼的螺钿在阳光下五光十色，厚重质朴的紫檀木雕刻成苍劲的松树，纹理细腻的象牙经过染色雕刻成嫩叶的模样……这些无不源自乾隆帝钟爱的江南风格。

这些装修的精美程度，尤其体现在珍稀材料的使用上，如玉片、玛瑙、琥珀、螺钿、绿松石、牛骨、玻璃画、竹黄、竹丝、珐琅、漆纱和双面绣等装饰，这些技艺在当时的宫廷中甚为罕见。其引进大大丰富了皇家装饰风格，拓展了宫廷造办处匠师们的设计视野，使皇家装修风格融入了江南的细腻与华丽，展现出典型的地方特色。

然而，与李质颖承办的几座建筑的精美装饰相对，乾隆花园其他部分的装修依旧由内务府造办处负责，风格较为简朴。这些不同风格的对比彰显出两种装饰手法的相对特点。材料哲学方面，南匠善用紫檀、沉香、鸡翅木等珍稀硬木构建隔罩，而宫廷匠师多以楠柏为基，仅以紫檀雕件点睛；工艺取向方面，南匠创造性地将薄螺钿、竹丝镶嵌、漆艺等脆弱工艺北迁，甚至突破性地运用瓷片拼嵌等手法装饰室内，而内务府装修则较少使用镶嵌技法，除了细节处的花纹外，风格更显简洁朴素。

无论是“南匠”北作，还是“南材”北用，这一地域间的技术交流与融合，极大促进了清中期传统手工业技术的发展。江南地区丰富的工艺技术与装饰材料被引入宫廷，进一步推动了皇家建筑内檐装修工艺的发展。这样的创新与变革，不仅使清中期内廷装修展现出清新、唯美的艺术风格，也使皇家建筑的传统样式传遍地方，影响了民间建筑的装饰手法。宫廷与地方装饰技术的相互影响与传播，为清中期装饰艺术注入了新的活力，成为一段难忘的历史篇章。

五　入世与出世：帝王心象中的异托邦与精神栖居

紫禁城深处，一方叠石理水的天地悄然铺展——乾隆花园以九曲回廊、玲珑假山与诗画亭榭，在森严宫阙间构筑起一座异质空间。这里既是帝王权柄的延伸，亦是心灵栖居的秘境，折射着乾隆帝对“入世”与“出世”这一永恒命题的深邃思考。

乾隆花园的营造，本质是帝王意志的物化。从乾清宫至太和殿的朝堂轨迹，在宁寿宫区转向了另一维度：通过“移天缩地”的造园手法，江南文人园林的隐逸意境与北方皇家园林的宏阔气象在此交融。六下江南携归的不仅是园林图卷，更将士大夫阶层的文化认同纳入皇权体系。

园林，作为“异托邦”的一部分，是远离世俗的理想空间。倪瓒以容膝斋的方寸天地浓缩着庄子“曳尾涂中”的哲学，抑斋以简素书房倒映着陶潜“心远地自偏”的意境。帝王在此暂卸冕旒，化身抚琴听松的文人。这种“大隐”的智慧，恰如白居易履道坊宅园的微缩山水——在朝堂与江湖的夹缝中，以象征性归隐维系着入世者的精神自治。

在乾隆帝的花园中，景观设计不仅展示了自然的瑰丽，更隐喻着他内心的哲思与帝王的责任。园林的每一角落、每一景致，都隐藏着乾隆帝对世界的理解，对权力的操控，以及对精神世界的超越与回归。他的园林，不仅是逃避政务的庇护所，更是他追求内心宁静与超脱的圣地。隐匿其中的“出世”与“入世”，如同曲水流觞，既是对自我矛盾的回应，也是他心灵的释放与转化。

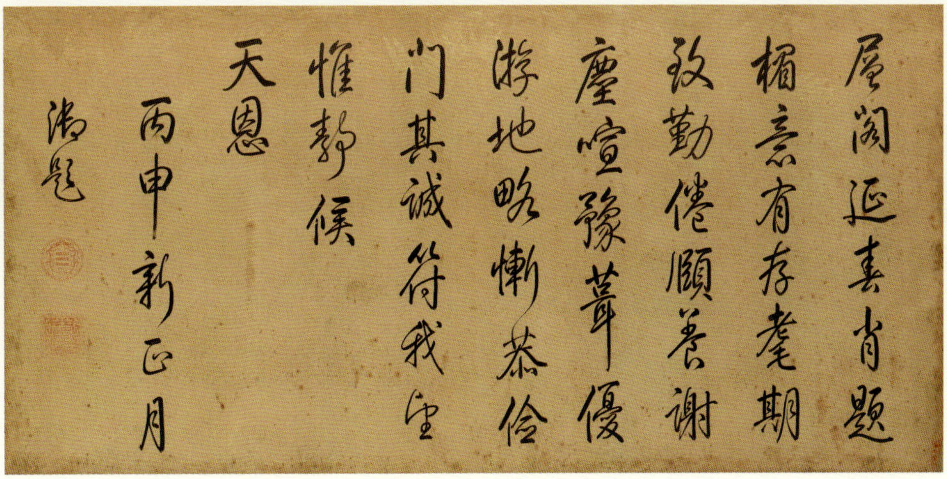

〔图 43〕
乾隆帝御笔《题符望阁》诗贴落

儒家“兼济独善”的进退之道，在叠山理水间获得空间转译；道家“无为而治”的哲学，化作曲径通幽的布局韵律；佛家“空明之境”的追求，沉淀为松风月影的禅意。王维辋川别业的诗画同构、李渔芥子园的心性书写，都可以在乾隆花园找到皇室版本。

“以出世的精神，做入世的事业。”这种超脱于尘世的精神，提醒人们在追求物质世界的同时，亦需保持内心的宁静与自持。而在政治与权力的漩涡中，乾隆帝始终未曾放下内心的渴求与追求。他的“出世精神”通过宁寿宫花园、圆明园等地的景致，悄然流露，正如他自己所言：“万物皆备于我，移天缩地在君怀。”在园林中寻找一方净土，他又通过权力和文化的交织，展现出“入世”的坚韧与智慧。

他“出世”与“入世”之间的张力，正体现在其园林设计中——一方面追寻内心的清宁与超脱，一方面又时刻不忘作为帝王的责任与威仪。这种微妙的平衡，让乾隆帝成为了一位矛盾交织的帝王，也使得他的园林成为了一个复杂而深刻的文化符号。

宁寿宫落成之后的每年腊月二十一日，乾隆帝都要在符望阁赏饭给王公大臣。之后的二十几年，每每游历花园并登临阁上，他都要在此抒情言志，为后人留下了很多诗文，不知这位皇帝心中的梦想与期盼是否真的遂其所愿、符其所望？归政后，乾隆帝真的能够“耄期致勤倦，颐养谢尘喧”吗？〔图 43〕

Weary from Diligence in Old Age, Retreat from Worldly Clamor:
A Return to Tranquility in the Qianlong Garden

Li Yue Department of Architectural Heritage, The Palace Museum

Abstract: In the 35th year of the Qianlong reign (1770), the sixty-year-old Qianlong Emperor initiated the construction of the Palace of Tranquil Longevity (*Ningshou gong*) as his future imperial retirement residence. Unlike the rudimentary retreat palaces of previous dynasties, the Palace of Tranquil Longevity (*Ningshou gong*) not only retained the ceremonial grandeur of imperial authority but also incorporated the aesthetic sensibilities of scholarly seclusion. The highlight of this complex, the Qianlong Garden, exemplifies the pinnacle of Chinese landscape artistry.

Through the meticulous arrangement of artificial mountains and waterways, winding paths, and poetic sceneries, the Qianlong Garden creates a uniquely immersive spatial experience. The seamless integration of architectural structures and landscape elements reflects Qianlong's refined aesthetic vision. The garden draws inspiration from the intimate charm of Jiangnan private gardens while maintaining the grandeur of northern imperial gardens. The interwoven presence of residential quarters, study rooms, Buddhist halls, and theatrical stages establishes an environment that is both reclusive and vibrant with daily life. Its carefully orchestrated spatial layout, fusion of northern and southern craftsmanship, and masterful use of Taihu stones collectively elevate it to the zenith of Qing Dynasty garden art.

Qianlong imbued this garden with his aspirations for retreat, striving to reconcile imperial authority with personal transcendence. However, whether this meticulously designed 'sanctuary of seclusion' truly fulfilled his vision of withdrawal from worldly affairs for peaceful cultivation (yiyang xie chenxuan) remains an open question.

Keywords: *Ningshou* Palace, Qianlong Garden, Imperial Garden, Scholarly Aesthetics, Rock and Water Landscaping, Philosophy of Seclusion

由萃赏楼望假山，陈敬哲摄

乾隆花园延趣楼，陈敬哲摄

乾隆六面与花园六事

刘仁皓　故宫博物院古建部

尽管时人常将乾隆花园、建福宫花园、慈宁宫花园和御花园并称为紫禁城内的四大花园，需要强调的是，今人所谓乾隆花园者，实不过乾隆帝为归政颐养而构筑的太上皇宫中的一隅。从传统造园角度审视，这四进正正方方的院子，连带其中二十七座充满皇家趣味的建筑，存在许多跳脱中国古典园林规制之外的设计。然而，与其规划者、使用者弘历之间紧密的联系，又使得这组院落在内在精神上有着高度的“园林性”。

目之所及，乾隆花园处处承载着乾隆帝个人的审美情趣、政治抱负，乃至人生经历。于是，在此仍沿用乾隆花园这一说法，并以其亭台楼阁、古木繁花为线索，稍窥乾隆帝其人于花园中的不同侧面〔图1〕。

一　文人与雅集

《园冶·相地》称：“市井不可园也；如园之，必向幽偏可筑，邻虽近俗，门掩无哗。”[1] 乾隆花园虽不在市井，但朝廷喧杂或许更甚，所以它的开篇正如《园冶》所言。踏入衍祺门，首先映入眼帘的便是一组山石屏障，其间有小径逶迤向前。绕过假山，视线豁然开朗，禊赏亭随即进入视线。雕刻竹枝的栏板望柱，天然纹理的大理石板，曲水蜿蜒的流杯石渠……禊赏亭中的诸多元素，对应的正是《兰亭序》所描绘的“崇山峻岭，茂林修竹”“清流激湍，映带左右”。看到此情此景，当年的乾隆帝想必会立刻将政务抛在脑后，剩下的，是对那场雅集无限向往的文人情怀〔图2〕。

乾隆帝的兰亭情结遐迩闻名。乾隆十六年（1751）第一次南巡，他便将兰亭列

1　（明）计成著，陈植注：《园冶注释》，中国建筑工业出版社，1988年，第60页。

〔图 1〕
乾隆花园远眺，陈敬哲摄

〔图 2〕
禊赏亭流杯渠，朱楷摄

〔图 3〕
禊赏亭竹纹望柱栏板，陈敬哲摄

在了行程中。一月到泰安，二月逛苏扬，三月初一在杭州已经按捺不住激动的心情，写下“即渐风光临上巳，便当觞咏赏兰亭”的句子[2]。可惜御船紧赶慢赶，到了兰亭还是已经到初八日下午，离上巳差了五天[3]。但这丝毫影响不了皇帝的好心情，站在兰亭，他接连赋诗六首，其中之一称：“即景还思晋永和，崇山真见绿嵯峨。斯人不只清谈辈，誓墓高风有足多。”[4]

憧憬的不止是雅集，更是文人誓墓不仕的清高风骨。

如此喜爱，以至年到六旬、筹划未来居所时，第一个想到的，还是十七年前造访过的那座兰亭。于是在花园的开端立起了禊赏亭，设置了流杯渠。紫禁城里难引活水，便在假山后藏个储水房，用铜管引水缸中的水入渠；北京城里不好种竹，就在大理石栏板上刻出摇曳的竹枝、竹叶〔图 3〕。甚至专门做了十一只银镀金杯和十一尊坐于莲花托盘上的五彩漆仙人[5]，让它们顺着弯曲的水道，自南向北缓缓流下。克服千难万险，那么，乾隆帝究竟是要与谁共饮呢？

可惜的是，现存的档案中没有留下明确记载。我们只知道光绪年间的流杯渠边，曾摆放有十二只绣墩[6]。它们可能是后人添改的，也可能是乾隆帝用来和古人、仙人[7]，甚至清风明月共饮的——两百年过去，或许只有禊赏亭的流杯渠还记得，当年的乾隆帝曾在这里有过怎样的雅集，曾怀过怎样的文人心事。

2 （清）弘历：《御制诗二集》卷二十五《三月朔日车驾至杭州驻跸之作》。

3 （清）弘历：《御制诗二集》卷二十五《兰亭即事》：“竹重春烟偏淡荡，花迟禊日尚敷荣（时三月八日）。”

4 （清）弘历：《御制诗二集》卷二十五《兰亭杂咏其二》。

5 《抑斋古华轩禊赏亭旭辉庭惬志舒怀陈设》光绪年抄本，第 6 页。

6 《抑斋古华轩禊赏亭旭辉庭惬志舒怀陈设》光绪年抄本，第 6 页。

7 徐斌：《与谁同坐——乾隆花园禊赏亭陈设考》，《美术学报》2023 年第 4 期。

〔图 4〕
乾隆花园第二进院落，陈敬哲摄

二　帝王与珍藏

路过禊赏亭，穿过古华轩，透过垂花门，便能看到乾隆花园的第二进院落〔图 4〕。不同于曲折婉转的一进院，这是一座平面相对简单的三合院，有一正两厢，中间以抄手游廊相连。外观看似寻常，这座院落却是乾隆花园的核心。目前已知唯一一张绘有乾隆与乾隆花园景象的清宫画作——《万国来朝图》中，他就端坐此处，等待着外国使臣的朝觐。

或许是因为它如此重要，所以乾隆帝在室内外都摆满了搜集的奇珍异宝。

室外是山石美玉。中央有青黄太湖石堆叠的影壁和“三阳开泰”主题的山子，四角则是四块造型独特的灵璧石，似山非山，似云非云。中国自古便有赏石的传统，乾隆帝也不能免俗。为了建造乾隆花园，他特意选用了北海琼岛的湖石[8]，传说那本是宋徽宗艮岳遗存的“花石纲”[9]。四块灵璧石也不简单，是乾隆帝专门挑选，从两淮千里迢迢运来的[10]——帝王权柄，可见一斑。

室内则置有书画古玩、奇技淫巧，内容包罗古今，贯穿中外。东配殿惬志舒怀有宫廷画师绘制的西洋风景主题通景画（trompe-l' oeil）[11]，正殿遂初堂室内摆有大量铜、玉、瓷器、钟表[12]，西稍间则是每年正月会安设一座巨大的鳌山灯，宽达 13 米，上面有山石、亭台、家具、牛羊、花草，乃至水法。当然，最重要的收藏还是书籍。“遂初”二字虽源自东晋孙绰作《遂初赋》，“遂初堂”这个名字却是南宋尤袤的首创。乾隆帝

8　乾隆三十七年六月初八日，《总管内务府奏为成造北海琼岛看画廊约估银两折》：“白塔西边旧有南太湖石，经宁寿宫拆运三百七十七块。”《清宫内务府奏销档》。

9　（清）弘历：《御制诗二集》卷二十九《燕山八景诗叠旧作韵》。

10　中国第一历史档案馆、香港中文大学文物馆编：《清宫内务府造办处档案总汇》第 43 册，乾隆四十五年十月二十一日，金玉作，人民出版社，2005 年，第 767 页。

11　中国第一历史档案馆、香港中文大学文物馆编：《清宫内务府造办处档案总汇》第 37 册，乾隆三十九年二月二十三日，如意馆，人民出版社，2005 年，第 112 页。

12　《遂初堂陈设档》光绪年抄本，陈 550。

沿用了这座藏书楼的名字，自然是想要在弆藏方面与尤袤一较高下的。于是我们在遂初堂陈设档中见到了《钦定天禄琳琅书目》《御制秘殿珠林石渠宝笈三编序》等书册的名字[13]。在《遂初堂口号》中，乾隆帝骄傲地宣称："芸编插架古香流，拟以他时伴课搜。别有全书贮三阁，底须录目效其尤。"[14]

皇帝下令纂修的《四库全书》有三万六千册，先期专门建了文渊、文源、文津三阁来存放——皇帝坐拥天下，可以为了珍藏的书画、刻帖、古墨，在建福宫设"四美具"，在含经堂建淳化轩，在养性殿立墨云室。如此丰功伟绩，古所未有，哪里需要像尤袤一样斤斤以书目自得？

三 隐士与山居

站在遂初堂北望，透过明亮的玻璃窗，嶙峋石壁如巨幅立轴轰然垂落——这便是乾隆花园的第三进院，一座以山为主角的院落。

中国古典园林中的山水，多是模仿山水画而成，讲究的是用"一卷代山，一勺代水"[15]。像乾隆花园第三进院假山这般充满所有角落的，几乎找不到第二个。是因为乾隆帝"俗而容拙"吗？所以即使"费累万金钱"，却依旧"山不成山，石不成石"[16]。可乾隆帝一生六次南巡，遍访江南名园，其中之佼佼者，如以掇山精妙著称的苏州狮子林、南京瞻园、杭州小有天园，更是被囫囵"搬"进了圆明园。有此过往，若说乾隆帝完全不懂工拙雅俗，恐怕是有些偏颇的。

于是，再细看这座假山。由遂初堂后的洞口向北，通过一处仅可容膝的蜿蜒石洞，走出几米，便有山石堆叠的岩壁拔地而起，形成幽谷〔图 5〕。秋冬之际，阳光透过谷口的"一线天"洒下，照亮半边峭壁——游走其间，仿佛身处林泉之中，正如乾隆帝诗中所言："假山真树友忘年，远隔红墙夏飒然。漫谓喜延山水趣，老来于此觉心偏。"[17] 心之所向，浑然不觉仍在禁宫大内的高堂邃宇之中。

郭熙在《林泉高致》中曾言："君子之所以爱夫山水者，其旨安在？丘园养素，所常处也。泉石啸傲，所常乐也。渔樵隐逸，所常适也。猿鹤飞鸣，所常观也。尘嚣缰锁，此人情所常厌也。烟霞仙圣，此人情所常愿而不得见也。"[18]

法国传教士王致诚（Jean Denis Attiret）曾坦言："中国皇帝要高度受其威严的约束……他什么也看不到，屋舍和店铺都对他关闭。人们到处都为防止他发现任何东西而支开了一张网……（他）就这样被迫生活在一种孤独之中。"[19] 被困在樊笼之中，

13 《遂初堂陈设档》光绪年抄本，第 3–5 页。

14 （清）弘历：《御制诗四集》卷五十五《遂初堂口号》。

15 （明）李渔：《闲情偶寄》清康熙刻本，第 518 页。

16 （明）李渔：《闲情偶寄》清康熙刻本，第 518 页。

17 （清）弘历：《御制诗五集》卷七十三《延趣楼》。

18 （宋）郭熙：《林泉高致》，《文渊阁四库全书》本。

19 ［法］杜赫德编，郑德弟译：《耶稣会士中国书简集：中国回忆录》第四卷，《耶稣会士和中国宫廷画师王致诚修士致达索先生的信》，大象出版社，2005 年，第 293 页。

〔图 5〕
乾隆花园第三进院落假山，陈敬哲摄

〔图 6〕
乾隆花园第三进院落假山，陈敬哲摄

于是以卷山勺水构建一方天地，将园林视作理想的载体，“不下堂筵，坐穷泉壑，猿声鸟啼依约在耳，山光水色滉漾夺目”[20]。

山青草盛，红墙高远，就算一再自警“归政以前不可耽静也”[21]，却还是忍不住“偷得浮生半日闲”〔图 6〕。

四　居士与静修

站在三进院的假山之上北眺，可以看到两栋楼阁式建筑。两栋之中，近些的是萃赏楼，远些的是养和精舍，它们室内均设有佛堂。

于园林中澡身浴德、养心养性并不罕见。东汉年间，仲长统《乐志论》称：“安神闺房，思老氏之玄虚。呼吸精和，求至人之仿佛。”[22] 明人文震亨《长物志》中也有专门讲述园林佛堂营造的章节。然乾隆花园的独特之处，在于这不到一公顷的院落中，有足足八处佛堂。[23] 八处佛堂中，养和精舍楼下和萃赏楼楼上的佛堂尤为特殊。

养和精舍的佛堂，位于曲尺形的南支东稍间，整体似仿毗诃罗（vihara）窟形式而建。想要进入其中，需越过精舍西面两组仕女、婴戏主题通景画，见过写着“此心时欲归于是，进步方超最上乘”的对联，跨过象征清净的莲花纹壶门。经历凡此种种后，站在佛堂再回首，便可看到壶门内侧高悬的乾隆帝御笔“西方极乐世界安养道场”横

20 （宋）郭熙：《林泉高致》，《文渊阁四库全书》本。

21 （清）弘历：《御制诗五集》卷五十一《延趣楼自警》。

22 （南朝宋）范晔：《后汉书》《仲长统传》，《文渊阁四库全书》卷七十九。

23 《抑斋萃赏楼养和精舍倦勤斋净尘心室玉粹轩佛堂》清抄本，陈 511。

〔图 7〕
乾隆帝御笔“西方极乐世界安养道场”横条

条〔图 7〕。整组空间体验就像是修行本身——超脱世俗，踏过莲花，最终抵达净土。此时此刻，再看墙上的腊梅天竺积雪贴落，或许就如乾隆帝《养和精舍有会》所言：“试看芽纽蕾含始，却胜嫣红姹紫罗。方寸有为及无欲，斯之未信敢言他。”[24]

萃赏楼的佛堂又不相同，设在二楼西稍间，这处佛堂内设高低床和两槽镶玻璃窗板墙。板墙与墙壁上，密密麻麻的佛画像间，有一件漆泥子佛挂龛[25]，其主尊正是文殊扮相的乾隆帝本人。他身下的坐垫上有两行藏文偈颂，译作：“睿智文殊人之主，游戏圣主法之王。金刚座上安奉足，意愿天成善福缘。”[26]

档案显示，藏传佛教中，至迟自元代起，就有赋予中原统治者“文殊菩萨（Manjusri）化身”“转轮王”称号的传统[27]。这么多皇帝，将自己装扮成了文殊模样，留影在唐卡、擦擦之中的却只有乾隆帝一人。有趣的是，他在御制诗《萃赏楼》中曾称：“层屋假山巅，翼然复洒然。不教陈御座，可以供金仙。”[28]

不能放龙椅，只能供佛像。本以为这是乾隆帝虔诚的表现，可诸多佛像之中又藏了个他自己——不是不懂“谁解能所忘，与入菩提境”的道理，只是到底无法做到“净尘心”[29]而已。

五　闲人与听戏

穿过萃赏楼，再过一座假山，便能看到乾隆花园的第四进院落。这座院落里坐落着乾隆花园除遂初堂外最重要的两座建筑——符望阁和倦勤斋。

从题名便可看出，这两座建筑寄托了乾隆帝怎样的期望：倦于勤政，渴望归政，愿符所望。如此重要，乾隆帝却选择以登基时建造的第一座园林建筑——建福宫花园为蓝本，将乾隆花园的第四进院落造得和它一模一样。就连敬胜斋中的风雅存戏台，也被原样搬来了倦勤斋，只可惜当年作画的郎世宁（Giuseppe Castiglione）已不在世，

24　（清）弘历：《御制诗五集》卷四十三《养和精舍有会》。

25　中国第一历史档案馆、香港中文大学文物馆编：《清宫内务府造办处档案总汇》第 39 册，乾隆四十一年二月五日，油木作，人民出版社，第 289 页。

26　故宫博物院、首都博物馆编：《长宜茀禄：乾隆花园的秘密》，紫檀雕画框极乐世界佛屏词条（罗文华撰），北京出版社，2014 年，第 208 页。

27　安海燕：《作为“转轮王”和“文殊菩萨”的清帝——兼论乾隆帝与藏传佛教的关系》，《清史研究》2020 年第 3 期。

28　（清）弘历：《御制诗五集》卷七十七《萃赏楼》。

29　（清）弘历：《御制诗四集》卷三十三《净尘心室》。

〔图 8〕
倦勤斋戏台及通景画，王时伟摄

只能叫王幼学照着稿子摹画，利用西方绘画的透视技法，在墙面、天花上满绘春日园景〔图 8〕。仿风雅存而建的戏台就位于通景画的西侧空间，据档案记载，当年乾隆帝会叫南府的小太监来，在此表演御制腔、太平歌词和岔曲。

与通俗作品中的“风流天子”不同，乾隆帝实际极为自律。法国传教士蒋友仁（Michel Benoist）曾在书信中提到“（接见时）皇帝总是与背后的靠垫保持一段距离，我们从未见他靠在上面或把臂肘支在上面”，“总是单独用膳，从未有任何人与他同席进餐”，“从不喝可使人极度兴奋的葡萄酒或其他甜烧酒”，“每餐饭用时从不超过一刻钟”，“证明皇帝是多么注意为鞑靼人作出榜样：必须避免一切耽于安乐的事情”。[30]

“内扰私欲纷，外引声色恣。小乃害身家，大必乱政治。”[31] 或许正因如此，明明已经归政颐养，明明这么喜爱听戏[32]，还是将戏台隐藏了起来。想要去到里面，必要通过位于西侧的寝宫，打开装作镜子的暗门，穿过一段昏暗无光的密道才能抵达。

“未臻归政称闲者，仍是循规蹈矩人。”[33] 想要做个富贵闲人，却又不敢放下帝王的责任——在乾隆花园中，我们时常能够体会到乾隆帝这般复杂的心态。

30 [法]杜赫德编，郑德弟译：《耶稣会士中国书简集：中国回忆录》第六卷《北京传教士蒋友仁神父致某先生的信》《蒋友仁神父的第三封信》，大象出版社，2005 年，第 24 页、第 60 页。

31 （清）弘历：《御制诗四集》卷四十一《题道宁斋》。

32 万依：《乾隆时期宫中文化措置》，见中国古都学会编：《中国古都研究（第四辑）——中国古都学会第四届年会论文集》，1986 年，第 177—186 页。

33 （清）弘历：《御制诗五集》卷三十五《矩亭口号》。

六　凡人与观花

追求风雅的文人、热爱收藏的皇帝、林泉高致的隐士、养心养性的居士、偶尔愉情的闲人，乾隆花园中的皇帝千千面面，但最终，他也不过是一届凡人。

是凡人，所以逃不过生离死别，阻止不了嫡子夭折、阻止不了发妻离世。

登基时，在寓意传位的密书中写下爱子名字时有多希冀，看见他去世时便有多痛苦。永琏去世的前一天，乾隆帝还在兢兢业业地工作，重审了逾百人的死刑犯人名册，只因“民命出入攸关……必须至详至慎”[34]。第二天，他便收到了爱子去世的消息。“永琏薨，辍朝五日”[35]，《清实录》仅有七字，却像用尽了所有气力。

可惜的是，就连为儿子默哀五日这样简单的愿望，乾隆帝最终也没有实现——时年江南歉收、安徽大旱、漕运堵塞，天下的臣民还在等着他们的皇帝，可怜的父亲只能“至撷芳殿奠酒，举哀毕，伫立门外，目送金棺良久”[36]，便要回宫继续他的工作。

九年之后，除夕之夜，乾隆帝在保和殿前赐宴外藩时，他的第二个嫡子永琮去世。这个孩子在佛诞日出生，曾被寄予厚望。然而再多宠爱，也抵不过命运——作为皇帝，甚至没有时间为他的死而悲痛，而是第一时间站出来，说明此事非上天责备，盖因“本朝……未有以元后正嫡绍承大统者……行先人所未曾行之事，邀先人所不能获之福，此乃朕过耶”[37]。第二天，他还要照常“御太和殿受朝，作乐宣表如仪”[38]。

若说永琏、永琮之死对乾隆帝造成了巨大打击，那么他的发妻——孝贤皇后的死，无疑永久地改变了他。

乾隆十三年（1748），在人生中第一次去封禅之地泰山的回程路上，富察氏抛下皇帝，溘然长逝。发妻的离开，让他陷入前所未有的悲痛。大批官僚因在皇后丧葬一事上的表现而被迁怒，超百人被训斥、降革、罚俸，甚至处死。就连皇帝的儿子也不例外，只因未表露哀伤，永璜和永璋就被斥责“此二人断不可承续大统”[39]。皇后棺椁离开紫禁城的那天，皇帝悲痛地写道：“因参生死俱归幻，毕竟恩情总是空……早知失子兼亡母，何必当初盼梦熊。”[40]

修身、齐家、治国、平天下，《大学》里说的这些乾隆帝八岁就能背诵。可在平了天下之后，皇帝却骤然发现，家没有了。从此，看花不再是花。

古华轩里精美的皮球花，遂初堂内小巧的樱花，延趣楼里瓷片上的西番莲，符望阁里沉香雕成的佛手、百宝镶嵌的牡丹，还有倦勤斋里双面绣上的水仙和小戏台隐天蔽日的藤萝……乾隆帝将这座退位用的花园装点得如此琳琅满目，或许就是想

34《清高宗纯皇帝实录》卷七十八，乾隆三年十月上。

35《清高宗纯皇帝实录》卷七十八，乾隆三年十月上。

36《清高宗起居注》卷三，乾隆三年十月。

37《清高宗起居注》卷六，乾隆十二年十二月。

38《清高宗纯皇帝实录》卷三百七，乾隆十三年正月上。

39 戴逸：《乾隆的家庭悲剧及有关的政治风波》，《清史研究通讯》1986年第1期。

40（清）弘历：《御制诗二集》卷三《大行皇后移殡观德殿感怀追旧情不自禁再成长律以志哀悼》。

向世人、向上天展现他的功绩、他的盛世，展现他是受到上天助佑的人，就像他的年号一样。

他是如此用力，以至人们每每走进乾隆花园，都能幻视到一位垂垂老矣的皇帝，在室外凛冽的寒风[41]、室内虚假的繁花中，捧着过去的珍宝——依旧自负、依旧自满，可逐渐老去的躯体，终于掩盖不住“乾隆”二字之下，名为弘历的凡人。

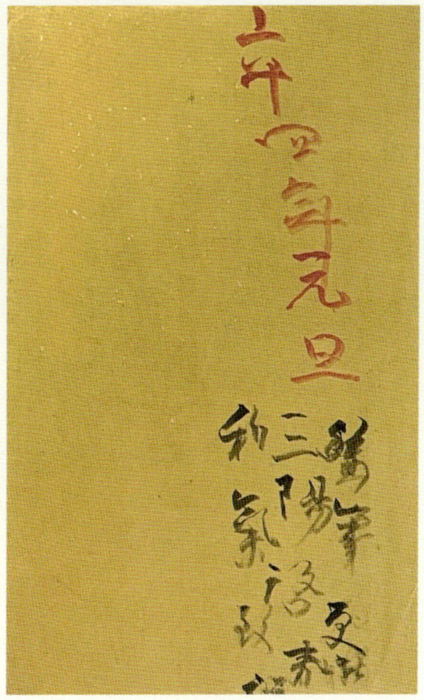

〔图9〕
乾隆帝最后的元旦开笔黄笺

太上皇宫落成时那么兴奋，可第二年，母亲就去世了。两位正妻，也早就已经撒手人寰。十七个儿子、十个女儿，大多比他先一步离开[42]。大臣们也走了，信任的、敌对的，曾那么憎恶的鄂尔泰、张廷玉，还有备受信赖的傅恒、阿桂，只剩下了和珅还在世——贪婪成性，但就这样吧。就连天下也不太平了，当年大小金川、准部、回部都不在话下，现在小小的白莲教竟然拖了这么久。若是孝贤皇后在，若是永琏在，若是老臣们在，若上天真的眷顾，一切是否会不一样？

“射鹄廿发中十九，抚茧蝇头书白手。射已前园泐以诗，书则西清贮之久。慢论臂病射弗为，书亦眼昏小字丑。文武臣中昔陪侍，经亲见者今稀有。无我质我弗重能，忆昔浑如滕诳口。”[43]——弓也挽不动了，字也写不好了，连记得我峥嵘岁月的人也没有了，不如归去罢〔图9〕。

退位的第四个年头，乾隆帝最后一次在乾清宫庆祝元旦。第二天，“执讯迅获丑，都同逆首来”[44]，伴随着这首《望捷》，皇帝在等待镇压白莲教起义的消息中，永久地合上了眼睛。《实录》称：“频望西南，似有遗憾。”[45]

“因参生死俱归幻，毕竟恩情总是空。”花开花谢，人生终不过一场幻梦。便是“层阁延春肖”，依旧“题楣意有存”[46]——乾隆花园的第四进院仿建福宫花园而建，可它永远成不了建福宫，它也不应当是建福宫。就像经过竹香馆漫长的、黑暗的通道，穿过倦勤斋的通景画，来到如梦如幻的戏台之中，返身回看——二十五岁的乾隆帝就站在那里，站在建福宫，站在发妻、嫡子身边，站在鲜花烂漫之中，对着上天祈求眷顾；而现在的皇帝，却只能在严冬之中，拖着年迈体弱的身体，孤独地坐在乾隆花园中，看着眼前虚幻的盛景，写下“此诚符我望，惟静候天恩”。

“凡所有相，皆属虚妄。一切有为法，皆如梦幻泡影，如露亦如电，应作如是观。”

41 ［法］杜赫德编，郑德弟译：《耶稣会士中国书简集：中国回忆录》第六卷，《蒋友仁神父的第三封信》：“皇帝一年中在北京只住三个月左右。他通常于冬至前几天去北京住一阵子……而在正月十五日前，皇帝及其随从就要去位于北京西北二法里处的别宫圆明园了。”大象出版社，2005年，第54页。

42 傅东光：《御笔〈林下戏题卷〉与乾隆晚年的“林下人”心境》，《故宫博物院院刊》2000年第4期。

43 （清）弘历：《御制诗四集》卷七十七《忆昔》。

44 （清）弘历：《御制诗余集》卷二十《望捷》。

45 《清高宗纯皇帝实录》卷三十八，嘉庆四年正月下。

46 （清）弘历：《御制诗四集》卷三十四《题符望阁》。

〔图 10〕
古华轩前盛开的花朵，朱楷摄

竹香馆的墙上刻着乾隆皇帝的《咏金刚六如》，这大概不是巧合。

不同于西方的花园，中国的园林从不是简单的景观罗列，而是园主人生的缩影。在我眼中，乾隆花园也正是这样一座关于乾隆帝的博物馆。

年轻时候许下的愿望实现了吗？上天对他真的有慈悲之心吗？他的一生究竟有意义吗？带着这些疑惑，让我们从乾隆花园入口进入，一步步走过遂初、符望、倦勤，来见证他的故事，认识他的面孔吧。

追求风雅的文人，意气风发的皇帝，心有丘壑的隐士，潜心修行的居士，陶情适性的闲人——到底哪个是乾隆帝？又或者都不是他。这个问题的答案只能由我们来回答。不是通过影视作品，不是通过文字图像，而是通过我们的眼，我们的心。

说起来又是春天，古华轩前那棵见证了乾隆帝流觞的楸树，又该开花了〔图 10〕。《兰亭序》说："虽世殊事异，所以兴怀，其致一也。"作为乾隆花园的保护工作者，我只愿"后之览者，亦将有感于斯文"。

The Six Personas of the Qianlong Emperor and the Six Narratives of His Garden

Liu Renhao Department of Architectural Heritage, The Palace Museum

Abstract: The Qianlong Garden, located in the Palace of Tranquil Longevity (*Ningshou gong*) within the Forbidden City, served not only as a retreat for Qianlong Emperor's retirement and leisure but also a physical manifestation of his inner world. From the floating cup channel by the Pavilion of the Purification Ceremony (*Xishang ting*) to the illusory grandeur of the Belvedere of Viewing Achievements (*Fuwang ge*), this garden is almost a personal museum of Qianlong, embodying his complex life trajectory in the architectural language. This article attempts to interpret the personality of Qianlong as reflected in the architecture of this garden, thereby digging into his spiritual pursuits.

Keywords: The Qianlong Garden, The Qianlong Emperor, Garden

“四美具”原状复原设想

张淑娴　故宫学研究院

“四美具，二难并”，出自王勃《滕王阁序》，意思是良辰、美景、赏心、乐事，贤主、嘉宾欢聚一堂[1]，是为文坛佳话。紫禁城内也有一个“四美具”，位于建福宫花园静怡轩西室。

乾隆十年（1745），因“内府秘笈王羲之快雪帖，王献之中秋帖，近又得王珣伯远帖，皆希世之珍也，因就养心殿温室易其名曰‘三希堂’”[2]。乾隆帝御笔亲书“三希堂”匾额，悬于御座上方，并下旨董邦达绘制“三希堂记山水画”一幅。乾隆十一年（1746），内府又收藏顾恺之《女史箴图》〔图1〕、李公麟《蜀江图》《九歌图》《潇湘卧游图》，按董其昌跋所言，此四图原为海上顾中含家所藏，后散佚，今先后俱入内府，因此仿“三希堂”之例将四图合而贮之于静怡轩西室，颜曰“四美具”，也命董邦达绘制《四美具合幅》并御笔题书[3]（著录于《石渠宝笈》三编，现藏于台北故宫博物院）。

静怡轩是紫禁城内廷西北建福宫区域的一座建筑。建福宫与建福宫花园建于乾隆五年（1740），后又陆续有些添建、改建工程。[4]建福宫区域建筑依次为抚辰殿、建福宫、惠风亭、静怡轩、吉云楼、慧曜楼，西为建福宫花园，花园内有延春阁、敬胜斋、凝晖堂、碧琳馆，玉壶冰等建筑，环山而建，幽然静谧。〔图2〕

静怡轩位于建福宫轴线上，惠风亭后，为建福宫寝宫。[5]面阔五间，进深三间，共十五间。前出抱厦三间，明三间前檐隔扇门。东西梢间前檐及静怡轩东西墙皆安支摘窗，周围游廊相连。室内明间三间相连为一大开间，设屏风、宝座，额曰“与物皆春”，联曰：“雨润湘帘，苑外青峦飞秀；风披锦幕，阶前红药翻香。”又有东梢间、东南梢间、西梢间、西南梢间，西室额曰“四美具”。后室五间并带仙楼，联曰：“墨壶琴荐相先后；旧咏新裁自品评。”又联曰：“兰殿蔼晴晖，炉烟结翠；芝宫呈秀采，

1　（清）弘历：《御制诗五集》卷二十九《六兼斋有会》：“昔人以良辰、美景、赏心、乐事为‘四美’，贤主、嘉宾为‘二难’，故王勃有‘四美具，二难并’之语。”

2　（清）弘历：《御制文初集》卷四《三希堂记》。

3　（清）董邦达绘《四美具合幅》跋文中写道：“今年（乾隆丙寅）孟夏，万几余闲，得龙眠潇湘图卷，按董题识，搜宝笈之藏，而彼三图者，适咸在焉。益深叹赏。亲制跋语，仍聚置一室，颜曰‘四美具’，并绘图以传之。”

4　张淑娴：《建福宫花园建筑历史沿革考》，《故宫博物院院刊》2005年第5期。

5　（清）弘历：《御制诗四集》卷六十一《题静怡轩》：“静怡轩，建福宫后之寝室也。”

〔图 1〕
东晋顾恺之《女史箴图》(局部)，大英博物馆藏

〔图 2〕
清丁观鹏《太簇始和图》(局部) 描绘的建福宫花园，台北故宫博物院藏

砌草舒芳。”[6] 静怡轩庭院种植梅树二株，乾嘉二帝留下多首咏梅诗。“江南应作子，冀北始舒花。以屈为信者，非冰即玉耶。满庭喷芳馥，不水故横斜。相对红梨朵，当谁宜月华。”[7] 赞咏梅花清秀典雅、柔美婉约的美丽和傲雪凌霜、冰清玉洁的精神。〔图 3〕

可惜 1923 年 6 月的一场大火，将静怡轩和建福宫花园焚毁殆尽，“四美具”的空间格局及原状陈设荡然无存。按照静怡轩仿建的宁寿宫景福宫内檐装修也于光绪二十九年（1903）全部撤除改换装修[8]，可依据的样本亦不复存在。乾隆十一年（1746）乾隆帝将四幅画合而收储于静怡轩西间，定名“四美具”，精心设计了一个雅室。室内安置洋漆格子宝座，靠背内放装有四幅画的插屏[9]，便于欣赏心爱的古画，后被撤换；二十三年（1758）又陆续更换了一些陈设[10]，“四美具”的原状也经历了不断变化的过程，至乾隆中期逐渐定型，之后则少有改变。本文就文献记载[11]、御制诗文[12]、内务府

6 （清）于敏中等编纂：《日下旧闻考》卷十七，北京古籍出版社，2001 年，第 224 页。

7 （清）弘历：《御制诗三集》卷七十二《静怡轩梅花》。

8 光绪二十九年八月，“奏为修理景福宫工程所需款项请由部库筹拨恭折仰祈圣鉴事。窃于光绪二十八年三月初三日奉懿旨：景福宫殿内装修著满撤去，按照现定式样成做，等因。钦此，钦遵，在案。因上年方向不宜，未能兴修。详查今岁合宜，即应遵旨修理。”参见故宫博物院内网，长编 69206，内务府新整杂件 377 卷。

9 中国第一历史档案馆、香港中文大学文物馆编：《清宫内务府造办处档案总汇》第 14 册，人民出版社，2006 年，第 617 页。

10 《清宫内务府造办处档案总汇》第 23 册，第 312 页。

11 参见《日下旧闻考》《国朝宫室》等。

12 参见《清高宗御制诗文全集》。

〔图 3〕
建福宫花园静怡轩，张淑娴摄

档案[13]、陈设档[14]和刘榕女士的文章《静怡轩建筑渊源及其复原设计》[15]，试图复原“四美具”的原貌。

一 “四美具”空间格局

“四美具”位于静怡轩西间，根据陈设档的描述占据两间，即西梢间和西南梢间（文中简称“西间”“西南间”）。西间与静怡轩明间有墙或隔扇相隔。由于档案的缺失，具体情况今人并不清楚，参照与静怡轩近似的建筑圆明园西峰秀色含韵斋的空间设计，隔断墙中安装嵌扇的可能性较大。

西间面阔 3.2 米，进深 6.38 米，面积约 20 平米；西南间进深 3.82 米，面积约 12 平米。两间之间应用槛窗相隔，中间小门相通，或仅开小门。西间与西北间之间，参照圆明园西峰秀色含韵斋的空间设计，应为一堵隔断墙，两间之间无通道，进西北间需从静怡轩后间辗转而入。〔图 4〕

“四美具”位于静怡轩的西南角，西南间的东面为实体墙。南面和西面安支摘窗，南面窗户两组上下四扇，西南间西窗三组上下六扇，西间西面窗户四组上下八扇。根据乾隆四十二年（1777）景福宫拟换安大玻璃窗的档案可知静怡轩“四美具”窗上安玻璃窗[16]。乾隆时期的玻璃窗均为玻璃窗户眼，即在窗户棂条中心安一长方形框，

13 参见中国第一历史档案馆藏《奏案》《奏销档》《活计档》等。

14 故宫博物院藏《静怡轩四美具百什件》，嘉庆七年十一月，设 31；《静怡轩现设陈设档》，道光十九年六月，设 177。

15 刘榕：《静怡轩建筑渊源及其复原设计》，《故宫博物院院刊》2005 年第 5 期。

16 乾隆四十二年七月，记事录：“于七月二十日，员外郎四德、五德将建福宫静怡轩四美具窗户三扇上现安玻璃，画得纸样一张，持进交太监鄂勒里呈览。奉旨：俱不必换安玻璃。钦此。”《清宫内务府造办处档案总汇》第 40 册，第 442 页。

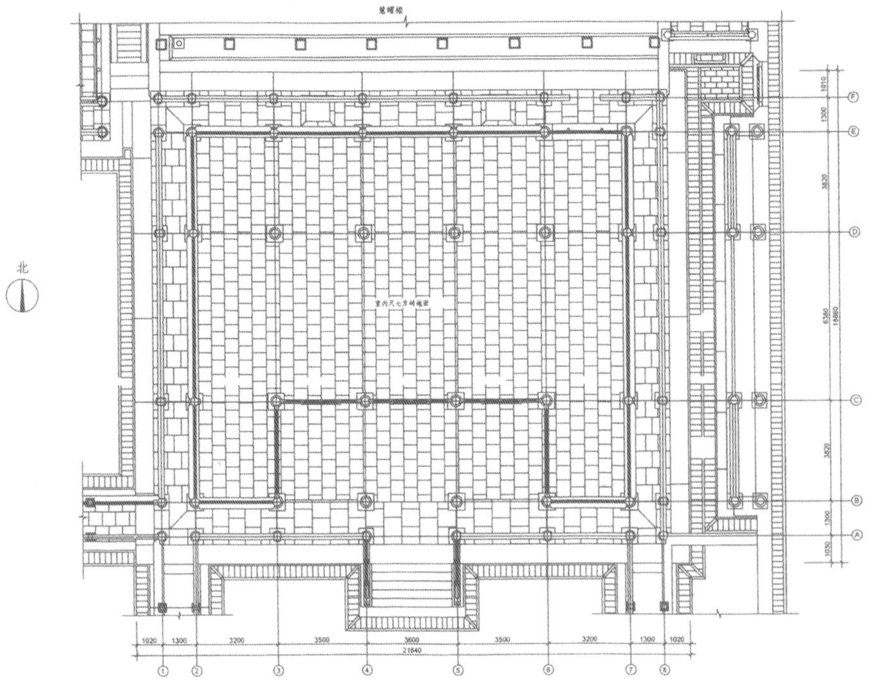

〔图 4〕
静怡轩平面图

内安小块玻璃，称为“玻璃窗户眼”。静怡轩使用的玻璃“长一尺四寸，宽九寸五分”[17]，安装在窗户中心的玻璃窗户眼。

二 “四美具”室内陈设

（一）西间陈设

“四美具”西间室内靠西墙正中面东设宝座，宝座上设衣素缎坐褥靠背迎手，后设五屏风、宫扇等常规陈设。宝座前左右各“南漆高低香几”一个，上面分别摆放“广珐琅水法嵌腰圆镜乐钟、青玉炉瓶盒、青玉双环洗、白地红花磁甘露瓶”等物品，香几下各摆放一个“洋磁巧工瓶”。宝座南北靠墙各放置一座紫檀木边座玻璃大穿衣镜插屏（背后贴御笔字二张）和广珐琅水法乐钟一座。西室室内陈设相对简单，是较为开放的礼仪性空间的陈设。

文献记载：“（静怡轩）西室，高宗纯皇帝御笔匾曰‘四美具’。屏间，则圣制‘视、听、言、动、思’五箴也。”（此处记载恐有误，根据乾隆帝御制诗《五事箴》应为“貌、言、视、听、思”五箴。）乾隆帝《五福颂有序》载：“轩有屏，尝撰五事箴，揭之以代铭座。”[18]

17 乾隆七年八月十五日：“司库白世秀、副催总达子将静怡轩向西窗户六扇需用玻璃尺寸折片一件，并贴得造办处无锡玻璃长一尺四寸、宽九寸五分六块，长三尺一寸二分、宽二寸二分一块，长四尺、宽一尺三寸一块，西洋玻璃门一件，持进交太监高玉等转奏。奉旨：准用造办处玻璃，如不足用向开其里要。钦此。”《清宫内务府造办处档案总汇》第 11 册，第 18—19 页。

18 （清）弘历：《御制文二集》卷三十七《五福颂有序》。

屏风为五扇，上贴落乾隆帝御笔《五事箴》。此后，将御笔揭下来，换上了缂丝《五事箴》。乾隆时期的档案记载了缂丝围屏心由苏州制作。档案具体如下：

乾隆十九年三月二十二日，员外郎白世秀将苏州织造安宁送到建福宫静宜（怡）轩缂丝围屏心五张随本文，持进交太监胡世杰呈览。奉旨：缂丝围屏心做的无空处，着他赔补，不准报销。所交活计安宁不用心看做，着德申饬。再照样做缂丝围屏心五张，边子不要花纹。钦此。

于四月二十日，太监董五经来说，首领桂元交御笔缂丝围屏心本文五张。传旨：着照本文另做缂丝围屏心五张，周围要绿色回文锦边一寸。钦此。

于本年十一月初五日，七品首领萨木哈将苏州织造安宁送到缂丝屏风心五张随本文，持进交太监胡世杰呈览。奉旨：着贴落。钦此。

于本月初六日，员外郎西宁、七品首领萨木哈将缂丝屏风心五张，持进赴静怡轩贴落讫。

于十一月初六日，监察御史西宁、员外郎达子、七品首领萨木哈将静怡轩换下缂丝五箴五张、御笔字条一张、对一付，持进交太监胡世杰呈览。奉旨：将缂丝五箴发往苏州，周围照花边宽窄接出素地杖，外仍缂做丝回纹边，其现有之花边裁下交边。再照静怡轩群板上坐龙捧圣一样缂做群板心五张，亦要绿回纹边，其挑山对上绫边起下做材料用，本文交懋勤殿归御笔收贮。钦此。

于十一月初七日，员外郎达子将画得绿回纹边放宽地杖五箴纸样一张，坐龙捧圣群板心纸样一张，持进交太监胡世杰呈览。奉旨：准照样缂做。钦此。

于二十年五月十五日，员外郎白世秀、副催总舒文将苏州织造安宁送到缂丝五箴五张补接出回纹边，随换下花边二十条，缂丝群板心五张，持进交太监胡世杰呈览。奉旨：将缂丝五箴、缂丝群板着收贮，俟指出地方时再贴，其换下旧花边交萨木哈镶边用，用时呈览，准时再用。钦此。[19]

乾隆十九年闰四月二十八日，监察御史西宁来说，太监胡世杰传旨：静怡轩现安五真（箴）屏风一座，将下边雕漆绦环拆去熔化，照上边的漆边一样黑漆画金夔龙边心子，交南边缂做，正面坐龙捧圣万字回文一寸，其做法要与先传五真（箴）字的边一样缂做，先画样呈览。钦此。

每扇外口高二尺八寸五分、宽二尺五寸二分，上边宽一寸二分五厘、下宽

19《清宫内务府造办处档案总汇》第 20 册，第 351—352 页。

一寸五分，两边各宽一寸二分五厘。里口高二尺六寸五分、宽二尺二寸七分，画边宽九分。

于五月初三日，监察御史西宁将画得静怡轩坐龙捧圣绿纹边一寸群板心子纸样一张，持进交太监胡世杰进呈览。奉旨：准交南边照样缂做。钦此。

于十一月初五日，员外郎达子将苏州织造安宁送到缂丝坐龙群板心五块随蓝册，持进交太监胡世杰呈览。奉旨：缂丝片贴落册子按例查核。钦此。

于十一月初六日，监察御史西宁、七品首领萨木哈将缂丝群板心五块持进静怡轩五屏上贴讫。[20]

苏州织造为静怡轩制作了缂丝《五事箴》围屏心和缂丝“坐龙捧圣”裙板，均为绿色回文锦边，裙板“外口高二尺八寸五分、宽二尺五寸二分，上边宽一寸二分五厘、下宽一寸五分，两边各宽一寸二分五厘。里口高二尺六寸五分（里口应高二尺五寸七分五厘）、宽二尺二寸七分，画边宽九分”。从档案中了解屏风为黑漆描金夔龙边框，屏风有屏心和裙板，但具体的式样并不清楚。

乾隆三十九年（1774）景福宫按照“四美具”制作屏风的档案给屏风勾勒了一个大概形状。

乾隆三十九年十月，行文，十二日，库掌四德、五德、笔帖式福庆来说，太监胡世杰传旨：新建景福宫着照静怡轩四美具现安设五屏罩壁并博古柜、闲余板、挂格一样画样呈览。钦此。随按新建景福宫地面、床尺寸照静怡轩四美具现设五屏罩壁并博古柜、闲余板、挂格样画得五件屏罩壁、博古柜、闲余板、挂格纸样一张，持进交太监胡世杰呈览。奉旨：罩壁背面绦环成做雕龙花纹，上面要通景落堂素心板雕龙搭脑，巴达马须泥座罩壁一座；柜子正面俱做雕龙花纹，两山板不必做雕龙，要素木板，另画大样呈览，准时发往粤海关成做。其柜下香几、闲余板、挂格，着造办处成做，亦另画大样呈览。钦此。

于本月二十六日，员外郎四德、库掌五德、笔帖式福庆为做新建景福宫五屏罩壁一座，做得雕龙搭脑、须泥座五屏罩壁小木样一件，前面画得大小柜三对、挂格一件，闲余板一件，并背面群板画得雕龙元光样通景落堂、素心板样，又画得五屏罩壁大纸样，持进交太监胡世杰呈览。奉旨：罩壁一座、大小柜三对，照样准发往粤海关成做，算伊贡内呈进，其背面通景心板不必配做，着造办处

20《清宫内务府造办处档案总汇》第20册，第464—465页。

配壁子。所做罩壁一座，得时不必成斗，即将搭脑、须泥座、群板、边框等送往京内成斗安挂。其挂格、闲余板并柜下香几仍留京内成做。钦此。

又于四十年三月初三日，员外郎四德、库掌五德、笔帖式福庆将景福宫五屏风罩壁后挂格一座，画得紫檀木雕龙闲余板、挂格纸样一张，素香几纸样一张，交太监胡世杰呈览。奉旨：照样准做，其香几照旧样成做。钦此。

于四十年九月二十三日，员外郎四德将粤海关送到五屏风罩壁一分，持进交宁寿宫讫。[21]

梳理档案可知，“四美具”屏风罩壁是五扇插屏，屏扇由屏心、绦环、裙板构成，屏心贴落缂丝《五事箴》，裙板贴落缂丝“坐龙捧圣”，绦环雕刻龙纹，屏扇背面群板为雕龙元光通景落堂，后改为素板。屏风边框是洋漆工艺，上有搭脑，下承巴达马须弥座。道光十九年（1839）《静怡轩现设陈设档》记载的是“洋漆边座镶床刻丝五屏峰（风）一座”，与档案记载相符。故宫博物院藏缂丝《五事箴》及龙额围屏十幅〔图5〕，字五幅，每幅长117厘米，宽64.2厘米，连边长132.4厘米，宽70.4厘米。分别缂丝“貌、言、视、听、思”五箴，绿色万字锦边；龙额五幅，缂丝“坐龙捧圣”，绿色万字锦边。此件《缂丝五箴及龙额围屏》与档案记载“四美具”屏风中所用《五事箴》及龙额尺寸、装裱锦纹皆相似，应为“四美具”屏风上的，屏心贴“五箴”五幅，裙板贴“龙额”五幅。

屏风前为宝座（或称为“宝座床”），宝座的工艺无记载，不过当时室内使用了大量黑漆描金的家具和装修，如黑漆描金插屏镜[22]、黑漆画金花炕案[23]等，屏风也为黑漆描金缂丝屏风，与家具和屏风相配，应为黑漆描金宝座，上铺衣素缎坐褥靠背迎手。宝座“靠背高一尺八寸五分、厚一寸九分、面宽二尺六寸三分，坐褥长四尺七寸、宽三尺一寸五分”[24]。

宝座前放置“南漆高低香几”各一对。南漆指南方制作的漆器，有雕漆、描金漆等品种。西间宝座及屏风均采用黑漆描金。前面左右应为描金漆高低香几，上置器物。南北各放置一座紫檀木插屏镜和水法乐钟。

21《清宫内务府造办处档案总汇》第37册，第436—437页。

22《清宫内务府造办处档案总汇》第11册，第96页。

23《清宫内务府造办处档案总汇》第23册，第312页。

24《清宫内务府造办处档案总汇》第23册，第255页。

（二）西南间陈设

西南间室内面积约12平米，南面和西面两面临窗。

靠西墙面东放置紫檀木罗汉床一张，足踏一件，上铺绣蓝红绸坐褥一件，红白

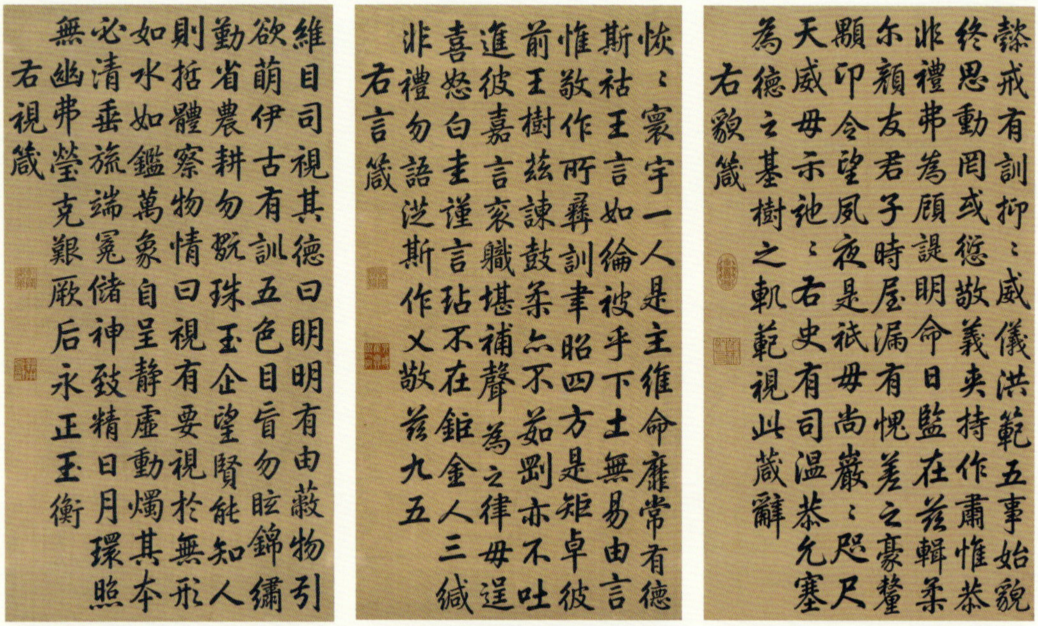

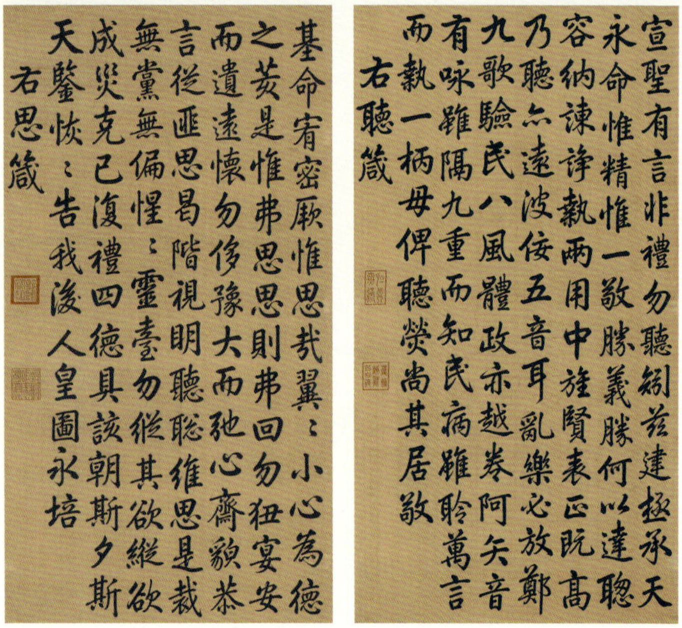

〔图 5〕
缂丝《五事箴》及龙额围屏，《五事箴》贴于屏风屏心上，“坐龙捧圣”贴于屏风裙板上，故宫博物院藏

毡二块，黄锦套一件，氆氇套一件，葛布套一件。

罗汉床南北各放木柜格一个。《陈设档》记载柜中各放置七件器物，柜门内放盒和笔筒两件小型器物。乾隆二十三年（1758）档案记载静怡轩博古柜一对，内装百什件 32 屉[25]。嘉庆七年（1802）十一月《静怡轩四美具百什件》中，百什件共有 32 屉，按照《千字文》排列，从“天”字到“阳”字，与档案相符，应放在博古柜内。乾隆

25 乾隆二十三年十月，百什件：“初九日，郎中白世秀、员外郎金辉来说，太监胡世杰传旨：静怡轩现安博古柜一对，着配装百什件，其古玩如不足向里边讨用。钦此。……于十二月十九日，郎中白世秀、员外郎金辉谨奏：今为配装静怡轩紫檀木古柜二对，内百什件三十二屉，查得现存古玩一千二百三十二件，今已配装得十二屉，用过古玩三百八十七件。其余二十屉，虽现有古玩八百四十五件，多系大件漆木铜磁葫芦盒碗等项，难以配装，请领小式玉器古玩等件以备陆续配装应用，缮写折片，持进交太监胡世杰呈奏。奉旨：知道了。钦此。”《清宫内务府造办处档案总汇》第 23 册，第 630—631 页。

三十九年（1774）档案记载景福宫“照静怡轩四美具现设五屏罩壁并博古柜、闲余板、挂格样画得五件屏罩壁、博古柜、闲余板、挂格纸样一张”“柜子正面俱做雕龙花纹，两山板不必做雕龙，要素木板”[26]。博古柜为紫檀木，上部为博古架，下部橱柜雕刻龙纹。柜格内放置各种古玩，格顶放《剿平三省邪匪方略》《九宫大成南北词宫谱》《御制诗》等书籍。

罗汉床南北各放一个紫檀木竹式香几，即紫檀木雕刻竹子式样的香几。“四美具”初设之时摆放的是黑漆描金香几，乾隆二十三年换成紫檀木香几[27]，上放“洋铜嵌冠架水法乐钟一对”，钟用玻璃罩罩着。香几下放带铜座的铜珐琅小缸一对。柱子上挂“御笔小挂屏一件”“文竹嵌铜字对一副”。

东部依东墙面西安设炕，高一尺四寸五分（46.4 厘米），炕一般宽 130—140 厘米，面西，正中铺坐褥迎手锦靠背，称宝座床。初期宝座床上用的是黑漆描金靠背，乾隆二十三年换成紫檀木靠背[28]，之后又换用衣素缎靠背。床上放置盘、盒、书籍等。

《陈设档》中有“紫檀木挂格一件”，没有明确位置，挂格应挂在宝座上方墙上。根据《陈设档》，挂格内盛放了各种古玩。挂格上有抽屉，抽屉内盛放铜表、象牙扇、鼻烟壶、玉器等小件器物，格顶上放置六方盒二对。从中分析挂格是一个立体的多宝格式挂柜，器物有些摆放在格子里，有些则是镶嵌在挂屏上的。这种多宝格式挂屏是清代中期较为盛行的装饰品，有雕漆、紫檀等不同材质的。故宫博物院现存多件紫檀木多宝格式挂屏，其中一件紫檀木边嵌玉石多宝格式挂屏〔图 6〕两件合为一组，现挂在养心殿东暖阁。其为紫檀木格，紫檀木回纹花牙，内嵌各种器物。右侧一小柜，铜荷叶把手，柜内放置器物。另一件剔红边嵌玉多宝格式挂屏〔图 7〕，为剔红缠枝花卉边框，剔绿夔龙纹花牙，内嵌玉瓶插折枝花枇杷、玉羊、仙人乘槎、珐琅炉、珐琅高足碗等器物。右侧一橱柜，柜门剔红回纹框，下层门框内嵌金边框，柜门安玻璃，安铜荷叶把手，内置画轴等器物。它们的样式和工艺做法与“四美具”紫檀木挂格异曲同工。

此外，西南间室内还有“四美具插屏一座”，其档案如下：

乾隆十一年四月，广木作，二十五日，七品首领萨木哈来说，太监胡世杰传旨：静宜（怡）轩洋漆格子靠背上着做夹堂插屏一座。钦此。

于本月二十六日，七品首领萨木哈将做得夹堂插屏合牌样一件，持进交太监胡世杰呈览。随交出李公麟九歌图一卷、李公麟蜀川图一卷、李公麟潇湘卧

26《清宫内务府造办处档案总汇》第 37 册，第 436 页。

27 乾隆二十三年，油木作：“十月初二日首领桂元传旨：着照四美具现安漆靠背、香几样另做紫檀木靠背、香几，先画样呈览。钦此。于十月初八日，郎中白世秀、员外郎金辉将画得四美具香几、靠背纸样二张，呈览。奉旨：照样准做素紫檀木的。钦此。于十一月初六日，副催总海升将做得紫檀木靠背一件、香几二件，持进安讫。于十一月十一日，郎中白世秀、员外郎金辉将四美具换下黑漆画金花炕案二张、漆靠背一件，持进安在养心殿呈览。奉旨：漆案几二件交敬事房，其靠背一件交造办处有用处用。钦此。”《清宫内务府造办处档案总汇》第 23 册，第 312—313 页。

28《清宫内务府造办处档案总汇》第 23 册，第 312—313 页。

游图一卷、顾恺之画女史箴并书真迹一卷。奉旨：照样准放大样呈览，准时再做紫檀木边、楠木心，抽屉要四个。钦此。

于五月十二日，司库白世秀将画得四美聚（具）插屏纸样二张，持进交太监胡世杰呈览。奉旨：夔龙绦环做商银丝的，其跕牙并丕水俱要雕做。再跕牙看面并大边、绦环地子、他萨俱要素的。钦此。

于六月十四日，司库白世秀将做得四美聚（具）插屏内抽屉样，持进交太监胡世杰呈览。奉旨：照样准做，手卷准垫起糊住，两头匣子准二色木的样款，用紫檀木做，要落堂，写字画画，将此匣子四件俱做成样呈览。钦此。

于十二年二月初一日，七品首领萨木哈将做得贴字画楠木心紫檀木边四美聚（具）夹堂插屏一座持进交太监胡世杰呈进讫。[29]

〔图 6〕
多宝阁式挂屏，故宫博物院藏

〔图 7〕
剔红边嵌玉多宝格式挂屏，故宫博物院藏

乾隆十四年十一月十八日，员外郎白世秀来说，太监胡世杰交四美具插屏一件，内盛抽屉四件，每件内匣子二件。一屉内盛手卷一卷，汉玉双鱼环佩一件（有璺），汉玉昭文带一件，匣上各嵌汉玉昭文带一件。一屉内盛手卷一卷，汉玉斧佩二件，匣上各嵌汉玉昭文带一件。一屉内盛汉玉双夔龙玦一件，汉玉连环一件，匣上各嵌汉玉昭文带一件。一屉内盛汉玉仙人一件，汉玉狮子一件，匣上各嵌汉玉昭文带一件。传旨：将字画绉处另糊手卷二卷，各换玉别子，其紫檀木边框有冽处着线逢收什。钦此。

于十二月初五日，七品首领萨木哈将四美具插屏一件，内盛抽屉玉器同持送安讫。[30]

29《清宫内务府造办处档案总汇》第 14 册，第 617—618 页。

30《清宫内务府造办处档案总汇》第 16 册，第 760 页。

〔图 8〕

紫檀三屉画箱插屏，故宫博物院藏

“四美具”插屏为紫檀木插屏，共有抽屉四层，每层抽屉内又放了两个匣子，配置各式玉器。但插屏的形式和雕刻的纹样不清。查阅档案，乾隆十六年（1751）《莲华经》按照“四美具”插屏制作插屏一座，紫檀木雕“佛样、坐龙牙子、火焰堂子、两边八吉祥小回纹边”[31]；永安寺的信炮册子也按“四美具”插屏套制作一分，“楠木扫金罩漆，前面板雕二龙捧签子，后面板要二龙捧珠”[32]；乾隆二十一年（1756）《五牛图》手卷亦照“四美具”插屏制作紫檀木插屏一件，“一边画松树，一边画竹子，拉道填金”[33]。这些插屏，从材质、工艺到纹饰都不一样，它们应仅仿照“四美具”插屏的式样，而后者的具体式样亦不明了。故宫博物院现存兰亭八柱插屏和梅竹三屉插屏〔图 8〕，推测“四美具”插屏式样与此两件相同。形制为紫檀木雕云龙纹，四层屉，每屉内又有两个匣子。《陈设档》载：

头层内盛李公麟九歌图一卷，紫檀木嵌玉匣二件，一件内盛青汉玉异兽一件，一件内盛汉玉连环一件；二层御咏李公麟蜀川图一卷，紫檀木嵌汉玉昭文

31 《清宫内务府造办处档案总汇》第 18 册，第 190 页。

32 《清宫内务府造办处档案总汇》第 18 册，第 276 页。

33 《清宫内务府造办处档案总汇》第 22 册，第 174 页。

带匣二件，每一件内盛青汉玉异兽一件；三层御咏顾恺之画女史箴并书真迹一卷，紫檀木嵌汉玉昭文带匣二件，一件内盛白玉带钩一件，一件内盛汉玉螃蟹一件；四层御咏李公麟潇湘卧游图一卷，紫檀木嵌汉玉昭文带匣二件，一件内盛汉白玉双鸠一件，一件内盛汉玉双鱼一件。

根据乾隆十一年（1746）档案记载，“四美具”插屏放置在“洋漆格子靠背上”，推测初设“四美具”时室内放置了一个带有格子的宝座，装着四幅画的插屏就放在靠背的格子里，以便展卷欣赏。之后室内装修则发生了变化，面西搭床，插屏应放在床上，放置的位置及式样不清，应该是放在床的最北头靠隔断的位置。

“四美具”插屏左右分别放雕紫檀木罩盖匣一对（内盛冯祥画写意人物一册，味余书室随笔一册，小玉器二件）、掐丝珐琅蟠龙瓶一件（内插班珠尔边股扇一柄，赵秉冲字、余省画，上拴汉玉葫芦扇器一件）、周环钮钟一件（上拴汉玉昭文带一件，汉玉结子三件，汉玉槌一件，紫檀木座上刻诗）。插屏下设红雕漆寿春宝盒四对、雕漆寿春宝盒一对。

宝座床南临窗放紫檀木小案，案上摆设青玉梅瓶、青白玉炉瓶盒、红雕漆四方三层盒。案下放黄公望《西山胜景册页》一册，紫檀木都盛盘和广珐琅亭式乐钟一对。

宝座的左边即南边（宝座面西，左边即南边）放紫檀木炕几一对，炕几上分别放置雕紫檀木顶竖柜，共二件，竖柜上放漆盒、瓷杯、青铜爵等器物。宝座的右边即北边放置了两件雕紫檀木柜格，上面摆放青铜、瓷器、珐琅、漆器等器物。

房间南窗户台下面放置箱、盒、匣、盘等器物，里面存储各色古玩。窗台上放置插屏、花觚、花瓶等各色摆件。

南北窗户见柱上挂铜镶玻璃边腰圆挂镜一件，西窗南北见柱上挂锡胎钧釉双管挂瓶一对。

三　总结

“四美具”为静怡轩西间和西南间两间，西间靠西窗正中面东摆放黑漆描金缂丝《五事箴》屏风，前为黑漆描金宝座，两边放置香几，南北有插屏镜和水法乐钟，为礼仪性的空间，整体风格黑漆描金，庄重华丽。西南间西墙面东放置紫檀木罗汉床，两边为紫檀木多宝格柜。东墙面西设宝座床，床上坐褥靠背迎手，宝座上方挂紫檀

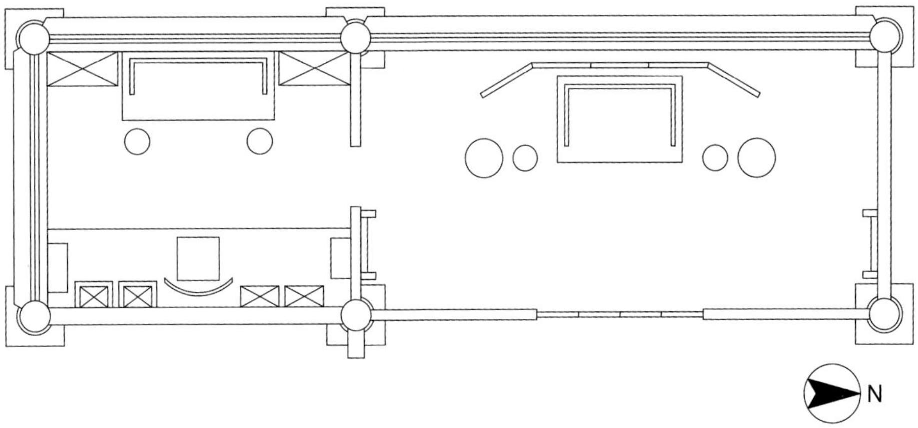

〔图9〕
“四美具”平面示意图

木多宝格式挂格。宝座南放置两个紫檀木顶竖柜，宝座北放置两个紫檀木柜格。床北头放“四美具”插屏，内存四幅“美人图”真迹。床的南头临窗摆一小案，床下放各种箱柜。室内柜格、挂格、多宝格内均放置各式珍玩，是乾隆帝鉴赏书画古玩的地方。室内均为紫檀木装修和家具，沉稳素朴〔图9〕。

“四美具”是乾隆帝收藏四件稀世珍品的地方，它“实与‘三希’互相辉映”。三希堂以三件稀世珍品寓意“士希贤、贤希圣、圣希天”。乾隆帝则在“四美具”中写道：“乃今四美具一室，赏心乐事无伦比。寓意于物不留意，咄咄是吾乃所以。”[34] 并让董邦达绘制图画、御笔亲书、大臣题跋，借此附会“良辰、美景、赏心、乐事，贤主、嘉宾欢聚一堂”的场景，以示君臣合作、共治国家。

34 （清）弘历：《御制诗初集》卷三十二《李伯时蜀江图歌》。

Concept for Restoring the Original State of Four Beauties Studio

Zhang Shuxian The Institute of Gugong Studies

Abstract: In the 11th year of the Qianlong reign (1746), the imperial collection of the four paintings–*Admonitions of the Instructress to the Court Ladies* by Gu Kaizhi and The Shu River, *The Nine Songs and Imaginary Tour of Xiaoxiang* by Li Gonglin–was housed in the western chamber of the Gallery of Tranquil Ease (*Jingyi xuan*) in the Palace of Established Happiness (*Jianfu gong*) which was thus named the Four Beauties Studio.

The Studio is situated in the westernmost chamber and the southwest corner chamber of the Gallery of Tranquil Ease. In the westernmost chamber, against the western wall stood a black lacquered and gold-painted throne, with a five-paneled screen inscribed with the *Five Admonitions* behind it. Palace fans and other conventional decorative items were placed on either side of the seat, creating a solemn and spacious ritual space.

Along the western wall of the southwest corner chamber stood an arhat bed with a throne facing west, upon which stood a four-panel screen holding the four masterpieces. The chamber was also furnished with various cabinets displaying an array of antiques, forming a dazzling collection–a space where Qianlong Emperor would leisurely appreciate his prized treasures. The Gallery of Tranquil Ease was destroyed by fire in 1923. This paper attempts to reconstruct the original state of the Four Beauties Studio based on archival research.

Keywords: Qianlong Emperor, Palace of Established Happiness, Gallery of Tranquil Ease, Four Beauties Studio

由延春阁望静怡轩，史宁昌摄

高阁临昆湖　玉蟾望琼阙
——清漪园望蟾阁写仿武昌黄鹤楼

贾珺　清华大学建筑学院

一　江楼胜迹

黄鹤楼旧址位于武昌长江东岸蛇山黄鹄矶（又名黄鹄山、黄鹤山、石城山），相传始建于三国东吴黄武二年（223），历代屡建屡毁，却一直是驰誉天下的名楼。关于"黄鹤"之名的来历说法不一，大多与古书记载的各种仙人驾鹤故事有关。唐宋以降，文人墨客吟咏不绝，以崔颢《黄鹤楼》诗最为著名，被严羽《沧浪诗话》推为"唐人七言律诗第一"[1]。一千多年来，又有多位名家为之作画，分别展现了不同时期的建筑形象。其中元代夏永笔下的黄鹤楼建于高台之上，由多座楼阁组合而成，最高一楼采用重檐歇山屋顶〔图1〕。明代安正文笔下的黄鹤楼共有三重屋檐，歇山屋顶的两面分别垂直插入一个小歇山顶〔图2〕。

乾隆《江夏县志·古迹》详载了不同版本的黄鹤楼典故：

> 黄鹤楼在黄鹄山。《南齐书》仙人王子安乘鹤过此，又世传费文伟登仙驾黄鹤憩此。《报恩录》载："鄂辛氏市酒，山头有道士数诣饮，辛不计酬。道士临别取橘皮画鹤于壁，曰：'客至，拍手引之，鹤当飞舞以侑觞。'遂致富。逾十年，道士复至，取所佩铁笛吹数弄，白云自空飞来，鹤亦下舞，道士飘然乘鹤去。辛氏即其地建楼，曰辛氏楼。"唐阎伯理作《郢记》，以文伟事为信。或又引梁任昉记，谓驾鹤者乃荀瑰，字叔祎，所遇虹裳羽衣客非文伟也。按《述异记》，叔祎憩楼上，西望有物飘然降自云端，乃驾鹤之宾也。宾主欢对，辞去，跨鹤腾空，渺然烟灭。后人误为文祎也。[2]

1　（宋）严羽：《沧浪诗话》，中华书局，1985年，第40页。

2　（清）陈元京纂修：《江夏县志》卷十五，清乾隆五十九年刊本。

〔图1〕
元夏永《黄鹤楼图》，云南省博物馆藏

〔图2〕
明安正文《黄鹤楼图》，上海博物馆藏

又载明嘉靖至清乾隆年间此楼的历史沿革：

按宋陆游记，当时楼已废，故址亦不复存，问之老吏，云在石镜亭、南楼之间，未知何年复建。嘉靖末火，隆庆五年都御史刘悫重建，汪道昆为之记，崇祯癸未毁于贼。国朝顺治中，御史上官鉝重建。康熙三年毁于火，总督张长庚、巡抚刘兆麟重建，布政刘显贵监造。甲申，总督喻成龙、巡抚刘殿衡修。壬寅，总督满丕、巡抚张连登修，布政张圣弼督修。乾隆□年御书“江汉仙踪”题额。

乾隆六十年（1795）前后，著名文士汪中应湖广总督毕沅之邀，作《黄鹤楼铭》详细描绘黄鹤楼周围环境及此楼概貌：

江出峡，东至于巴邱，沅、湘二水入焉；又东至于夏口，汉水入焉。于是西自岷山，西南自牂牁，南自桂岭，西北自嶓冢，五水所经半天下，皆汇于是，以注于海。而江夏黄鹄山当其冲，江环其三面，再折而后东，故地形称险焉。县因山为城，山之西有矶，起于江中，石立如植，激水逆行，恒数里，于形为尤险。其上为楼，咸取于山，以为名。始自孙吴，郦氏著之；《齐》《梁》二书，并载其迹。于后，楼之兴废，史莫能纪。乾隆元年，大学士史文靖总督湖广，乃更其制。自山以上，直立十有八丈，其形正方，四望如一，高壮闳丽，称其山川。历年六十，坚密如新。[3]

3 （清）汪中：《黄鹤楼铭》，见纪宝成主编：《清代诗文集汇编》第410册，上海古籍出版社，2010年，第59页。

〔图 3〕
清关槐《黄鹤楼图》，台北故宫博物院藏

另一位文士陈木立《黄鹄山名胜记》载：

> 黄鹄山名石城山，长竟里，高十寻有奇，东连高冠，绵亘郡城。先是圭土为城者置此山阛阓中，首瞰大江，头陀寺显蔽之，又黄鹤楼扼其吭。……更上即黄鹤楼，楼上高百尺、八窗洞达者三层，嵌空玲珑，胜甲三楚。[4]

乾隆时期宫廷画家关槐曾经绘有一幅《黄鹤楼图》〔图 3〕，现藏于台北故宫博物院，图上表现了当时黄鹤楼的准确形象和周围的山水景致，可与同时期的文献相互印证。楼坐落于小山之麓，西临江岸，以一圈不规则的短垣环绕。前设拱门，后有殿堂亭轩，自成一院。楼高三层，主体部分平面呈正方形，四面各出三间抱厦，屋顶为攒尖式样，颇显繁复，与元明时期画卷上的形象有所不同。建筑的二层和三层均略有收分，比例和谐。底层似设有外廊，上部两层则以门窗围合。

此楼四个立面完全一致，被汪中概括为“四望如一”。有一处细节需要辨析：抱厦外凸的侧面并非垂直于主体部分，而是呈斜抹状态，使得平面轮廓更加复杂，相当于在四个正向之外又增加了四个斜向的观景视角，各个斜面与正面都设有槛窗，

4 （清）陈本立：《黄鹄山名胜记》，见（清）陈元京纂修：《江夏县志》卷十二。

被陈本立称为“八窗洞达”。

关于乾隆时期黄鹤楼的高度，汪中称“十有八丈”，陈本立称“百尺”（即十丈）。清制一丈合今制 3.2 米，则十八丈为 57.6 米，十丈为 32 米。现存中国古代最高的木构建筑为辽代所建的山西应县佛光寺释迦塔，高约 67.31 米；其次为清代乾隆年间所建承德普宁寺大乘阁，高 36.65 米；第三为光绪年间重建的颐和园佛香阁，高 36.44 米。武昌黄鹤楼仅有三层，不可能达到 57.6 米之高，而 32 米左右则是相对合理的尺度。据张子安先生推算，同治时期基本按照原样重建的黄鹤楼总高约“九丈七尺五寸”[5]，合 31.2 米，也大致符合“百尺”的说法，可为旁证。

中国古代水景名胜区常常在关键地段修建高大楼阁，成为标志性的中心建筑，同时又是登临观赏周边风景的最佳场所，如洞庭湖畔的岳阳楼、赣江东岸的滕王阁、黄河之滨的鹳雀楼，均为流传千古的名楼。黄鹤楼属于同样的情况，扼守长江、汉水两条大河的交汇口，巍峨雄壮，耸出山际，极为醒目。登楼一观，四面八方、远近高低的山水、林木、亭台全部历历在目——隔江西岸有龟山与蛇山相对，延伸至江面的禹功矶上建有晴川阁，江中有鹦鹉洲，而蛇山之南的洪山上有一座明初所建的宝通寺塔，北面有重峦叠嶂的大别山，武昌、汉阳、汉口三镇的街市闾巷也尽收眼底。

清代诗人延续前朝传统，经常登黄鹤楼赏景、宴饮，为此楼写下很多诗篇，不吝赞美之辞。如清初曾任湖广提学道佥事的王孙蔚《黄鹤楼》诗云：“独立飞楼尺五天，窗环平野入樽前。长江晓结千峰雨，大别晴开万树烟。紫雁北来迷楚浦，白云西去认秦川。凭栏愁看陶公柳，舞却春风又一年。”[6]祖籍安徽歙县、久居武昌的程封《喜黄鹤楼新成》诗云：“楼际重听汉水声，四窗依旧白云生。当年玉笛重吹彻，此日朱甍又落成。江上渺茫仙子路，山前荒废楚王城。登临不禁悲今古，犹见晴川树影横。”[7]

乾隆帝平生足迹未履湖北，没有亲眼见过黄鹤楼，但非常熟悉历代相关咏楼诗文，曾经通过画卷领略其风貌，并为此楼御笔题写“江汉仙踪”匾额，还先后作过三首御制诗，其中为宫廷画家董邦达所绘《楚南名胜图册》中的第十幅《黄鹤楼》题诗云：“迴临雉堞瞰江流，崔颢题诗楼上头。太白顾而不复作，卓哉此意足千秋。”[8]另外两首分别为明代仇英和清代邹一桂各自所绘《黄鹤楼图》而作，前诗曰：“两个地仙谁跨鹤，千秋佳话空传楼。高楼江夏今好在，我欲寻之叹路悠。莫谓昔人传昔事，试看春水生春洲。依然别馆拈吟处，不知何事惹闲愁。”[9]后诗曰：“黄鹤楼高古名迹，一桂惨淡图成之。汉阳树影连空翠，鹦鹉芳草仍离披。自言奉使偶经过，追摹烟景曾无遗。

5 张子安，张声著：《清同治黄鹤楼的构思与形制》，《华中建筑》1985 年第 2 期，第 40 页。

6 （清）王孙蔚：《黄鹤楼》，见（清）陈元京纂修：《江夏县志》卷十四。

7 （清）程封：《喜黄鹤楼新成》，见（清）陈元京纂修：《江夏县志》卷十四。

8 （清）弘历：《御制诗三集》卷九四《黄鹤楼》，清光绪二年刊本。

9 （清）弘历：《御制诗三集》卷九四《黄鹤楼》。

〔图 4〕
题明仇英《江汉揽胜图》(局部)，武汉博物馆藏

我虽未登快骋望，楚江坐对奚然疑。即今一桂亦如我，向所俯仰徒存思。”[10] 仰慕之意，跃然纸上。

武汉博物馆藏有一幅《江汉揽胜图》〔图 4〕，图上绘有黄鹤楼，且带有仇英落款，但未盖清宫印玺，应该不是当年乾隆帝所见的那幅仇英《黄鹤楼图》。另有学者考证此图为清代画家托名而绘，图上的黄鹤楼实为清代形象而非明代原貌[11]，与关槐图上的黄鹤楼较为相近。

二　万寿献礼

乾隆十六年（1751），正逢乾隆帝生母崇庆皇太后六十圣寿之年，乾隆帝在当年十一月为其在北京举行了盛大的万寿庆典，王公大臣、各地官员、外藩首领、外国使节齐集，为此特意在紫禁城西华门至西郊清漪园东宫门沿途道路两侧张灯结彩，寺庙、店铺修饰一新，还搭建了大量的楼阁、亭轩、戏台、彩棚、牌楼、园林，其间唱戏奏乐，罗列仪仗，场面极为热闹。

仪典期间所搭楼台都属于临时的布景性质，分别由内务府、王公贵胄和地方官员负责起造，庆典过后即予拆除。其中各地供奉的彩楼往往竭力表现本地的特色，

10　（清）弘历：《御制诗二集》卷十《黄鹤楼歌题邹一桂画》。

11　徐望生：《黄鹤楼建筑形式漫议》，《档案记忆》2018 年第 10 期，第 13—16 页。

穷极工巧，其情形有些类似今天的彩车巡游。

乾隆年间著名学者赵翼在《檐曝杂记》中记录了当时亲眼所见的景象：

> 皇太后寿辰在十一月二十五日。乾隆十六年，届六十慈寿，中外臣僚纷集京师，举行大庆。自西华门至西直门外之高梁桥，十余里中，各有分地，张设灯彩，结撰楼阁。天街本广阔，两旁遂不见市廛。锦绣山河，金银宫阙，剪彩为花，铺锦为屋，九华之灯，七宝之座，丹碧相映，不可名状。每数十步间一戏台，南腔北调，备四方之乐。侲童妙伎，歌扇舞衫，后部未歇，前部已迎，左顾方惊，右盼复眩，游者如入蓬莱仙岛，在琼楼玉宇中，听《霓裳曲》，观羽衣舞也。其景物之工，亦有巧于点缀而不甚费者。或以色绢为山岳形，锡箔为波涛纹，甚至一蟠桃大数间屋，此皆粗略不足道。至如广东所构翡翠亭，广二三丈，全以孔雀尾作屋瓦，一亭不啻万眼。楚省之黄鹤楼，重檐三层，墙壁皆用玻璃，高七八尺者。浙省出湖镜，则为广榭，中以大圆镜嵌藻井之上，四旁则小镜数万，鳞砌成墙，人一入其中，即一身化千百亿身，如左慈之无处不在，真天下之奇观也。时街衢惟听妇女乘舆，士民则骑而过，否则步行。绣毂雕鞍，填溢终日。余凡两游焉。此等胜会，千百年不可一遇，而余得亲身见之，岂非厚幸哉！京师长至月已多风雪，寒侵肌骨，而是年自初十日至二十五日，无一阵风，无一丝雨，晴和暄暖，如春三月光景，谓非天心协应，助此庆会乎？二十四日，皇太后銮舆自郊园进城，上亲骑而导，金根所过，纤尘不兴。文武千官以至大臣命妇、京师士女，簪缨冠帔，跪伏满途。皇太后见景色巨丽，殊嫌繁费，甫入宫即命撤去……[12]

文中特别提到“楚省”进贡了一座模仿黄鹤楼的楼阁。这座楼阁由曾任湖广总督的阿里衮进献——所谓“楚省”即湖北、湖南两省的代称。

据《清史稿》记载，阿里衮字松崖，钮祜禄氏，满洲正白旗人，为内大臣尹德第四子、大学士讷亲之弟，乾隆初年出任总管内务府大臣，乾隆十五年（1750）任湖广总督，次年改任两广总督。后官至户部尚书、协办大学士。乾隆三十四年（1769）十二月在征伐缅甸的军营中去世，谥号襄壮，入祀贤良祠。[13] 阿里衮与崇庆皇太后同为钮祜禄氏一族，深受乾隆帝宠信，此次为万寿盛典所献楼阁最为恢宏壮丽，冠绝各省。

乾隆十六年至二十六年（1751—1761），宫廷画家张廷彦等奉旨工笔绘制一套由

12 （清）赵翼：《檐曝杂记》卷一，上海古籍出版社，2012年，第14—15页。

13 赵尔巽等编：《清史稿》卷三一三，上海古籍出版社，1986年。

〔图5〕
清张廷彦等《万寿图》中的黄鹤楼形象，故宫博物院藏

四幅长卷组成的《万寿图》，全面展现了崇庆皇太后万寿盛典的绚丽场景。在第四卷《兰殿延禧》图上可见皇城西安门内大街上新建一座楼阁，造型酷似武昌黄鹤楼，正是阿里衮所进的贺寿贡品〔图5〕。

这座楼阁共设三层，体量硕大，在正方形平面的主体四面各出一间抱厦，底层、二层抱厦与主体转角位置均布置45度斜边。各层分别设有一圈开敞的外廊，围以木质栏杆，地面铺设花毯。二层与三层均以斗栱架设平坐层。屋面采用攒尖顶，每面各置一座悬山小顶，二层抱厦则采用独立的歇山顶。各层檐均铺绿色琉璃瓦，屋脊及兽件、宝顶则为黄色琉璃所制。门窗花格采用宫廷建筑最高规格“三交六椀菱花”样式。各层外面一圈廊柱刷深绿色油漆，内柱刷红色油漆，柱头、额枋、斗栱、栏板均绘有彩画。三层檐下分别挂满花灯，玲珑剔透，充满喜庆色彩。按赵翼所记，其墙壁安装了高达七八尺的玻璃，十分昂贵。

此楼的平面与立面形式与关槐《黄鹤楼图》上的黄鹤楼如出一辙，均有“四望如一”“八窗洞达”的特点，区别在于上部两层的收分更加明显，二层抱厦的歇山顶更为别致。此外，檐角、门窗、兽件、斗栱、彩画等细部做法可能更近于北方官式建筑，但总体上明显就是武昌黄鹤楼的翻版。

万寿盛典结束后，各地进献的彩楼陆续拆除，这座黄鹤楼也不例外。但乾隆帝认为此楼耗资巨大，不宜浪费，便赏赐阿里衮万两白银，将所有材料留下，并运至新建的御苑清漪园南湖岛上重新搭建，更名为“望蟾阁”。对此，乾隆帝在《题望蟾阁》诗中有明确强调：“庆典本缘学武昌，构移湖上恰相当。倒看山景入纨影，远挹天光接镜光。”自注：“是阁式肖黄鹤楼为之，盖圣母六旬万寿，阿里衮为湖广总督，庆典所备，嫌其所费多，因赐以万金，而留材木构阁于此云。”[14]

三　仙阁望月

清漪园昆明湖中筑有南湖岛、藻鉴堂、治镜阁三座岛屿，象征神话中的东海三仙山。南湖岛靠近东岸，以一座十七孔桥连通。《日下旧闻考》对此岛格局有简略记载：“廓如亭西，度长桥为广润祠，祠西为鉴远堂，东北为望蟾阁。”[15]岛上保留一座祭祀龙王的广润祠，共有三路建筑，中路为正殿所在，山门前建三座牌坊，竖立旗杆；东路为跨院；西路分为两进庭院，从南至北依次建鉴远堂、澹会轩和月波楼，其中月波楼为两层歇山顶建筑。岛北部堆叠土山，宏伟的望蟾阁高踞土山之上，北临湖面。

乾隆帝非常喜欢这座望蟾阁，几十年间以之为题一共写了16首诗，另有4首其他诗篇也与此阁相关。乾隆十九年（1754）所作《湖上杂咏》诗的第二首首次提到望蟾阁：“望蟾阁外放烟舟，澄照欣看镜里游。绿柳红桥堤那畔，驾鹅鸥鹭满汀洲。”[16]次年《昆明湖泛舟》诗再一次提及此阁[17]，同年稍晚时候又作《题望蟾阁》诗：“高阁湖心号望蟾，每来小坐未曾淹。”[18]另在乾隆五十一年（1786）所作《登望蟾阁极顶作歌》诗注中声明：“乙亥年因圣母大庆，阿里衮所备仿武昌黄鹤楼三层，木料给值，筑于此。”[19]在嘉庆元年（1796）《望蟾阁叠昨岁韵》诗注中则称：“是阁乾隆壬申年所建。”[20]所谓“壬申年”和“乙亥年”分别指乾隆十七年（1752）和二十年（1755）。由此推断，望蟾阁可能于乾隆十七年开始动工兴建，乾隆二十年建成，但在乾隆十九年已经大致成形，可以从湖上欣赏其高大的身影。

在《万寿图》上清晰可见望蟾阁形象〔图6〕，与西安门的彩楼几乎一模一样，主要差别在于屋面除了铺设绿琉璃瓦之外，还增加了黄琉璃剪边，并且不再悬挂花灯。

14　（清）弘历：《御制诗四集》卷二一《题望蟾阁》。

15　（清）于敏中等编纂：《日下旧闻考》，北京古籍出版社，1981年，第1406页。

16　（清）弘历：《御制诗二集》卷四七《湖上杂咏》。

17　（清）弘历：《御制诗二集》卷五四《昆明湖泛舟》。

18　（清）弘历：《御制诗二集》卷五九《题望蟾阁》。

19　（清）弘历：《御制诗五集》卷二四《登望蟾阁极顶作歌》。

20　（清）弘历：《御制诗余集》卷四《望蟾阁叠昨岁韵》。

〔图 6〕
清张廷彦等《万寿图》中的望蟾阁与南湖岛，故宫博物院藏

望蟾阁下的假山磴道有百级之多，加上楼内陡峭的楼梯，要登到最上层，需要一定的体力。乾隆二十五年（1760），乾隆帝年近五十，在诗中写道："高阁临湖迥且虚，玲珑石洞负阶除。丹梯先陟百十级，历历白玉栏杆扶。升堂香扆略憩息，其上更有二层庐。却笑九年三到吾，三层跻乃一次无。似此构筑缘底事，自问不解况人乎。今朝佳景可再孤，索须极顶登临诸。"[21] 感慨自己九年来三次登阁，之前两次都没有上到第三层，这一次终于登顶。二十六年后的乾隆五十一年（1786），虚岁七十六的乾隆帝又在诗注中说："历年以来虽四登，惟庚辰、甲申两登其最上层。"并自夸："古稀有六步尚强，拾百十级消俄顷。"[22] 他退位为太上皇之后，于嘉庆元年（1796）最后一次为望蟾阁题诗，总结自己四十多年来一共登阁 12 次。阁上风大，夏日尤为凉爽，乾隆帝的感受是："朱明伏暑最炎朝，拾级乘凉正斯候。扶栏未出曲折幽，已觉白汗辞肤腠。"[23]

清代乾隆年间宗室弘旿所绘《都畿水利图》以相对写意的笔法展现了望蟾阁及其周边的景致〔图 7〕，可见此阁位处昆明湖东部的中心位置，不但是湖上最醒目的景观建筑，同时也宜于眺赏四周景物。

21 （清）弘历：《御制诗三集》卷六《登望蟾阁极顶放歌》。

22 （清）弘历：《御制诗五集》卷二四《登望蟾阁极顶作歌》。

23 （清）弘历：《御制诗三集》卷四〇《登望蟾阁作歌》。

〔图7〕
清弘昨《都畿水利图》（局部），中国国家博物馆藏

望蟾阁北侧临湖处设有码头，乾隆帝经常由此乘船往来于南湖岛和北岸乐寿堂之间，其《乐寿堂得句》诗曾咏："望蟾阁下放烟航，径渡山阳乐寿堂。"[24]《昆明湖泛舟》诗描写了从湖上乘船观赏望蟾阁的景象："凤池春水碧溶溶，雁已回翔鱼未喁。却见湖心望蟾阁，晶盘擎出玉芙蓉。"[25]更多御制诗描写自己登阁赏景的感受，如："岧峣杰构俯昆明，金粟银漪映座清。"[26]"高阁临昆湖，眺远不计顷。峰态及林姿，四时供日骋。"[27]《登望蟾阁极顶放歌》诗还将楼阁四面风光一一点明："畅春东望神仙区，西山真是西竺如。北屏万寿南明湖，就中最胜耕织图。"[28]东南西北各有妙景，可谓绝胜。

除了点景、观景功能之外，望蟾阁还蕴含着双重文化主题，意境深远。

首先，南湖岛作为蓬莱仙境的化身，望蟾阁是其重要的组成元素。乾隆帝在香山静宜园静室一带远眺此阁，作诗云："昆明湖上望蟾阁，疑是蓬莱驾海涛。"[29]武昌黄鹤楼有许多关于神仙的传说，望蟾阁仿其形制，也间接强化了自身的仙境主题。

阁名源自东汉时期的志怪小说《洞冥记》。这部书记载汉武帝在钓影山上建望蟾阁，阁上悬挂特殊的金镜："钓影山去昭河三万里，有云气望之如山影，丹藿生于影中，叶浮水上。有紫河万里，深十丈，中有寒荷，霜下方香盛。有降灵坛，养灵池，分光殿五间，奔雷室七间，望蟾阁十二丈，上有金镜，广四尺。元封中，有祇国献此镜，照见魑魅，不获隐形。"[30]清漪园望蟾阁上没有悬挂类似的宝镜，但楼前湖平如镜，可以产生类似的联想，乾隆帝咏阁诗也不止一次提到"镜光"，如"倒看山景入纨影，

24（清）弘历：《御制诗三集》卷一三《乐寿堂得句》。

25（清）弘历：《御制诗二集》卷五四《昆明湖泛舟》。

26（清）弘历：《御制诗二集》卷六四《望蟾阁》。

27（清）弘历：《御制诗五集》卷八六《望蟾阁有会》。

28（清）弘历：《御制诗三集》卷六《登望蟾阁极顶放歌》。

29（清）弘历：《御制诗三集》卷一三《静室口号》。

30（汉）郭宪：《汉武帝别国洞冥记》，中华书局，1991年，第2页。

远挹天光接镜光。”[31]

乾隆帝从湖对面看望蟾阁，作诗云:“隔湖飞睇者，望此作蟾宫。”[32] 所谓“蟾宫”，指的是月亮上的广寒宫。南湖岛平面轮廓近于圆形，本身也是月亮的象征，岛中有月波楼，岛东侧的十七孔桥上刻有“偃月”二字，加上望蟾阁，构成一个完整的月中仙境模式。

中国素有“近水楼台先得月”的谚语，以临水楼阁为赏月的最佳场所，看天上之月与水中倒影相互映衬，方为赏心乐事，著名者如杭州西湖之平湖秋月楼、扬州瘦西湖之二分明月楼。黄鹤楼同样是赏月的胜地，宋代范成大《水调歌头》词曾咏:“今年新梦，忽到黄鹤旧山头。……想见姮娥冷眼，应笑归来霜鬓，空敝黑貂裘。酾酒问蟾兔，肯去伴沧洲。”[33] 清代王守正《黄鹤楼玩月歌》云:“黄鹄山头高硉矹，黄鹤楼头尤峭崒。楼头秋夜寂不喧，碧空高挂一轮月。乘兴来游静领之，皎如水镜悬贝阙。旁无浮云三五点，中有桂花影突兀。月白晴烟横大江，月明山壁森见骨。江月白兮渔火青，山月明兮鹊声发。倚栏惟有清风生，当年长笛吹已歇。兴酣摇笔饮且歌，凌虚一啸神仙窟。”[34] 与此类似，赏月也是望蟾阁的另一重主题，其名兼取“远望蟾月”之意。

乾隆九年（1744）乾隆帝曾经写过一篇《圆明园后记》，告诫后世子孙“必不舍此而重费民力以创建苑囿”[35]，但不久却自食其言，大建清漪园，故而作《万寿山清漪园记》辩白，似有检讨之意，并声称:“园虽成，过辰而往逮午而返，未尝度宵，犹初志也，或亦有以谅予矣。”[36] 终乾隆帝一生，确实从未在清漪园过夜，算是践行诺言，但由此带来一个很大的遗憾：没有机会在夜晚登望蟾阁欣赏月色。

清晨时分太阳东升，西边垂落的月亮还没有完全消失，但此刻皇帝一般都正在圆明园中处理政务，难得亲临清漪园观赏，只能偶尔一到，如乾隆帝《望蟾阁》诗所云:“望蟾最宜夕，而我曾未到。偶来每值晨，汉边尚堪眺。我闻太阴光，实藉太阳照。金乌已腾辉，玉兔奚能耀。游目适斯闲，因心观厥徼。”[37]“不负佳名是今度，尚看晓魄挂西楹。”[38] 他在另一首诗中吟道:“望蟾宿所名，而却未一试。晓蟾值寅卯，正为理政际。晚蟾当酉戌，游罢早归跸。设如期副名，耽玩非美事。责实纵孤其，无逸恒勉自。”[39] 一方面对辜负美景表示遗憾，一方面又标榜自己勤政无逸。

乾隆帝内心对于登阁赏夜月一事始终念念不忘，多次在御制诗及诗注中表达向往之情，如“何尝座里披金镜，为忆阶前响玉签（率在圆明园驻跸，尚未登此望月也）。”[40]“徘徊有所思，孤负是良宵（是阁虽云‘望蟾’，而从未于此赏月，故戏及之

31 （清）弘历：《御制诗四集》卷二一《题望蟾阁》。

32 （清）弘历：《御制诗三集》卷八二《题望蟾阁》。

33 （宋）范成大：《范石湖集》，上海古籍出版社，1981年，第467页。

34 （清）王守正:《黄鹤楼玩月歌》，见（清）陈元京纂修:《江夏县志》卷一四。

35 （清）弘历：《御制文初集》卷四《圆明园后记》。

36 （清）弘历：《御制文二集》卷十《万寿山清漪园记》。

37 （清）弘历：《御制诗三集》卷五《望蟾阁》。

38 （清）弘历：《御制诗二集》卷六四《望蟾阁》。

39 （清）弘历：《御制诗五集》卷四四《望蟾阁》。

40 （清）弘历：《御制诗二集》卷五九《题望蟾阁》。

云)。”[41]“所谓‘望蟾’，亦虚有其名耳。”[42]

在《望蟾阁有会》诗中，乾隆帝总结道：“高阁临昆湖，眺远不计顷。峰态及林姿，四时供日骋。而吾额望蟾，会意别有领。峰林具实质，蟾魄标虚景。筹其虚实间，斯应托怀永。”[43]坦承“望蟾”二字题额表现的主要是虚景，但能够让自己体会相关的意境，寄托情怀，就算达到目的。

望蟾阁南侧是供奉龙王的广润祠，乾隆六十年（1795）乾隆帝来祠中祈雨，顺便最后一次登上望蟾阁，作诗云：“迩来望捷并望雨，逸兴都无懒作诗。兹因请雨灵祠叩，祠侧高阁聊临之。”[44]次年作《望蟾阁叠昨岁韵》诗，又一次提到祈雨，将望蟾阁视为广润祠的外围扩展，亦与春雨相关联，赋予其景观之外的特殊寓意，也使得这座楼阁的内涵更加丰富。

四　写仿再造

乾隆帝本人在诗中反复强调此楼样式全盘写仿黄鹤楼，所言非虚。

“写仿”取“摹写＋仿建”之意，是中国古代园林史上一种特殊的艺术创作手法，即以经典名园胜景为原型，先做图写，再在异地进行仿建，最早的记载见于《史记·秦始皇本纪》：“秦每破诸侯，写放[45]其宫室，作之咸阳北阪上，南临渭，自雍门以东至泾、渭，殿屋复道周阁相属，所得诸侯美人、钟鼓以充入之。”[46]

秦代之后的历代皇家宫殿和园林建设也不断会出现类似的例子，但写仿的对象以自然界的名山为主，例如东汉洛阳西园以大假山模仿关中少华山，东魏邺城仙都苑中堆山模仿五岳，北宋东京艮岳中筑山以仿杭州凤凰山和万松岭，南宋临安德寿宫叠石以仿灵隐飞来峰，均以相对较小的尺度概括其形，勾画其神，具有类似立体山水画的意趣。

清代皇家园林将这一手法进一步发扬光大，使其成为御苑造景最主要的方式。圆明三园、清漪园、避暑山庄、静明园、静宜园、静寄山庄等各大御苑中都有若干写仿式的景观，而且写仿的对象除了名山、名水之外，还扩大到私家园林、公共景观建筑和佛寺、祠庙，尤其以江南地区的名胜风景所仿最多，被晚清诗人王闿运形容为“谁道江南风景佳，移天缩地在君怀”[47]。

从具体的造园手段来看，清代皇家园林的写仿现象非常复杂，粗略梳理一下，大概有六种不同的情况。

41 （清）弘历：《御制诗三集》卷九九《登望蟾阁》。

42 （清）弘历：《御制诗五集》卷九七《望蟾阁作歌》。

43 （清）弘历：《御制诗五集》卷八六《望蟾阁有会》。

44 （清）弘历：《御制诗五集》卷九七《望蟾阁作歌》。

45 此处“放”是“仿”的通假字。

46 （汉）司马迁：《史记》卷六，上海古籍出版社，1986年。

47 （清）王闿运著，马积高主编：《湘绮楼诗文集》，岳麓书社，2008年，第1405页。

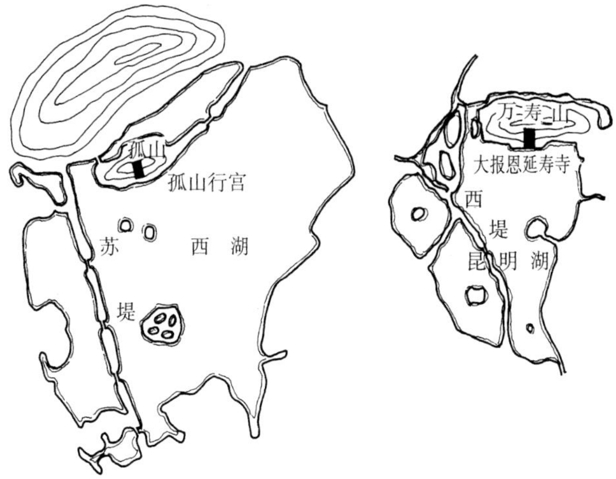

〔图 8〕
杭州西湖与清漪园平面比较图（引自《颐和园》）

第一种写仿是通过地形改造，对某处风景区的山水形态进行大规模的模拟，气魄最大，以仿杭州西湖的清漪园为典型代表。昆明湖虽然尺度略小于西湖，但两处水面的平面轮廓基本一致；昆明湖上筑有西堤，和西湖苏堤一样均呈东南—西北方向布置，而且堤上都以六座石桥串联；万寿山与孤山一样都位于湖北岸，万寿山中部位置的大报恩延寿寺也与孤山中部的康熙行宫一样居于主导地位；而清漪园以西的玉泉山、香山等山峰也如同西湖西侧的群山一样，成为可以凭眺的远景；万寿山山北有一湾曲折狭长的后河，也正相对于孤山北面的里湖。二者对照，几如孪生姐妹〔图 8〕。如此高度相似的大规模仿建，也是清代皇家园林中绝无仅有的一个例子。

第二种写仿是以园仿园，即以某一座江南名园为原型，在御苑中仿建一座相对独立的园中之园，比如圆明三园中先后建有仿海宁陈氏园的安澜园、仿江宁瞻园的如园、仿扬州趣园的鉴园、仿苏州狮子林的长春园狮子林〔图 9、图 10〕，清漪园中建有仿无锡寄畅园的惠山园（后改称谐趣园）。这些写仿园林的重点在于模仿原型的布局结构。

第三种写仿是同主题的景观再现。这种情况并不强调对原型的布局结构或建筑形制的模仿，而是以写意的方式再现原型的山水植物景貌以及场所意境，属于“得其意而忘其形”的变体创作，如圆明园中仿设的西湖十景大多如此。这十景除了三潭印月逼真地仿建了三座小石塔而外，其余诸景实际上与西湖十景原型的差异很大，

〔图 9〕

清钱维城《苏州狮子林全图》，加拿大阿尔伯特博物馆藏

〔图 10〕

清人绘《弘历雪景行乐图》（长春园狮子林），故宫博物院藏

〔图 11〕
圆明园廓然大公假山遗址与无锡寄畅园假山比较，贾珺摄

像平湖秋月、曲院风荷、柳浪闻莺、花港观鱼等景区的取景重点分别在于湖月、荷花、柳荫、花港和游鱼等主题，而非原型的园林格局或建筑形式；圆明园中的苏堤春晓只是岸边的一小段河堤，断桥残雪是一座简单的木板桥，远非西湖原貌，乃是以符号化的手段进行点题；而两峰插云、雷峰夕照、南屏晚钟则与原型差距更大，仅是环境风貌略有几分相似，因此借用西湖旧景之名而已。

〔图 12〕
紫禁城宁寿宫花园第三进院假山叠石，贾珺摄

第四种写仿是对某些名园的假山片段进行模仿。中国古典园林最重视掇山，皇家园林中常常花费大量的人力物力堆叠假山。江南地区盛产湖石、黄石，假山艺术甲于天下，广受推崇。乾隆帝曾经专门征召江南叠石匠师为长春园狮子林堆叠奇巧玲珑的湖石山景，也曾经模仿无锡寄畅园对圆明园廓然大公景区北部的假山进行改造，山径盘旋，石峰崎岖，还参照寄畅园八音涧，在西北端引来一股涧流，至今遗址犹可见其嶙峋之态〔图 11〕。紫禁城宁寿宫花园第三进院堆满山石，几无空地，里面掩藏着幽深的洞穴和盘旋的磴道，人行其间，如入深山幽谷〔图 12〕。这种独特的处理手法很可能也借鉴了苏州狮子林的假山形式——乾隆帝曾经为苏州狮子林的一座亭子题“真趣”匾额，后来将宁寿宫花园第三进院西侧的楼阁题曰“延趣”，似乎也暗示了二者相似的趣味。

第五种写仿是对一些著名佛寺或祠庙的仿建。清代皇家园林中颇多宗教性建筑，分属儒释道三家以及杂祀性质，反映了清帝复杂的信仰倾向，其中有一些实例以其他地区的著名同类寺庙为写仿对象。典型者如清漪园后山建有一座佛寺须弥灵境，其前半部分和一般汉族寺院一样设有山门、天王殿、大雄宝殿，后半部分则直接以西藏地区著名的古寺桑耶寺为蓝本，中央修筑大型楼阁，周围按照佛经中关于“四大部洲”的说法设置大小不同的塔、殿。这组建筑造型奇特，色彩艳丽，底座采用雄伟的大红台造型，体现了浓烈的藏式风格。清代帝王通过这里特殊的宗教建筑形式表现出对国家版图内不同民族文化兼收并蓄、充分尊重的态度，同时也是对园林景观不可替代的重要点缀。又如圆明园中的汇万总春之庙，模仿杭州的花神庙而建。嘉庆帝曾经专门派遣官员测绘江苏淮安府运河边的惠济祠与河神庙，并在绮春园中加以仿建，以此寄托平息淮河水患的愿望。

第六种写仿侧重于仿建某种特殊形式的景观建筑，其中具有代表性的实例是圆明园中仿宁波天一阁的文源阁、仿绍兴兰亭的坐石临流、仿杭州龙泓亭的飞睇亭，避暑山庄中仿镇江金山寺的小金山、仿嘉兴烟雨楼的烟雨楼，清漪园中仿湖南岳阳楼的景明楼、仿杭州蕉石鸣琴的睇佳榭等等，仿武昌黄鹤楼的望蟾阁也属于这种情况。

这种写仿手法的重点仍在于景区的整体格局和山水环境，对于具体的建筑造型往往只略取其制，很少具象模仿，与原型存在明显的差异。但黄鹤楼是一个罕见的例外，因为直接利用湖广地区进献的彩楼翻建而成，与原型相似程度很高。

望蟾阁的四面抱厦各带倾斜侧面，忠实再现了黄鹤楼独有的“四望如一”“八窗洞达”的特征，对此乾隆帝深有领悟，在诗中吟道：“九霄飒爽座间披，四面画图镜中斗。”[48]“一径石桥通，崇台迥据中。四时延座景，八面纳窗风。”[49]“却欣八面珠帘卷，水色山光取次拈。”[50]

两座楼阁不但造型雷同，而且所在景区的建筑格局也有相近之处——黄鹤楼与一组亭台轩馆共同构成相对独立的庭院，望蟾阁同样与相邻的祠宇殿堂形成完整的组群，两座高楼均居于临水一侧，而望蟾阁南侧的月波楼与黄鹤楼东侧的南楼地位相当，都属于陪衬的附楼性质。

如果将视野放到昆明湖周边，会发现望蟾阁与黄鹤楼各自所在的山水环境较为相似，正如清华大学建筑学院编著的《颐和园》一书所言：“南湖岛上的望蟾阁，弘历的《题望蟾阁》诗注已明白道出其‘盖仿式武昌黄鹤楼之制’，就其环境而言，望蟾阁前临辽阔的湖面隔水遥对万寿山的态势，与黄鹤楼前临长江隔水遥对龟山的态势却也颇有相似之处。”[51]

具体而言，昆明湖的水面堪比辽阔的长江，南湖岛比拟黄鹄矶、北岸的万寿山

48 （清）弘历：《御制诗三集》卷四〇《登望蟾阁作歌》。

49 （清）弘历：《御制诗三集》卷八二《题望蟾阁》。

50 （清）弘历：《御制诗二集》卷五九《题望蟾阁》。

51 清华大学建筑学院编：《颐和园》，中国建筑工业出版社，2000 年，第 108 页。

对应龟山，临湖的三层文昌阁对应晴川阁，湖上的知春亭小岛对应江上的鹦鹉洲，而昆明湖西侧玉泉山上的玉峰塔对应洪山上的宝通寺塔。据此推断，望蟾阁的这次写仿兼顾到更大尺度的园林风光，在一定程度上再现了楚天江山的形势与神韵。

乾隆帝在咏望蟾阁诗中多次提及与黄鹤楼相关的诗文典故，尤其对崔颢的诗句推崇备至，如“翠屿湖中紫阁翔，迥堪揽胜且维航。玉蟾最可望琼阙，黄鹤由来肖武昌。山色茏葱带云色，波光浩渺接烟光。何须古迹追江夏，文祎荒唐太白狂。”[52]“兴寄白榆宇，制规黄鹤楼（是阁制如武昌黄鹤楼）。设云武昌是，崔句孰能俦。”[53]“因之即景思崔咏，衹合于斯笔砚藏。”[54]这些诗句进一步表明了二者的亲缘关系。

五　故景依稀

清代宫廷档案记载，清漪园望蟾阁自建成以后，经常出现破损情况，屡次维修。对此张龙先生《颐和园样式雷图档综合研究》提出此阁“位于南湖岛北侧，直接面对西北红山口，春冬两季此处风大，三层的望蟾阁更是首当其冲，翼角极易被吹坏。在现存档案材料中，该阁的维修次数最多，相关工程人员也受此牵连。”[55]并依据中国第一历史档案馆藏《鉴远堂等处陈设清册》等文献考证嘉庆十五年至十七年间（1810—1812），望蟾阁被彻底拆除，原址另建了一座单层的涵虚堂。

咸丰十年（1860）英法联军入侵北京，焚掠三山五园，清漪园同遭劫难，南湖岛上的建筑均被焚毁。光绪时期重建颐和园，在原址上按照嘉庆年间的涵虚堂样式复建五间殿堂，北出三间抱厦，景观效果远逊当年的望蟾阁〔图13〕。堂南悬有“晴川藻景”匾额，能让人联想起崔颢《黄鹤楼》诗中的名句“晴川历历汉阳树”，似乎与昔日的写仿历史依然有微妙的呼应关系〔图14〕。

武昌黄鹤楼的经历更为坎坷。据李华先生《清代黄鹤楼建筑考》考证，咸丰五年（1855）太平军攻占武昌，放火焚毁黄鹤楼。同治四年至六年（1865—1867）在湖广总督李瀚章、湖北巡抚郭柏荫的主持下进行重建。[56]同治、光绪年间有几幅黄鹤楼老照片留存至今，可以清晰辨别其造型仍为正方形平面的主体四面出抱厦，抱厦侧面斜45度，屋顶为攒尖式，每面另加歇山小顶，底层留有一圈通透的外廊。与乾隆时期相比，主要差异在于二三层均无收分，外观保持直上直下〔图15〕。

光绪十年（1884）武昌黄鹤楼再次失火被毁。光绪三十三年（1907）湖广总督张之洞调任军机大臣，属下官员和门生筹资在黄鹤楼故址附近建了一座风度楼以纪念

52 （清）弘历：《御制诗三集》卷八二《题望蟾阁》。

53 （清）弘历：《御制诗三集》卷九〇《望蟾阁》。

54 （清）弘历：《御制诗四集》卷二一《题望蟾阁》。

55 张龙：《颐和园样式雷图档综合研究》，天津大学博士学位论文，2009年。

56 李华：《清代黄鹤楼建筑考》，《武汉理工大学学报（社会科学版）》2004年第3期。

其在任期间的政绩，后更名为奥略楼。此楼共有三层，采用歇山屋顶，体量比黄鹤楼小得多，造型也有很大差异，但仍被视为黄鹤楼的替身。1957 年建长江大桥武昌引桥时，占用了黄鹤楼的原址，将奥略楼拆除。1981 年在距离旧址约 1000 米的蛇山峰岭上重建复古风格的新黄鹤楼，1985 年落成，内部采用钢筋混凝土结构，五层总高 51.4 米，造型参考了清代的样式〔图 16〕而更显巍峨，与颐和园现存涵虚堂南北遥望。

1998 年颐和园被联合国教科文组织列入世界文化遗产名录，被誉为“中国皇家园林的传世绝响”，其前身清漪园的艺术成就更胜一筹，特别在写仿创作上极有巧思。望蟾阁利用皇太后万寿盛典的木料修造而成，忠实模仿武昌黄鹤楼，兼顾群体格局和山水环境，景致独胜，视野广阔，同时蕴含深刻的文化内涵，堪称古代造园的优秀范例，虽然原构早已不存，却依然值得今人品赏追忆。（感谢徐家宁先生和杨中一先生为本文的撰写提供资料。）

图 13	图 14
图 15	图 16

〔图 13〕
颐和园南湖岛上涵虚堂东北侧景象，贾珺摄

〔图 14〕
颐和园涵虚堂南面景象，贾珺摄

〔图 15〕
武汉黄鹤楼，［英国］约翰 · 汤姆逊（John Thomson）摄于 1871 年（旧影志工作室提供）

〔图 16〕
武汉黄鹤楼今景，陈刚摄

参考文献

（汉）郭宪：《汉武帝别国洞冥记》，中华书局，1991 年。

（宋）范仲淹：《范文正公文集》，中华书局，1985 年。

（宋）范成大：《范石湖集》，上海古籍出版社，1981 年。

（宋）严羽：《沧浪诗话》，中华书局，1985 年。

（清）赵翼：《檐曝杂记》，上海古籍出版社，2012 年。

（清）于敏中等编纂：《日下旧闻考》，北京古籍出版社，1981 年。

（清）陈元京纂修：《江夏县志》，清乾隆五十九年刊本。

（清）奕䜣等编：《清六朝御制诗文集》，清光绪二年刊本。

（清）吴友如：《点石斋画报大全》乙集，上海集成图书公司宣统二年刊本。

赵尔巽等编：《清史稿》，上海古籍出版社，1986 年。

纪宝成主编：《清代诗文集汇编》第 410 册，上海古籍出版社，2010 年。

孟白、刘托、周奕扬主编：《中国古典风景园林图汇》第 5 册，学苑出版社，2000 年。

清华大学建筑学院编：《颐和园》，中国建筑工业出版社，2000 年。

刘潞主编：《十八世纪京华盛景图——清乾隆皇太后〈万寿图〉全览》，故宫出版社，2019 年。

朱诚如主编：《清史图典》，紫禁城出版社，2001 年。

张龙：《颐和园样式雷图档综合研究》，天津大学博士学位论文，2009 年。

张子安、张声著：《清同治黄鹤楼的构思与形制》，《华中建筑》1985 年第 2 期。

李华：《清代黄鹤楼建筑考》，《武汉理工大学学报（社会科学版）》2004 年第 3 期。

徐望生：《黄鹤楼建筑形式漫议》，《档案记忆》2018 年第 10 期。

A Study on the Imitation of Yellow Crane Tower in Wuchang by Moon Watching Tower (*Wangchan ge*) in *Qingyi* Garden

Jia Jun School of Architecture, Tsinghua University

Abstract: Yellow Crane Tower located on the bank of the Yangtze River in Wuchang has been repeatedly destroyed and built throughout history, and is very famous with numerous poems and writings. During the reign of Qianlong Emperor in the Qing Dynasty, Moon Watching Tower (*Wangchan ge*) was built on Nanhu Island in *Qingyi* Garden, a imperial garden in the northwest suburbs of Beijing. It fully imitated the shape of the Yellow Crane Tower, thus became an iconic landscape in the garden and was deeply loved by Qianlong Emperor, who wrote many poems for it. Based on historical documents and visual materials, the author tries to research the construction process of Yellow Crane Tower and *Wangchan ge*, summarize the different imitation models of imperial gardens in the Qing Dynasty, and further analyze the scenic features and cultural connotations of *Wangchan ge*.

Keywords: *Qingyi* Garden, Moon Watching Tower (*Wangchan ge*), Yellow Crane Tower, imitation, Qianlong Emperor

武汉黄鹤楼旧影，宝记照相馆摄于1890年代

景山五亭旧影，黎芳摄于 1880 年代

浅谈艮岳造园影响的二元性

朱育帆　清华大学建筑学院

去年12月的一天，因展览事宜特意造访了一下故宫博物院的乾隆花园，一座声名远扬又充满神秘色彩的帝王花园，静静地隐藏在故宫中轴线的东侧。当登顶符望阁，南出环廊周匝，很自然地转向了西面，再一转便框见了那熟悉的景山，夕阳余晖斜下，很是迷人。这个视角下，那个曾经威严森森的紫禁城却彰显出些许如画的园林色彩。景山上的那五座亭子颇具标志性，由东往西分别是周赏亭、观妙亭、万春亭、辑芳亭和富览亭，万春亭居于山顶，傲然凌绝于北京城中轴线之上，使得这座本来就居中对称的镇山更加具有仪式感，而茂密植被覆盖下的巨大山体又难掩园林的自然气息。

不禁浮想，当年乾隆帝晚年游赏此地时会不会知道：景山与六百年前的一代名苑密切相关，而那里也有一座人造大山，上面也有五座亭子，分别是半山亭、麓云亭、介亭、极目亭、萧森亭，介亭居首，它就是北宋东京汴梁城的皇家苑囿——艮岳〔图1〕。

艮岳之“艮”字并不常用，一定程度影响了它的传播。艮岳有许多别称，如凤凰山[1]、万岁山[2]、寿岳[3]、华阳宫[4]等等，倒是后来民间将万岁山与名著《水浒传》所讲述的“花石纲”和“生辰纲”相关联的说法为这座神秘园林增添了更多的普及色彩。

“艮”字的含义，指东北。简而言之，“艮岳”就是指位于东北方向的山。因为宋徽宗即位初年子嗣少，道士进言如果少阳艮位有山便可子嗣广衍。艮岳始建后，子嗣问题果然大有改观，皇帝便更加崇信道教。[5]艮岳的名声并不好，连同这位亡国之君，背负了亡国之园的骂名，不少人将北宋灭亡归结于赵佶等无节制地兴建以艮岳为代表的园林。艮岳是个传说中的园林“聚宝盆”，是奇花异木、珍禽猛兽、怪石灵泉等无穷财富的聚集之地。

1　（宋）蔡絛《史补·宫室苑囿篇》、（宋）赵彦卫《云麓漫钞》、（宋）袁翁《枫窗小牍》均记艮岳初名“凤凰山”，原因在于其取象余杭之凤凰山。

2　（宋）赵彦卫《云麓漫钞》记“且曰万岁山”；《宋史·地理志》记“万岁山艮岳”。

3　（宋）李焘撰，（清）黄以周等辑补：《续资治通鉴长编附拾补》卷三十四“（宣和六年九月庚寅）手诏以金芝产于艮岳万寿峰，宜改名寿岳”，上海古籍出版社，1986年。

4　（宋）祖秀《华阳宫记》记“工已落成，上名之曰华阳宫”。

5　（宋）张淏之《艮岳记》记“徽宗登极之初，皇嗣未广。有方士言京城东北隅，地协堪舆，但形势稍下，倘少增高之，则皇繁衍矣。上遂命工倍其冈阜，使稍加于旧，已而果有多男之应……政和间，遂即其地，大兴工役，筑山号寿山艮岳”。参见《丛书集成》卷一五〇八。

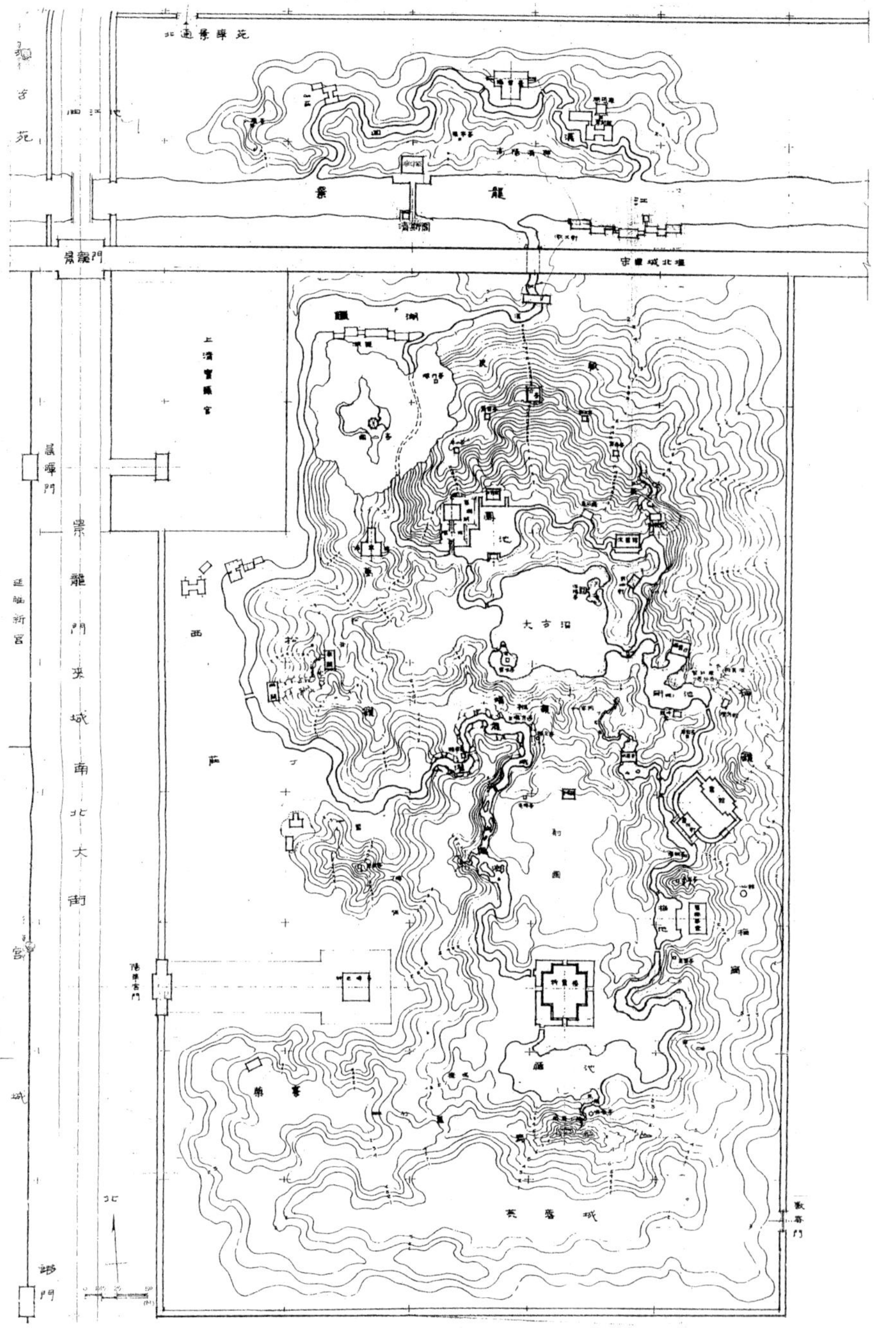

〔图1〕
艮岳想象平面图，朱育帆绘

根据记载，艮岳的位置和范围大致是清晰的。北界与宋内城北城墙有关，西界与景龙门和大内东墙有关，东界与内城北封丘门有关，南界与大内东墙謻门有关，其中景龙门、謻门和内城北城垣均有考古挖掘成果支撑。而就艮岳具体的园林样貌，至今还只能根据核心文献记载进行推测，总体尚停留在猜想层面〔图2〕。与王维的辋川别业和谢灵运的山居等类似，艮岳的传播早已摆脱了其作为园林物质空间的真实性。本文主要谈谈艮岳造园影响后世的二元属性：自然山水园与都城营国形制。

一　艮岳的后期沿革与艮岳遗石

金灭北宋，强占东京城建皇宫，金皇宫北部地望原系宋延福宫及宝箓宫和部分艮岳址。根据宋邹伸之《使燕日录》和元扬奂《汴故宫记》的描述进行推测分析，金皇宫仁智殿北之“百泉山”与宋延福新宫五位后“杏岗”[6]位置相同，很可能是在此基础上改建的。仁智殿与百泉山均在大内中轴线上，这部分区域属金皇宫后苑，苑东部即原宝箓宫和艮岳位，亦有水石但并不为胜。可以说自北宋灭亡之后，宋大内以北已由原东北山势居高演变为金及金以后的中轴线山势居高的格局。被划入金大内的部分艮岳，其余土石尚可被利用，但原格局已完全被破坏而不复存在。因此，自金大内、金里城建成后，再探讨艮岳是否作为园林存在已全然没有意义了。金以后感怀吟咏艮岳的诗文不断涌现，但由于受金大内与宋大内城址变迁，而诸条件又不足以明察其变的局限，诗文作者多在对艮岳基址的判断上出现了偏差。如明李梦阳作《艮岳十六韵》就误以城西北龙德宫前三山为艮岳；还有一种将明周藩王府旧山（金大内旧山所改）误作艮岳，如清杨恩圣《周藩故宫行》等。这些诗文若以题咏艮岳记，显然感怀伤古，寻错了方位，而若以题咏艮岳遗迹或遗物记，倒还是可以的。

自靖康之变后，艮岳园中流出之物都能成为话题。其中园林遗石作为造园活动可传承的流通品，似乎成为艮岳留给后世的一种托孤，于是便有了“艮岳遗石”的说法，后人往往以获得一鳞半爪的艮岳遗石为荣，足见艮岳已成为优秀的代名词。

艮岳之石除被用作当年靖康守城炮石之外，因其藏量巨大，仍有大量花石被遗弃于旧址。宋范成大于乾道六年（1170）至汴梁城，此时的金大内已扩至九里十三步，但尚无展筑里城，范称“出旧封丘门，即安远门也，金改为元武门。门西金水河，旧夹城曲江之处，河中卧石魂，皆艮岳所遗”[7]。艮岳中上佳之巨石如“玉京独秀太平岩”及“敕赐卿云万态奇峰”被移至金大内后苑仁智殿前，“殿后用怪石叠成山”称“百泉山”，又有“涌翠峰”等，这些石山之石材极有可能直接取自艮岳遗石。

明洪武二十年（1387）以后，黄河淹及开封城。在此之前，明周藩府及金元之宋金大内均在平面上有过改动，艮岳遗石在此期间损失的可能性不大。但随着黄河灌城次数的增加，艮岳遗石便大多日渐淤塞于河泥之下了。

清钱仪吉在《留云峰记》中记述其在嘉庆十四年（1809）过大梁，慕名访城南馆舍之“赵宋遗迹，或曰艮岳数峰”[8]。观后二十余年，再造大梁至史致昌卷石山房，见

6 《宋史·地理志》记“延福宫”：“东西配大内，南北稍劣。其东直景龙门，西抵天波门。”按此延福宫应是大内东西序北延至里城北垣的区域，进而推证金水门（即天波门）应位于大内西序南北一线与里城北垣的交汇点附近。宋李焘《续资治通鉴长编》卷七十一“大中祥符二年四月己亥”条下记真宗欲修玉清昭应宫，丁谓献媚称：“建宫于宫城之乾地，正可以祈福。”玉清昭应宫位于天波门以东至旧酸枣门以西的区域，若天波门位于大内西序南北一线，玉清昭应宫正好位于宫之坎地（正北）。丁谓所称乾地（西北）断难解释，而且《宋史·地理志》延福宫东西均配大内一说，尚无其他文献可对合之。因此笔者认为延福宫东序配大内而西序不配，天波门应位于大内西序南北一线更西的区域。由于延福宫东门晨晖门与上清宝箓宫宫门相对（注：《宋史·地理志》误作“景晖门”），而宝箓宫又紧邻艮岳，因此研究廷福宫布局对研究艮岳园址形状具有一定的意义。但是由于史料的局限，尚不能确定晨晖门的具体方位，所以笔者并没有将延福宫的研究内容加入正文，特此注之。延福宫的史料据《宋九朝编年备要》卷二十八、《东京梦华录》及《北道刊误志》。

7 （宋）范成大撰：《揽辔录》，参见《丛书集成》卷三一一〇。

8 （清）沈传义等修，（清）黄舒昺纂：《祥符县志·丽藻·记》卷二十一。

〔图 2〕
艮岳想象鸟瞰图（原题：艮岳图想），朱育帆绘

拉奥孔群雕，娄玮摄

中西方园林中雕塑的收藏和展示
——以西汉与文艺复兴时期为例

郭欣悦　故宫博物院数字与信息部

西汉时期的中国正处于一个空前的文化繁荣期，丝绸之路的开辟促进了东西方[1]之间的联系。随着国家实力不断增强，民众有余力营建居室陵墓，帝王也通过兴造宫苑来满足军事和娱乐需求。园林开始兼具实用和观赏功能。一系列规划和装饰手法应运而生，用以表达文化立场和对天地自然法则的理解。作为景观配置的大型宫苑雕塑在这一时期也开始出现，可惜留存下来的极少。而在西方，文艺复兴时期被普遍认为是园林里收藏和展示雕塑极繁荣的时期。随着与东方各国的贸易往来，新兴资产阶级的阵营不断发展壮大，以文艺复兴运动的发祥地佛罗伦萨和政治中心罗马为代表，人文主义思潮下对田园情趣的追求加速了庄园兴造之风的盛行。文艺复兴园林不仅从古代继承了多种造园要素，复兴古典文化的风尚也促进了对古代希腊罗马雕塑的挖掘和复制，并最终收藏展示在庭园之中。可见，无论是东方还是西方，雕塑在园林中的收藏和展示都是各自历史文化背景下审美追求与哲学观念相互作用的结果。马克·特雷布（Marc Treib）在《雕塑与花园：历史概述》（*Sculpture and Garden：A Historical Overview*）中介绍了园林与雕塑两者之间可能存在的关系，包括园林作为雕塑的收藏空间，雕塑作为园林的装饰艺术以及在两者的发展中形成的园林雕塑和雕塑花园的概念。园林作为雕塑的收藏空间在诸如普林尼（Pliny the Elder，23 或 24—79）的《博物志》（*Naturalis Historia*，又译《自然史》）等亚文学材料中清晰可见。雕塑作为园林的装饰艺术，一个典型案例是古希腊列柱中庭（peristylium）中的赤陶雕像替代花草作为装饰物，其作用与盆栽或喷泉并无二致。无论是收藏行为还是装饰行为，雕塑本体的艺术性是承续的，但是其与特定园林空间共同构建起的体验与象征意义，以及作为收藏品或装饰艺术品的角色与当时的文化语境与实践存在的关联与互动作

1　指广义上的东方与西方，非本文所讨论西方园林中现代意义的欧洲或“西方世界”。

用会有差异。本文选取西汉上林苑的牛郎、织女石刻像作为中国园林雕塑的典型案例，引申至受其影响的清乾隆时期清漪园镀金铜牛与耕织图景区，对比文艺复兴时期梵蒂冈望景楼庭院的拉奥孔群雕，探讨在各自的辉煌时期，中西园林中的大型雕塑在美学形态和象征匠艺上的表达，雕塑和园林的互动关系，收藏和展示的目的和实践，以及它们所承载的更广泛的历史文化信息。

一　中西方雕塑艺术的背景

雕塑在西方园林中屡见不鲜，在中国数量却极少。梁思成先生曾有言，“艺术之始，雕塑为先”然“我国言艺术者每以书画并提”[2]。中国的雕塑艺术从石器时代开始发展，至今也有几千年的历史，其中不乏辉煌时期，铜像、俑、画像砖、瓦当、造像，不胜枚举。然而以故宫博物院举例，故宫 1925 年建院，直到 1958 年才在奉先殿设雕塑馆，竟是我国博物馆历史上第一个雕塑专馆。雕塑这一极其重要的艺术形式，长久以来被国人所忽略，谈及雕塑大师作品，往往只能想到西方雕塑。[3] 其中一个关键的社会历史因素在于，工匠群体长期以来在主流观念中未能获得应有的尊重与认可，其身份定位被普遍排除在“艺术家”之外。另一个原因是，中国雕塑作品常做墓葬礼器，属于“封闭性雕塑”，起到象征、寄托意义，却欠缺观赏性。再者，佛像于魏明帝时传入中国，魏晋南北朝开始兴起造像，雕塑逐渐成为宗教崇拜的载体，被庄严供奉于石窟、佛堂之内，成为一种宗教艺术。雕塑作品与世俗生活长久以来保持着一定的距离，这就导致雕塑在古代中国较少出现在园林中。园林里虽常有赏石、假山，但不能全然作为雕塑去一并而论，故不在本文讨论范畴之内。

二　中国古典园林中的大型雕塑

（一）牛郎、织女石刻像与上林苑昆明池（西汉）

本文的第一个案例——牛郎、织女石刻像，是中国雕塑历史上一件重要的、少有的秦汉苑囿圆雕石刻作品。牛郎、织女石刻〔图 1〕按左牛郎、右织女设置在汉代长安上林苑昆明池东西两岸，学界普遍认为这是迄今所知最早的一对汉代园林大型装饰雕塑。牛郎像面部五官轮廓清晰，双眼圆睁，向右前方凝视，身体呈踞坐状，右手置胸前，左手贴腹；织女像面部饱满，鼻头比例略显宽大，下颌处残毁严重，

2　梁思成：《中国雕塑史》，百花文艺出版社，2003 年。

3　前揭梁思成《中国雕塑史》。

〔图 1〕
牛郎、织女石刻像，西安市周秦都城遗址保护管理中心藏，屈铁军摄

已被后人以水泥修补，坐姿与其他汉墓出土的跽坐女俑相近，双手作笼袖姿态置于身前。[4]石刻材质为陕西秦岭花岗岩，雕塑技艺上灵活地结合了圆雕、浮雕和线刻，虽艺术风格古拙质朴，但细节刻画不失精细，形象不失真趣，反映了汉代早期石雕艺术既写实又写意的风格。

上林苑是汉武帝刘彻于建元三年（公元前 138）下令在秦代旧苑址上扩建而成的皇家宫苑，规模雄伟，兼具多种功能，是皇家观赏和游猎的胜地。其中的池沼之一昆明池是具备综合用途的人工湖沼。关于昆明池的初次修建时间，学术界虽存在不同看法，但基于《汉书》和《史记》等史料推导，基本形成武帝元狩三年（公元前 120）为讨昆明夷而初建和元鼎二年（公元前 115）为伐南越再修的主流观点。[5]昆明池在中国园林史上的一个深远意义，在于开创了我国以大型水体为核心来构建园林景观的先河，是为中国古典园林设计中“借景”“对景”等造园理论的重要实践基础。据《汉书·武帝纪》记载：“（元狩）三年春……发谪吏穿昆明池。”其后颜师古撰《汉书注》引西晋臣瓒：“西南夷传有越嶲、昆明国，有滇池，方三百里。汉使求身毒国，而为昆明所闭。今欲伐之，故作昆明池象之，以习水战，在长安西南，周回四十里。”[6]《史记·平准书》载：“是时越欲与汉用船战逐，乃大修昆明池，列观环之。治楼船，高十余丈，旗帜加其上，甚壮。”[7]可见，修凿昆明池的初衷并不是游山玩水，而是用

4 田军、陈雪华：《试析长安斗门汉牛郎织女石刻像的艺术价值》，《美术观察》2007 年第 10 期，第 116 页。关于两座石雕中孰为牛郎、孰为织女的问题，存在两种不同的看法。另一说法与本文所引材料相反，与现西安市石婆庙、石爷庙对应一致。鉴于辨明孰为牛郎、孰为织女与本文所论主题无直接关联，不作进一步探讨，可参考汤池：《西汉石雕牵牛织女辨》，《文物》1979 年第 2 期；顾铁符：《西安附近所见的西汉石雕艺术》，《文物参考资料》1955 年第 11 期。

5 冯晓多：《汉代长安昆明池修建初衷再审视》，《秦汉研究》2022 年第 1 期，第 134—144 页。

6 （东汉）班固：《汉书·武帝纪》，清乾隆四年武英殿校刻本，第 325 页。

7 （西汉）司马迁：《史记》卷三十《平准书》，清乾隆四年武英殿校刻本，第 1436 页。

来水师操练。后昆明池兼具了为长安供水和调节漕运的功能。又因昆明池水面浩渺清澈，水产丰饶，以其为中心环绕岸边的亭台楼阁、植物花卉倒映其中，影影绰绰，逐渐变成了一处“一池三山”般的游船泛舟、歌舞嬉戏之地。《历代赋汇》卷五中就有汉武帝夜游昆明池的记载，与司马迁、司马相如歌颂星辰之《星赋》。而汉武帝对星象的痴迷，是牛郎、织女石刻的重要背景。《三辅黄图》卷四引《关辅古语》曰：“昆明池中有二石人，立牵牛、织女于池之东西，以象天河。”杜甫《秋兴八首·其七》中的“织女机丝虚夜月，石鲸鳞甲动秋风”都说明武帝以昆明池象征天上的银河，而立于池东、西两侧的牛郎和织女石像对应河鼓、牵牛、牛宿以及织女、婺女、女宿这些星宿，旨在构建一个地上人间与天上星辰交相辉映的宏观宇宙模型，传达中国早期造园追求“天人合一”的艺术境界。在秦汉时期，天象被视为上天意志的直接体现，帝王通过观天象、顺天意来治国理政，是巩固政权的关键手段之一。汉赋中的“天人之符”、帝王宫室园池的布局形制映射天帝所辖的“三垣二十八宿”等，都说明汉代多以星象之学来论证帝国的顺天应时，宣扬帝王的丰功伟绩。汉武帝尤其迷信神仙方术：“台观”“东海三仙山”“是时上求神君，舍之上林中蹄氏观”[8]，许多与方术相关的建设和活动都说明在上林苑这样一个方士们描绘的神仙乐园里，牛郎、织女石刻也是顺应了“好仙道”的心理。“神仙乐园”并不仅为东方苑囿所追求，在西方语境下，往往可联系到“乐园”（paradise）一词，如《旧约圣经》中的伊甸园。古埃及法老们亦崇敬诸神，大力兴造“神苑”“圣林”“墓园”以求与神灵交流，通往来世乐土。

符号学视角下的牛郎、织女石刻，是一个天象深入人心到民间情感的实证。睡虎地秦简《日书》甲种一五五简有记：“戊申、己酉，牵牛以取织女，不果，三弃。”吴小强《秦简〈日书〉集释》翻译作：“戊申日、己酉日，是牵牛迎娶织女而失败的日子，假如在这个日子中结婚，婚后不到三年，妻子就会被丈夫休弃，或者妻子离开丈夫而逃走。”[9]可见，星宿拟人化的开端，是一个充满吉凶宜忌和警示意味的天文象征，虽与人的婚姻命运紧密相连，却无完整的情节叙述。而从先秦《日书》的书写时代至汉代记载牛郎、织女石刻以前，这段时期相关史料稀缺，这使牛郎、织女石刻成为其故事从星宿发展至神话传说的链条中的一个关键引子，也是研究星象与中国古代民间信仰、社会风俗与艺术表现不可多得的里程碑，更是汉代不少诗歌文学的灵感源泉。随着时间推移，昆明池早已不复存在，但昆明池畔的牛郎、织女像仍然矗立着，两者间距约 3 公里，如班固《西都赋》所云：“集乎豫章之宇，临乎昆明之池，左牵牛而右织女，似云汉之无涯。”

8 （西汉）司马迁：《史记》卷十二《孝武帝本纪》，清乾隆四年武英殿校刻本，第 819 页。“是时上求神君，舍之上林中蹄氏观。神君者，长陵女子，以子死悲哀故，见神于先后宛若。”

9 吴小强：《秦简〈日书〉集释》，岳麓书社，2000 年，第 108 页。

〔图 5〕
拉奥孔版画，《罗马辉煌之镜》系列，大都会艺术博物馆藏

〔图 6〕
拉奥孔群雕，乌菲兹美术馆藏

〔图 7〕
拉奥孔版画，《罗马辉煌之镜》系列，大都会艺术博物馆藏

异教的政策愈加严酷，又将一些异教雕塑和艺术品移出，或被禁止公开展示，或限制民众接触。因与多任继任者密切相关，在此不再详谈，相关探讨可参考艺术史学家弗朗西斯·哈斯克尔（Francis Haskell）和英国国家美术馆前馆长尼古拉斯·潘尼（Nicholas Penny）的合著《趣味与古物》（*Taste and the Antique: The Lure of Classical Sculpture, 1500-1900*）。

在文艺复兴时期，关于各庄园及其中珍稀雕塑收藏的信息通过版画在欧洲广泛传播。这些版画还成为现代研究园林史学与艺术史学交叉领域的重要图像资料。本文将通过《罗马壮丽之镜》（*Speculum Romanae Magnificentiae*）收录的版画，简要追溯拉奥孔和其重要复制品的沿革。由于出土时拉奥孔的右臂、大儿子和小儿子的右手缺失〔图 5〕，根据瓦萨里（Giorgio Vasari，1511—1574）的说法，大约在 1510 年望景楼的建筑师伯拉孟特举办了一场关于拉奥孔右臂修复方案的比赛，由拉斐尔（Raphael，1483—1520）担任评委。比赛以雅各布·桑索维诺（Jacopo Sansovino，1486—1570）主张的伸展手臂的方案获胜，但此方案只停留在复制品实验阶段。1520—1525 年间，巴乔·班迪内利（Baccio Bandinelli，1493—1560）受命于利奥十世（Pope Leo X，Giovanni di Lorenzo de' Medici，1513—1521 在位）以半伸展的手臂为蓝本制作了拉奥孔的早期大理石复制品，目的是将其作为外交礼物赠与法国国王弗朗西斯一世。这座雕塑后被摆放在美第奇花园的壁龛中〔图 6〕。约 1532 年，米开朗基罗的门生蒙托尔索里（Giovanni Angelo Montorsoli，1507—1563 年）才真正意义上主持了拉奥孔手臂的修复工作，拉奥孔进入“上举手臂抗争”的形象时期〔图 7〕。1798 至 1816 年间，根据法兰西第一共和国与教皇国签署的《托伦蒂诺条约》（Treaty of Tolentino），拉奥孔从望景楼移至巴黎。这一时期见证了古物收藏在政治与战争背景下的动荡命运。

1903年，犹太学者路德维希·波拉克（Ludwig Pollak，1868—1943）在拉比卡纳大街（Via Labicana）附近的考古发掘中购买了拉奥孔缺失的右臂，并于1906年捐赠，使之得以重归拉奥孔雕像。[25] 遗憾的是，波拉克没有机会亲眼见证发生在1957至1959年间的修复。1943年，他死于奥斯维辛集中营。拉奥孔手臂方向的改变，至关重要的原因在于其情感表达发生了根本性的转变：他从来不是一个伸着手臂抗争死亡的英雄，而是一个在恐惧与绝望中沉默而无力地放弃与命运抗争的形象〔图2〕。正如同集中营中的波拉克。

〔图8〕
教皇尤利乌斯二世肖像，拉斐尔及工作室绘于1511—1512年，德国施泰德博物馆藏

可见，无论是雕塑还是园林，它们的命运都与人息息相关。尤利乌斯二世〔图8〕时期的重要雕塑收藏，之所以能在望景楼完好地保存至今，其中一个关键原因是他不将艺术收藏视为家族敛财的工具，而把主要精力放到复兴教皇国上。体现在他的艺术政策上，就是要使教廷内的艺术收藏超群，成为整个欧洲的艺术中心。通过不朽的艺术来追求教廷永恒的荣耀。除了望景楼的宏伟蓝图外，他的艺术赞助先后支持了百余位文艺复兴时期艺术家的创作。经过尤利乌斯二世的努力，16世纪初，罗马取代了15世纪的佛罗伦萨，成为百家争鸣的艺术殿堂，而教廷也成为亚平宁半岛的政治重心，文艺复兴进入全盛时期。但是，对艺术的资助和对教会的维系不能仅依赖精神基础，无论是兴造园林还是收藏雕塑都需要大量资金支持。彼时筹措经费的一个主要方式就是出售赎罪券及其他替代性的大赦手段。为了资助这些工程，尤利乌斯二世时期的教廷继续依赖赎罪券的销售，继位的利奥十世更无节制。当他们对艺术的慷慨建立在人民的忏悔与信仰之上时，慷慨便不再虔诚，文艺复兴的行为不再与它的思想基础相统一。这从某种程度上，揭开了宗教改革运动的序幕。

25 Ondarza,Paolo. "Vatican Museums:Laocoön and 'The Arm of Memory.'" Vatican News, 2024. https://www.vaticannews.va/en/vatican-city/news/2024-05/segreti-musei-vaticani-braccio-laocoonte-ludwig-pollak-nogara.html.

四　结论

拉奥孔群雕与西汉时期的牛郎、织女石刻像，在艺术风格与表现手法上展现了显著的差异。前者通过精致的肌肉雕刻和动态的表现，融合了理想化美学与戏剧性的情感张力。人物虽处于极端痛苦状态，但肌肉线条和身体姿态仍体现出古希腊雕刻对人体美学和解剖学精度的追求，呈现出极致的视觉美感。相比之下，牛郎、织女石刻则呈现出一种原始、粗犷的艺术风格，注重的是象征意义和文化传递，非单纯的形态完美。在纪念碑性方面，拉奥孔群雕人物的姿态传达出接近于悲悯的严肃情感，使观者产生敬畏、肃穆的心理反应。主题对命运、生命的表达，使得雕塑超越了对某一历史人物的纪念，而是一种普遍性的象征，体现出典型的纪念碑性。相较之下，牛郎、织女石刻实际上并不具备典型的纪念碑性特征，其更贴近于文化符号，承载了特定的历史信息，但与之关联的民族记忆与文化历史象征赋予了它一定程度的纪念意义。普遍来说，中国园林中的雕塑作品相对含蓄，以传统文化为内核，承载了个人记忆并逐渐转化为集体记忆，具有民族性；同时承载文化寓意与社会理想，所以具有亲和力和感染力。西方园林中的雕塑收藏和展示，遵循古代文化影响下视觉语言的图像学规范（iconographical conventions），以神的崇拜为出发点，更注重审美和装饰性。这逐渐演化成权力和财力的象征，服务于政治、精神和美学的象征意义，以其稀有性和历史积淀展示主人的深厚文化修养与高尚品位，进一步彰显或提升自己或所在群体的社会地位与声望。但是中西相同的是，雕塑与园林主人（或委托制作人）密切相关，是主人意志的强烈体现。甚至西方偶尔作为自我崇拜和偶像崇拜的载体进行个人形象的塑造和表达——如古希腊时期亚里士多德的弟子提奥弗拉斯特（Theophrastus，公元前 372—前 287）的“利西乌姆园”中立有亚里士多德塑像，以及奥林匹亚附近圣林中放置的体育比赛优胜者的半身或全身塑像。其次，西方雕塑在园林中多可独自成立，且立于独立空间或显著位置；中方则多与景观联动，作为园林小品或山水景观的一部分，服务于造景，更多表现气势，营造氛围。但毋庸置疑的是，无论中西，都存在雕塑和园林的整体展示语境。

The Collection and Display of Sculptures in Chinese and Western Gardens:

A Comparative Case Study of the Western Han Dynasty and the Renaissance Period

Guo Xinyue Digital and Information Department, The Palace Museum

Abstract: Sculptures such as the Cowherd and Weaver Girl (*Niulang and Zhinv*) by *Kunming* Pool in *Shanglin* Garden enhanced garden aesthetics while embodying cultural symbolism and national aspirations. In the west, sculptures like the *Laocoön* in the Cortile del Belvedere played a pivotal role in reviving classical art and culture. These works differ significantly in artistic expression: *Laocoön* combines idealized aesthetics with emotional intensity through intricate muscle detailing and dynamic composition, whereas the *Niulang and Zhinv* features a primitive and rugged style, focused less on formal perfection. This paper offers a comparative analysis of Chinese and Western cases, noting that in Chinese gardens, sculptures are deeply rooted in traditional culture and convey personal and collective memories, though their primary purpose is to enhance the landscape. In the West, on the other hand, they typically symbolize power and wealth, as well as express spiritual and aesthetic ideals. Despite differences in style, purpose, and practice, both traditions reflect the strong influence and will of their owners.

Keywords: Garden Sculpture, *Niulang and Zhinv*, Laocoön, Cortile del Belvedere, Bronze Ox

乾隆花园禊赏亭流杯渠，陈敬哲摄

编后记

故宫博物院举办的展览，因其历史背景和地理位置等因素，观展者是丰富且多元的。因此，关注大众诉求，力求呈现传统对于当下的价值，是近年来故宫博物院策展的重要理念之一。本次“乐林泉：中外园林文化展”，聚焦承载人类理想环境的古典园林主题。面对受众的广泛与主题的超然，策展首先要解决的问题是：在时代变革之下，如何将稍显遥远的古典园林与当代大众的需要相连接？尤其是中国古典园林相较于世界其他园林体系所具有的独树一帜的山水精神，如何与当下的心灵同频共振，并兼容中外？需要找到打破地域和文化限制的共通之处。

我们把“人”作为解题的关键。园林景观因人的活动需要而营造，虽然文化不同，景观各异，但古往今来人们的栖居活动大体相似。展览以六类“园事”，即雅集、鉴藏、游山、静修、观花、畅音为线索，并联起中外园林景观的展品。在展品的选择上，以紫禁城中的园事为起点，延伸至更广泛的中国园林，同时观看外国园林，进而阐释来自不同地域的人们在园林中的生活方式、思想情感和文化追求。

作为同时包含中国与外国园林文化内容的展览，此次策展并未采用经典的对比展或交流展的形式，没有把中国与外国按照时间、地域或事件等叙事方式整体对照，而是采取六种中国园事分别对应不止一个地域和时代的外国园事的方式。其中以中国园事为参考系，组织与之关联的更广泛的外国园事，试图在更多元的连接中展现人与园林的丰富互动。我们希望展览作为一盏烛台、一把钥匙，能点亮人们心中的光，开启探索未知的门，而不止步于文物与文献的对照，不强求对历史认知的机械铺排。中国园林独具的“游园”之妙，便是这一策展思路的灵感源泉。

漫步中国古典园林，我们看到的总是一个局部，总有着对全景的渴求和移步的窥想。渐渐地，不同层次的风物，叠加出变幻无穷的眼前美景。中国园林的观看方式，充满着个人的选择性和随机性，可谓妙趣横生。我们也由此找到了展览逻辑的合理性和观看展览的最佳方式。

当看似独立的六个“园事”被放进曲径通幽的展厅设计，一场以“游园”为打开方式的展览被逐渐塑成。从东方士人的雅集，到法国贵妇的沙龙；从中国的拳石洞天，到意大利的人造洞窟。在包容中外的园林展览中，“游园”变得不期而遇，不断有新奇发现。铺陈六种“园事”的展览明线，在松弛的观展环境中又投射出不同的“暗线”——展品中涉及园林、文学、雕塑、宗教、音乐、戏剧等多领域的中外对比和交流，希望让每位观众都如沉浸式地游园一般，所见、所思、所感，全凭一己之兴。

策展逻辑的创新，需要与之契合的展陈设计。展览立足中国古典园林的文化语境，这就要求展厅的空间设计能够营造出东方园林的“游园”感受。而并置中外园林展品，更需要贯通中外与古今的设计语汇。最终，午门的三个展厅以水、山、花设计元素建构“游园”的曲折路径，有意让两种园事在同一个展厅的设计语汇中相得益彰，将对应主题的展品铺陈其中。情境先于展品一步抵达观众的视线，这是展览主题想更贴近观展者的一步，也是策展者通过设计在内容的整体性和局部性之间所做出的平衡。

此次具有突破性的综合型艺术展览得以举办，背后凝聚着不同专业领域同仁的智慧。饱含了故宫内外的专家学者，对于现代人如何美好栖居的关切和如何以时代需要的形式展示历史文化的不竭探索。故宫博物院长期以来的广泛国际交流，为跨文明的研究和展示提供了源源不断的能量。基于策略和理念的相互协调、真诚合作，贯穿整个展览的策划和执行，共同构成着这场游园展览。每一位沉浸于展览中的观众，都是游园人，自此园林可常驻心间。

谨以上述文字，作为本书的编后记。

尚倩宇

2025 年 4 月

Postscript

Exhibitions held by the Palace Museum attract a wide and varied group of visitors, drawn by the institution's historical significance and unique location. This diversity has urged the museum to explore how traditional culture can forge meaningful connections with contemporary society. The exhibition, 'Rejoicing in Woods and Springs: A Journey through Garden Cultures in China and the Wider World' celebrates classical gardens as representations of humanity's vision of an ideal environment. Given the exhibition's wide-reaching visitors and the elevated nature of its theme, the key challenge for the curators lies in bridging the gap between the seemingly remote world of classical gardens and the needs of contemporary society amidst the changes of the modern era. In particular, the distinctive spirit of landscape in Chinese classical gardens must find resonance with today's public, in a manner that blend both Chinese and global perspectives. This requires identifying common ground that transcends regional and cultural boundaries.

We focus on human experience as the key to solving the challenge. Gardens have always been created to meet human needs. While cultures and designs may differ, the way people live and interact with their surroundings has remained largely consistent throughout history. The exhibition is organized around six themes of garden-related human experience–Gardens and Enlightenment, Gardens and Antiquarianism, Gardens and Nature, Gardens and Retreats, Gardens and Inspiration, and Gardens and Recreation–to weave together exhibits from Chinese and international garden landscapes. The exhibition starts with the traditions of the Imperial Gardens in the Forbidden City, before extending to a broader exploration of Chinese gardens and then gardens of other countries. Through this journey, it delves into how people from different regions and eras have expressed their way of life, emotions, and cultural aspirations through gardens.

To explore garden cultures of China and other countries, this exhibition takes an unconventional approach. Rather than following the traditional approach of comparing Chinese and foreign gardens by timelines or events, it adopts the experience of Chinese gardens as its central framework. Each of the six themes about Chinese garden experiences is linked to corresponding garden experiences from different regions and historical periods around the world, showcasing the diverse and meaningful connections between humans and gardens across cultures and eras. The exhibition aims to serve as a guiding light and a key, sparking curiosity and inspiring visitors to explore new horizons. It moves beyond simply juxtaposing artifacts and texts or rigidly reconstructing historical narratives. Instead, the curatorial approach is inspired by the unique charm of 'strolling and wandering' in Chinese gardens.

When strolling through Chinese classical gardens, one is always catching glimpses of partial views, which sparks a desire to see the entire landscape and a curiosity to discover what lies ahead with each step. The scenery unfolds layer by layer, creating an endlessly changing and captivating visual experience. The way these gardens are observed is highly personal and spontaneous, brimming with charm and delight. This has helped us discover the logical structure of the exhibition and the most effective way to experience it.

As the six seemingly independent themes of 'garden experiences' are brought together within the winding and immersive design of the exhibition space, an exhibition inspired by the concept of 'strolling through gardens' gradually takes shape. From the gardens and enlightenment of the East to the salons of French noblewomen, and from China's grottoes and rock formations to Italy's artificial caves, the exhibition bridges Chinese garden and gardens from around the world. The exhibition presents the six themes of 'garden experiences' with a distinctly structured narrative, while subtly revealing hidden layers within its relaxed viewing atmosphere. This way, it contrasts the obvious and with the implied and blends moments of activities and moments reflection. Encompassing a rich array of content–from gardens, literature, and art to religion, music, and theater–the exhibition immerses each visitor in an experience akin to wandering through a garden, where what they see, think, and feel is guided entirely by their personal interest and imagination.

Innovative curatorial logic must be matched by equally thoughtful exhibition design. Rooted in the context of Chinese gardens, the exhibition requires a spatial design that recreates the experience of 'strolling through a garden.' The juxtaposition of Chinese gardens and gardens of other countries in the exhibition relies on the language of design to connect them seamlessly. In the final draft of the design, the three exhibition halls at the Meridian Gate are structured around the elements of water, mountains, and flowers, creating a winding path that evokes the experience of 'strolling through a garden.' The design carefully integrates two garden themes within a single landscape, ensuring they complement each other and align with the showcased themes. The atmosphere of the exhibition reaches the visitor before the exhibits themselves, creating a stronger engagement between the theme of the exhibition and the visitor's experience. This reflects the curator's thoughtful balance between overarching cohesion and thematic details.

This groundbreaking comprehensive art exhibition is the result of collaboration and expertise from professionals across diverse fields. It reflects the dedication of scholars, both from within and beyond the Forbidden City, to exploring how modern people can create beautiful lives and how historical culture can be presented in ways that meet contemporary needs. The Forbidden City's tradition of international exchange has continuously fueled cross-cultural research and exhibitions. Through carefully coordinated ideas and strategies alongside genuine teamwork, this exhibition brings to life the essence of 'strolling through gardens.' Every visitor who immerses themselves in the experience becomes a wanderer in the garden and will bring home its spirit with them long after their visit.

I humbly present the above thoughts as the editor's postscript for this book.

Shang Qianyu
April 2025

图书在版编目(CIP)数据

乐林泉 ：中外园林文化展 / 故宫博物院编 .
北京 ：故宫出版社，2025．4．-- ISBN 978-7-5134
-1677-1
Ⅰ．TU986.1
中国国家版本馆 CIP 数据核字第 2025AJ8290 号

乐林泉：中外园林文化展

故宫博物院 编

出 版 人：果美侠
责任编辑：王志伟
英文编辑：方 妍 朱芮菡
图片提供：故宫博物院数字与信息部
乾隆花园特约摄影：陈敬哲
责任印制：常晓辉 顾从辉
出版发行：故宫出版社
地址：北京市东城区景山前街 4 号 邮编：100009
电话：010-85007800 010-85007817
邮箱：ggcb@culturefc.cn
制版印刷：北京雅昌艺术印刷有限公司
开 本：889 毫米 ×1194 毫米 1/16
印 张：29.25
版 次：2025 年 4 月第 1 版
2025 年 4 月第 1 次印刷
书 号：ISBN 978-7-5134-1677-1
定 价：560.00 元